Emerging Technologies in Wastewater Treatment

Emerging technologies in wastewater treatment plant is an ecological, profitable, and natural technology designed to eliminate heavy metals, radionuclides, xenobiotic compounds, organic waste, pesticides, etc. from contaminated sites or industrial downloads through biological means. Since this technology is used in conditions on site, it does not physically disturb the site unlike conventional methods, that is, chemical or mechanical methods. In this technology, higher plants or microbes are used alone or in combination for the phytoextraction of heavy metals from sites contaminated with metals. Through microbial interventions, metals are immobilized or mobilized through redox conversions in contaminated sites. If they are mobilized, accumulating metal plants are placed to accumulate metals in their bodies. Next, metal-loaded plants are collected and recycled to reduce the volume of waste and then, disposed of as hazardous materials or used for the recovery of precious metals, if possible. In the case of immobilization, metals are no longer available to be toxic to organisms. There are very few books published on the proposed theme. A good number of books have been published on environmental bioremediation, but the proposed book is a new and innovative proposal specifically in wastewater treatment. Looking into the importance of emerging technologies in wastewater treatment research, the book will have a high and applicable value in industrial wastewater treatment research.

Features:

- The book highlights the importance of emerging technologies in the wastewater treatment plant to clean up the environment from pollution caused by human activities.
- It assesses the potential application of several existing bioremediation techniques and introduces new emerging technologies.
- It is an updated vision of the existing emerging technologies in environmental bioremediation strategies with their limitations and challenges and their potential application to remove environmental pollutants.
- It also introduces the new trends and advances in environmental bioremediation with a thorough discussion of recent developments in this field.
- Highlights the importance of bioremediation to deal with the ever-increasing number of environmental pollutants.

Wastewater Treatment and Research

Series Editor:
Maulin P. Shah

Wastewater Treatment: Molecular Tools, Techniques, and Applications
Maulin P. Shah

Advanced Oxidation Processes for Wastewater Treatment: An Innovative Approach
Maulin P. Shah, Sweta Parimita Bera, and Gunay Yildiz Tore

Emerging Technologies in Wastewater Treatment
Maulin P. Shah

Bio-Nano Filtration in Industrial Effluent Treatment: Advanced and Innovative Approaches
Maulin P. Shah

Membrane and Membrane-Based Processes for Wastewater Treatment
Maulin P. Shah

For more information, please visit: https://www.routledge.com/Wastewater-Treatment-and-Research/book-series/WASTEWATER

Emerging Technologies in Wastewater Treatment

Edited by

Maulin P. Shah

CRC Press
Taylor & Francis Group
Boca Raton London New York

CRC Press is an imprint of the
Taylor & Francis Group, an **informa** business

First edition published 2023
by CRC Press
6000 Broken Sound Parkway NW, Suite 300, Boca Raton, FL 33487-2742

and by CRC Press
4 Park Square, Milton Park, Abingdon, Oxon, OX14 4RN

CRC Press is an imprint of Taylor & Francis Group, LLC

Library of Congress Cataloging-in-Publication Data
Names: Shah, Maulin P., editor.
Title: Emerging technologies in wastewater treatment / edited by Maulin P. Shah.
Other titles: Emerging technologies in wastewater treatment (CRC Press)
Description: First edition. | Boca Raton : CRC Press, [2023] |
Series: Wastewater treatment and research | Includes bibliographical references.
Identifiers: LCCN 2022041255 (print) | LCCN 2022041256 (ebook) |
ISBN 9780367759780 (hbk) | ISBN 9780367759810 (pbk) | ISBN 9781003164982 (ebk)
Subjects: LCSH: Sewage–Purification–Technological innovations.
Classification: LCC TD745 .E4798 2023 (print) | LCC TD745 (ebook) |
DDC 628.3—dc23/eng/20221115
LC record available at https://lccn.loc.gov/2022041255
LC ebook record available at https://lccn.loc.gov/2022041256

ISBN: 9780367759780 (hbk)
ISBN: 9780367759810 (pbk)
ISBN: 9781003164982 (ebk)

DOI: 10.1201/9781003164982

Typeset in Times
by codeMantra

Contents

Editor

Maulin P. Shah is very interested in genetic adaptation processes in bacteria, the mechanisms by which they deal with toxic substances, how they react to pollution in general and how we can apply microbial processes in a useful way (like bacterial bioreporters). One of his major interests is to study how bacteria evolve and adapt to use organic pollutants as novel growth substrates. Bacteria with new degradation capabilities are often selected in polluted environments and have accumulated small (mutations) and large genetic changes (transpositions, recombination, and horizontally transferred elements). His work has been focused to assess the impact of industrial pollution on the microbial diversity of wastewater following cultivation-dependent and cultivation – independent analysis. He has more than 280 research publications in highly reputed national and international journals. He is an Editorial Board Member in CLEAN-Soil, Air, Water (Wiley), Editor in Current Pollution Reports (Springer Nature), Editor in Environmental Technology & Innovation (Elsevier), *Journal of Biotechnology & Biotechnological Equipment* (Taylor & Francis), Current Microbiology (Springer Nature), Ecotoxicology (Microbial Ecotoxicology) (Springer Nature), Geo Microbiology (Taylor & Francis), Applied Water Science (Springer Nature), *Archives of Microbiology* (Springer), *Journal of Applied Microbiology* (Wiley), Letters in Applied Microbiology (Wiley), Green Technology, Resilience and Sustainability (Springer), Biomass Conversion & Biorefinery (Springer), *Journal of Basic Microbiology* (Wiley), Energy Nexus (Elsevier), e Prime (Elsevier), IET Nano biotechnology (Wiley), Cleaner and Circular Bioeconomy (Elsevier), and International Microbiology (Springer).

He has edited 150 books in wastewater microbiology, industrial wastewater treatment. He has edited 25 special issues on Industrial Wastewater Treatment & Research theme in high-impact factor journals with Elsevier, Springer, Wiley, and Taylor & Francis.

Contributors

A. Luís Amaral
CEB – Centre of Biological Engineering
University of Minho
Braga, Portugal
and
Instituto Politécnico de Coimbra
Instituto Superior de Engenharia de Coimbra
Coimbra, Portugal
and
Instituto de Investigação Aplicada
Laboratório SiSus
Coimbra, Portugal.

Rachna Bhateria
Department of Environmental Science
Maharshi Dayanand University
Rohtak, India

Anjali Bishnoi
Department of Chemical Engineering
LD College of Engineering
Gujarat, India

Orhan Taner Can
Department of Environmental Engineering
Bursa Technical University
Bursa, Turkey

Mamta Chahar
Nalanda College of Engineering
Bihar, India

Sohel Das
Structural Biology and Nanomedicine Laboratory
Department of Biotechnology and Medical Engineering
National Institute of Technology
Odisha, India

Deepti
Center for the Environment
Indian Institute of Technology
Assam, India

Erhan Demirbas
Department of Chemistry
Gebze Technical University
Gebze, Turkey

Bandita Dutta
Department of Biotechnology
Maulana Abul Kalam Azad University of Technology
West Bengal, India

Eugénio Campos Ferreira
CEB – Centre of Biological Engineering
University of Minho
Campus de Gualtar, Portugal

Erhan Gengec
Department of Environmental Protection
Kocaeli University
Kocaeli, Turkey

Sougata Ghosh
Department of Microbiology
School of Science
RK University
Rajkot, India

Risha Hazarika
Department of Biosciences and Bioengineering
Indian Institute of Technology
Assam, India

Nagireddi Jagadeesh
Department of Civil Engineering
National Institute of Technology
Andhra Pradesh, India

Alireza Khataee
Department of Environmental Engineering
Gebze Technical University
Gebze, Turkey
and
Department of Applied Chemistry
University of Tabriz
Tabriz, Iran

Sarita Khaturia
Mody University of Science & Technology
Rajasthan, India

Mehmet Kobya
Department of Environmental Engineering
Gebze Technical University
Gebze, Turkey

Dibyajit Lahiri
Department of Biotechnology
University of Engineering & Management
West Bengal, India

Jaya Lakkakula
Amity Institute of Biotechnology
Amity University
Mumbai, India

B. Yamini Lakshmi
Department of Civil Engineering
National Institute of Technology
Andhra Pradesh, India

Jyoti S. Mahale
DBT-ICT Centre for Energy Biosciences
Institute of Chemical Technology
Matunga, Mumbai

Njabulo S. Mdluli
Department of Chemistry
College of Science, Engineering and Technology (CSET)
University of South Africa
Johannesburg, South Africa

Daniela Mesquita
CEB – Centre of Biological Engineering
University of Minho
Campus de Gualtar, Portugal

Nomvano Mketo
Department of Chemistry
College of Science, Engineering and Technology (CSET)
University of South Africa
Johannesburg, South Africa

Sharma Mona
Department of Environmental Science & Engineering
Jambheshwar University of Science & Technology
Hisar, India

Piyal Mondal
Chemical Engineering Department
Indian Institute of Technology
Assam, India

Uma Sankar Mondal
Structural Biology and Nanomedicine Laboratory
Department of Biotechnology and Medical Engineering
National Institute of Technology
Odisha, India

S. Muthukumar
Department of Bioengineering
National Institute of Technology
Andhra Pradesh, India
and
BIT Mesra
Ranchi, India

Moupriya Nag
Department of Biotechnology
University of Engineering & Management
West Bengal, India

Philiswa N. Nomngongo
Department of Chemical Sciences
University of Johannesburg
Johannesburg, South Africa

Preeti H. Pandey
DBT-ICT Centre for Energy Biosciences
Institute of Chemical Technology
Mumbai, India

Sanjukta Patra
Department of Biosciences and Bioengineering
Indian Institute of Technology
Assam, India

Subhankar Paul
Structural Biology and Nanomedicine Laboratory
Department of Biotechnology and Medical Engineering
National Institute of Technology
Odisha, India

Hitesh S. Pawar
DBT-ICT Centre for Energy Biosciences
Institute of Chemical Technology
Matunga, Mumbai

Mihir Kumar Purkait
Chemical Engineering Department
Indian Institute of Technology
Assam, India

Cristina Quintelas
CEB – Centre of Biological Engineering
University of Minho
Campus de Gualtar, Portugal

Rina Rani Ray
Department of Biotechnology
Maulana Abul Kalam Azad University of Technology
West Bengal, India

Patra Sanjukta
Department of Biosciences and Bioengineering
Indian Institute of Technology
Assam, India

Ramya Sankar
Department of Chemical Engineering
National Institute of Technology
Calicut, India

Satyam
Department of Biosciences and Bioengineering
Indian Institute of Technology
Assam, India

Pankaj Shinde
DBT-ICT Centre for Energy Biosciences
Institute of Chemical Technology
Mumbai, India

Har Lal Singh
Mody University of Science & Technology
Rajasthan, India

Rimmy Singh
Department of Environmental Science
Maharshi Dayanand University
Rohtak, India

V. Sivasubramanian
Department of Chemical Engineering
National Institute of Technology
Calicut, India

Baranidharan Sundaram
Department of Civil Engineering
National Institute of Technology
Andhra Pradesh, India
and
BIT Mesra
Ranchi, India

Dalui Sushovan
Department of Biosciences and Bioengineering
Indian Institute of Technology
Assam, India

Gangar Tarun
Department of Biosciences and Bioengineering
Indian Institute of Technology
Assam, India

Tejas M. Ukarde
DBT-ICT Centre for Energy Biosciences
Institute of Chemical Technology
Matunga, Mumbai

Ayush Vasishta
DBT-ICT Centre for Energy Biosciences
Institute of Chemical Technology
Matunga, Mumbai

Nilesh S. Wagh
Amity Institute of Biotechnology
Amity University
Mumbai, India

1 Emerging Nanotechnologies for Detection and Removal of Metal Ions from Aquatic Environment

Sohel Das, Uma Sankar Mondal, and Subhankar Paul
National Institute of Technology

CONTENTS

1.1 INTRODUCTION

Heavy metal contamination and toxicity by dyes in water is a global problem caused by numerous human activities, which include the exponential development of industries, an increase in urban waste, inappropriate agricultural activities, etc. High concentrations of heavy metals such as cadmium, arsenic, lead, chromium, mercury, etc., in water are another form of toxicity in the aquatic environment and biological systems that are affecting the ecosystem and humans tremendously

DOI: 10.1201/9781003164982-1

through the food chain. Due to high toxicity and detrimental health effects, heavy metal ions such as lead, cadmium, chromium, and mercury, which even at very low levels could cause cancer, diseases causing heart condition, liver diseases, kidney diseases, and reproductive disorders, could also adversely affect the central nervous system, and for children, it can cause more serious problems. Atomic absorption spectroscopy (AAS), X-ray photoelectron spectroscopy (XPS), and inductively coupled plasma atomic emission spectrometry (ICP-AES) are a few techniques that have been largely used for the detection of heavy metals in the environment. However, their typical μM detection limit, lengthy functioning time, and inability to use in online mode render them unsuitable for fast detection and removal of heavy metal ions. Apart from these techniques, a colorimetric approach using different nanoparticles has also gained interest, but the need for expensive organic reagents and urbane systems is disadvantageous. Though being most reported for heavy metals detection, different voltammetric, potentiometric, and electrochemical techniques suffer from ion selectivity, long-term steadiness, compatibility with different aqueous environments (even in the biological environment), and sensitive on-site detection (in-vitro and in-vivo).

Nanoparticle and quantum dot-based nanotechnology are creating a fuss for easy detection and removal of heavy metal ions from wastewater. Simple synthesis approaches and size-dependent optical-electronic properties make them a very good candidate for metal sensor production and removal. Indeed, there has been significant growth in the last decade in the development of novel methods based on nanoparticles (NPs), specifically quantum dots, for their efficient optical properties for sensing metal ions. Due to simplicity, very low detection limits, and sensitivity even in biological environments, quantum dots have become a great optical sensor for metal detection in environmental sites or even in living cells and organisms.

1.2 SOURCES OF WATER CONTAMINATION

Water, which is also known as a "universal solvent," can dissolve more substances than any other liquid present on earth, making it undeniably exposed to contamination. Toxic matters from different agricultural sources, industrial sources, and households are readily disposed and easily mixed with them, causing water pollution. When pollution starts from a single source, it is then called Point source pollution. Examples of this kind of contamination are different manufacturing units, oil factories, or different wastewater treatment facilities.

Furthermore, nonpoint source pollution is when water gets contaminated by diverse sources. These include agricultural or stormwater runoff or debris blown into waterways from land. Due to the difficulties toward regulation, nonpoint source pollution is the leading cause of water pollution as there's no single detectable source. These toxic chemicals, nutrients, and heavy metals are carried from different industrial sectors and cities through torrents and rivers into the bays and the sea, which incidentally get into the food chain, thus affecting the environment and human health.

1.3 CONVENTIONAL TREATMENT FOR THE REMOVAL OF METALS FROM WASTEWATER

There are several conventional methods, such as adsorption, chemical precipitation; membrane filtration, ion exchange, and electrolysis reverse osmosis, to remove heavy metal ions from the aquatic environment. Such methods are mostly expensive and insignificant for the detection and removal of low concentrations of heavy metal ions present in a high volume of solution. Adsorption is one of the most common methods which is considered as an effective, reliable, and economic method for the removal of metal ions from water among all the above-mentioned methodologies. Clay crystals, zeolites, and activated carbon are among various materials which are used as conventional adsorbents for heavy metals in water purification system. Natural zeolites are becoming widely popular in different environmental applications because of their properties and significant availability in nature. Natural zeolites (hydrated aluminosilicate minerals) are porous minerals that have properties such as molecular sieving, adsorption, cation exchange, and catalysis, hence making them a great tool for removing contaminants from water. Over the past years, ion exchange property of natural zeolites has been used to remove heavy metal ions from water. Recently, activated carbon is also being widely used as an adsorbent for different toxic inorganic and organic materials from water. The presence of micropores and mesopores in large volume makes activated carbon to provide a large surface area for the adsorption and removal of heavy metal ions. Activated carbon can be obtained easily from coal, wood, and a number of other agricultural wastes. Activated carbon can be used as a good adsorbent for the adsorption and removal of metal ions such as Pb(II), Fe(II), Cu(II), Zn(II), Cd(II), and Ni(II) from wastewater. Modified Activated carbon modified by immobilizing tetrabutyl ammonium iodide (TBAI) and sodium diethyl dithiocarbamate (SDDC) at their surface was investigated for the adsorption of metal ions, such as copper, zinc, chromium, and cyanide from wastewater body [1].

1.4 EMERGING NANOTECHNOLOGY-BASED WATER TREATMENT AND ITS ADVANTAGES

Nanotechnology is the study and production of materials with a size of 1–100 nm with various ranges of applications. Due to the nano-sized particles, nanotechnology provides insight on the fundamental atomic or supramolecular properties and characteristics of the material. At the nanometric level, the property of the material highly differs from the property when it is present as bulk. The exclusive size and shape-dependent property of a material in nano scale provide an exciting viewpoint to scientists to develop new materials having potential use in different fields of science and technology. The challenge comes in the synthesis and modification of nanoparticles according to different applications. Top-down and bottom-up approaches are the main two synthesis procedures for nanoparticle production. For controlled size, shape, and electronic properties, bottom-up approaches are mainly followed. Quantum dot is when the nanoparticle is of size less than 10 nm.

Since their encounter in the early 1980s, quantum dots (QDs) have attracted increasing attention. From the literature, most QDs' core is a product of an inorganic material, usually from groups II–VI, III–V, or IV–VI elements. Then the surface of the QDs is covered with organic surfactant molecules (ligands), which provide functionalization of the quantum dots for specific binding. QDs are semiconductor nanocrystals that have many unique properties such as size-tunable fluorescence, narrow emission peak, and wide range of excitation. The quantum confinement effect [2] occurs when the physical dimensions of the QDs are reduced to a size less than the material's exciton Bohr radius or the physical size is brought down to less than 10 nm, the energy levels get discretized, thus the bandgap of the material increases along with an inclined Coulomb interaction between the charge carriers. Due to this, an optical change in its properties is observed, with an increase in the band gap energy with the decreasing size. Thus, the corresponding photophysical and chemical properties changes due to the significant modification of the intraband and interband relaxation pathways. Thus, small interactions between core or surface ligands and the surrounding molecules can highly affect the QD's photoluminescence (PL) properties, which makes them very sensitive and selective for the application in chemical sensing. There are a few properties which enhance the potential of QDs as chemical sensors, which are versatility due to size and shape-controlled fluorescence property, elevated quantum yields (QY), high resistive nature against photobleaching, and large Stokes shifts with narrow emission bands. For chemical recognition, the coating layer of QDs can be very useful as an interface. Semiconductor QDs show photoluminescence after being excited by a specific wavelength and when recombination of that exciton occurs, that is, band gap recombination (the carrier recombines before they are trapped) or near band gap recombination (when they are trapped in between recombination region). Having this property if in any case this core electron–hole recombination can be interrupted, then it is expected that the changes affect the luminescence efficiency as well. Thus, the charges on the QD surface and the ligand groups present on the QD surface play an important role for the luminescence change of the QD, thus it can be used as the binding or sensing site of the QD. Therefore, the fluorescence of QDs gets affected due to physical adsorption or chelation of analyte ions on the surface of the QD, which indirectly signifies sensing or recognition of targeted molecule, and further process can be followed to remove metal ions. Many QDs have recently been used and are being commercialized as chemical sensors. CdSe, CdTe, ZnSe, ZnS, CdS, ZnO, carbon, graphene, and carbon nitride are the few QDs that have gained a very good interest in the field of heavy metal sensing. Though among most QDs (II–VI) group QDs are having larger stokes shifts and high QY, and carbon and its derivative dots are gaining interest for their better optical and electronic properties.

There are many nanomaterials which already found great potential in different applications such as electronics, catalysis, and drug delivery, sensing and thus reaching the industrial scale for production. Hence, their unique optoelectronic and physicochemical properties have developed the interest to discover new materials and their high-end synthesis methods to produce environment-friendly and efficient nanoparticles.

The above said diversity in different physical and chemical properties has made the production of nanoparticle and QDs attractive in the field of nanomaterial and nanotechnology. As mentioned previously nanomaterials possess diversity in surface modification and large surface area, hence providing dynamic surface sites for adsorption. Therefore, nanoparticles-based nanotechnology solutions prove to be a promising approach for the removal of metal ions from water.

1.5 FLUORESCENCE-BASED STRATEGIES FOR SENSING OF METAL IONS

1.5.1 Turn-Off Strategy

Previously we have defined how nanoparticles and quantum dots are optically sensitive, now as we are concerned about quantum dots as metal sensors so here, we all elaborate strategy where QDs fluorescence change with the presence of targeting molecule or metal ions. To design a QD based metal sensor the QD used and the detected metal ion is of great concern. If after the addition of analyte ions, the fluorescence of the reduces that its control value then it is said to be quenching the fluorescence, thus the strategy of sensing the analyte becomes a "turn off" strategy.

The quenching in the fluorescence can happen through charge transfer (CT) or energy transfer. Further energy transfer can be divided into two: frequency resonance energy transfers (FRET) and Dexter energy transfer. For CT the turn-off strategy is defined with respect to QD's fluorescence, if the fluorescence of the QD reduces in the presence of the analyte then the sensor will be called a turn-off sensor.

For FRET, there is a donor (QD) and an acceptor, the fluorescence energy from donor excites the acceptor, thus FRET-based sensors are called as turn-off sensor when the whole system, donor-acceptor together is showing a quenching in the fluorescence. As acceptor's fluorescence is dependent on QD, so if QD's fluorescence quenches, the fluorescence from the acceptor will quench as well.

There are many ways that can induce fluorescence quenching in QD, such as (a) through a PET (photoinduced electron transfer) process, when the analyte binds onto the ligands or QD surface, facilitating the non-radiative electron transfer, or (b) through the displacement of the surface ligand from QD surface, due to its strong affinity toward analyte ions which makes QD surface imperfect and induces agglomeration, hence leading to the ineffective charge transfer process.

1.5.2 Turn-On Strategy

As mentioned earlier Turn-on sensor is based on the fluorescence enhancement of QD only after the sensing of targeted analyte, even if there is occurrence of FRET, then also if the fluorescence from the QD or rather say emission at the wavelength from the QD is enhanced the sensor is taken as a turn-on sensor. Here based on the strategies depending on the charge transfer mechanism or energy transfer mechanism four categories have been defined (a) when the binding of analyte on the QD surface enhances QD's fluorescence, (b) first turning off the QD by chemical etching, (c) host-guest-based turn-on sensing, and (d) when the active FRET gets broken due to the presence of analyte.

1.6 DIFFERENT METAL SENSING MECHANISM

1.6.1 Charge Transfer

Optical sensing ability of semiconductor NPs/QDs originates from the excitation of electrons from valence band to the conduction band through photo-excitation, hence creating holes in the valence band. The electron–hole pair created after the excitation of NP is called an exciton. These excited electrons, lose energy in the form of vibration and radiating photons (fluorescence) to come back to the valence band and recombine with photogenerated holes. Sometimes the presence of different trap sites on QD core surface makes the recombination non-radiative in nature. Therefore, these two processes, radiative recombination and non-radiative recombination happen simultaneously and to get an efficient quantum yield from NPs radiative recombination rate must higher than the non-radiative recombination. Sometimes in the presence of other specific molecules excited electrons from NP's conduction band gets transferred and recombined to that of the other molecule's electronic states, which suppresses the fluorescence from NPs. Hence, electron–hole pair separation in the NP electronic states induces an effective quenching of the luminescence of the NP which is also named as the photoinduced electron transfer (PET) mechanism. This mechanism provides a scheme to either switch on the fluorescence of the NP or switch it off by quenching (Figure 1.1a).

The interaction between the QD surfaces/capping groups and targeted metal ions is the key to interrupting the normal radiative recombination process, which is the main idea for designing sensors based on this mechanism. Displacement of the surface ligand from the NP surface is another strategic way of affecting the NPs fluorescence. The strong affinity between the NP surface ligand and the targeted metal ions causes the imperfection on the NP surface, which induces agglomeration of the NP. This leads to quenching of the fluorescence of the NP.

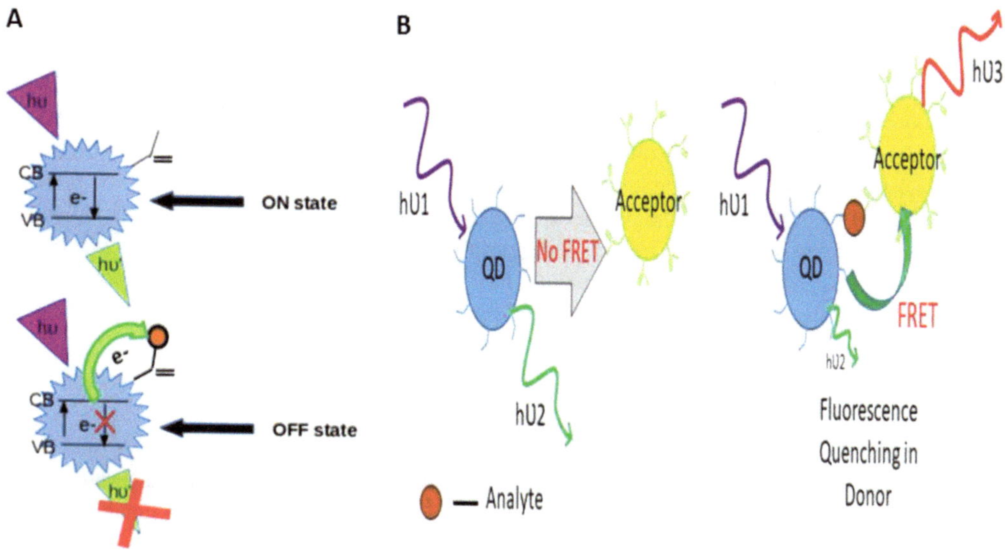

FIGURE 1.1 (a) Charge transfer mechanism, (b) energy transfer mechanism.

1.6.2 Energy Transfer

Another mechanistic way based on fluorescent metal sensor is energy transfer (ET)-based sensing system. Resonance energy transfer (RET) depends on two different species of material; one is donor (mostly NP/QD) which must transfer its energy through photoluminescence to the other species, i.e. acceptor (organic fluorophore), then the acceptor will show fluorescence.

According to Forster's theory, this transfer of energy is possible when there is an occurrence of resonance between the emission spectra of the donor collides with the excitation wavelength of the acceptor. There are other factors as well that induces energy transfer besides the spectral overlap between the emission of the donor and the absorption of the acceptor. The molecular closeness between the donor-acceptor, donor's quantum yield, and the relative orientation of the transition dipoles of donor and acceptor [3] highly affect the energy transfer between donor and acceptor. Therefore, high quantum yield of the donor is very expected for a better energy transfer.

Hence FRET occurs due to long-range dipole–dipole interactions between a donor with a high quantum yield (excited state) and an acceptor (ground state), in their proximity (Figure 1.1b).

Recent studies suggested that metal having a d10 electronic configuration shows a strong affinity toward other metal ions configured in similar electronically closed-shell nature [4]. Metallophilic interaction associated with strong d10–d10 interaction of metals is an interesting phenomenon which is also known as Dexter energy transfer (DET). This phenomenon is used to sense metal ions through d10 interactions and facilitating the DET process.

In Ref. [5] authors used (BSA)-capped HgS QDs for the detection of Hg^{2+} ions. The metallophilic interaction between Hg^{2+} present in HgS QDs and Hg^{2+} ions (i.e., 5d10—5d10) had led to the luminescence quenching of HgS QD, which demonstrated the sensing of Hg^{2+} ions.

Inner filter effect (IFE) is another mechanistic way of making the NP/QD system a sensory system. IFE initially was considered as an error in the fluorometric system due to the high concentration of fluorophore. Researchers have tried to minimize the inaccuracy by IFE by restoring the linear relationship between NP's fluorescence and its concentration. There are two types of IFE, when different material's (chromophores) absorbance wavelength range falls in the range of excitation wavelength of the NP or QD then the excitation of the NP never occurs, thus change in the fluorescence is observed, which is called as Primary IFE. In secondary IFE chromophore's absorbance wavelength range collides with the emission wavelength of the NP, hence luminescence quenching is observed. Unlike FRET, IFE-based sensory applications are more flexible, as it does not directly propose the need for a pair of an acceptor and a donor (NP).

For a sensor designed based on the IFE mechanism, the absorber must not reduce the emission of the donor or NP through any of the ways mentioned above. Moreover, the emission must be affected in the presence of the analyte by the absorber. In the presence of the analyte or metal ions, the absorber must possess an adequate overlap area with the excitation and/or emission spectrum of donor, here NPs/QDs [6]. In the presence of analyte, the absorber could exist in a complex form which must have an absorbance range overlapping with the NPs excitation/emission wavelength. The complex formed after addition of analyte must unveil a sensitive effect on the fluorescence to the deviation of the metal ion concentration. Table 1.1 provided a comparison of different works based on different nanoparticles with different sensing mechanisms.

1.7 ADSORPTION-BASED METAL REMOVAL TECHNIQUE

The basic idea of adsorption is when different molecules or atoms get attached to a particular surface of another material, also known as adsorbent. The molecules of the material which get adsorbed on the surface of the adsorbent are known as adsorbate. Adsorption is a phenomenon connected to surface which is a consequence of surface energy. The surface tension between adsorbent and adsorbate causes particles to adhere onto the adsorbent's surface. Normally in

TABLE 1.1
Different Work on Metal Detections Based on Turn-Off Strategy

Paper Ref.	QD	Mechanism	Ions	Range	LOD
[7]	Carbon	CT	Cu^{2+}	5–100 nM	1.53 nM
[8]	CdTe	CT	Cu^{2+}	0.01–20 μM	7.6 nM
[9]	CdSe/ZnS	CT	Cu^{2+}	0–10 μM	0.9 μM
[10]	CdTe	CT	Cu^{2+}	0.02–1.0 μM	8 nM
[11]	CdTe	CT	Cu^{2+}	$5.16 \pm 0.07 \times 10^{-8}$ to $1.50 \pm 0.03 \times 10^{-5}$ mol/L	$1.55 \pm 0.05 \times 10^{-8}$ mol/L
[12]	$CdTe/Fe_3O_4$	CT	Cu^{2+}, Ag^{+}	10^{-7} to 10^{-3} M	4×10^{-3} M
[13]	CdTe/CdS	CT	Cu^{2+}	2.4×10^{-2} and 28 μg/mL	1.3×10^{-3} μg/mL
[14]	CdSe	CT	Cu^{2+}, Zn^{2+}	4–160 μM	4 μM
[15]	CdTe	ET	Ag^{+}	0.01–8.96 μM	4.2 nM
[16]	Mn-doped CdS/ZnS	ET	Hg^{2+}	1–10 nM	0.49 nM
[17]	CdTe	ET	Hg^{2+}	0–100 μM	260 nM
[18]	Graphene	ET	Ag^{+}	0–115.2 μM	250 nM

bulk, adsorbent attracts adsorbate molecules through its open ionic, metallic or covalent bonds, as they are not fully acquired. The physical and chemical properties of the adsorbent and adsorbate highly affect the exact nature of the attachment between each other. Adsorption mainly follows two approaches, (a) physisorption, which involves weak Van der Waals forces and (b) chemisorption involving different covalent bonding. Electrostatic attraction forces can also sometimes cause adsorption.

Adsorption of adsorbate on the surface of the adsorbent causes change in the size of the adsorbent, leading to agglomeration of the adsorbent molecules. The size change of adsorbent can cause its adsorption intensity and peak to shift or reduce. The surface attachment of adsorbate on adsorbent causes reduction in absorbance intensity of adsorbent, which is used as the presentation of the sensing of materials or metal ions. As the adsorbed molecules can be taken out with the adsorbent molecule, thus it is a way for removal as well of the adsorbate or the toxic metal ions. Therefore, adsorption is a good way for sensing and removal of toxic metal ions from aquatic environment. Recently this principle is used in many sensing and purification systems to remove heavy metal ions dissolved in it.

In Ref. [19], the authors developed a sensor for the detection of Cu^{2+} metal ions using the principle of adsorption, which further can remove Cu^{2+} metal ions as well. The nano-conjugation of reduced graphene oxide and Ag NPs synthesized from pine leaf extracts was used as the adsorbent. The sensing of the Cu^{2+} ions was represented through the color change of the adsorbent solution. The color change was observed due to the agglomeration of rGO@AgNP in the presence of copper ions.

1.8 CARBON AND ITS DERIVATIVE NANOPARTICLES FOR SENSING AND REMOVAL OF METALS

Recent investigations are widely oriented around QDs of carbon and its variability as its presence in nature is enormous. Different electronic, electrical, mechanical, and optical properties are what making a great and interesting area of research. These nanoparticles have good conductivity, high chemical stability, high quantum yield, large bandwidth and a large absorption spectrum. The recent advancements on carbon, graphene, fullerene, and carbon nanotubes are the opening doors toward new discoveries in science and nanotechnology.

1.8.1 Carbon and Graphene Quantum Dots

Carbonaceous nanoparticles have drawn much attention in the field of environment and pollution management, as it offers suitable surface and optical properties with low cost and easy availability. Among all carbon-based nanoparticles, graphene quantum dots (GQDs) have attracted attention in the field of nanotechnology due to their versatile properties and divergent applications.

As it is mentioned earlier, fluorescence spectroscopy has become a powerful tool for the selective detection of metal ions. There have been many mechanisms discovered such as chromogenic and fluorogenic reagents-based sensing, FRET and surface plasmon resonance (SPR) mechanism-based sensing. In regard to fluorescence-based sensors, carbon-based fluorescent nanoparticles like GQDs and carbon QDs is considered as an excellent candidate for high luminescent metal sensors. Due to several electronic and chemical properties such as water solubility, chemical inertness, high biocompatibility, high quantum yield, lower level of photobleaching and cytotoxicity, carbon QDs have gained a great interest in the field of optical sensors.

Compared to carbon dots, GQDs show a better crystalline structure and also demonstrated unique electrical, mechanical and optical properties. GQDs are also very stable, show less toxicity and are highly soluble in multiple solvents just like CQDs. Graphene sheet is single layer of graphite, which is then converted to particle with layers on one another having size of less than 10 nm, which is then called GQDs. It is a variable of carbon, it is normally synthesized from graphene or graphene oxide. GQDs exhibit similar physical and chemical properties like graphene. The difference between graphene and graphene QDs is that in graphene QD the confined sp^2-carbon conjugation and the edge effect open their band gap, which leads to bright and photostable tuneable emissions. Biocompatibility is also an important property for its biomedical uses. Compared to CdTe, CdSe, CdS, ZnSe, and ZnS QDs, the other well-investigated semiconductor quantum dots, CQDs and GQDs exhibit higher quantum yield, superior photostability against blinking and photobleaching, lower toxicity, lower cost of production, less pollution, availability for large-scale production and greater biocompatibility.

There has been a great amount of research in detecting metal ions such as Al^{3+}, K^+, Be^{2+}, Au^{3+}, Co^{2+}, Ni^{2+}, Sn^{2+}, Cr^{6+}, Pb^{2+}, Mn^{2+}, Bi^{3+}, Fe^{3+}, Zn^{2+}, Cd^{2+}, Cu^{2+}, Au^{3+}, Co^{2+}, Ni^{2+}, Ag^+, and Hg^{2+} [20], using CQDs or GQDs in the last decade. Many of these developments follow the same basic idea, i.e. quenching of the photoluminescence in the presence of the analyte.

In many cases innovation comes from the simple synthesis methodologies of CQDs and GQDs, green synthesis is also gaining interest due to its simple low-cost synthesis procedures.

An example is, Ref. [21] has found that Hg^{2+} changed the surface states of nitrogen-doped CQDs (N-CQDs), thus, quenching the fluorescence of N-doped CQDs. The data through the investigation showed that the sensor was greatly selective toward Hg^{2+}, even in the presence of other metal ions in high concentrations.

1.8.2 Carbon Nanotube (SWCNT and MWCNT)

Since their discovery, carbon nanotubes (CNTs) have attracted increased attention due to their excellent electrochemical properties, large surface area, high rate of electron transfers and wide potential window. Just like carbon's other variables, CNTs also have a wide range of applications, such as in nanoelectronics, biomedicine, environmental and pollution engineering and electrochemical sensing.

There are mainly two classes of CNTs single-walled carbon nanotubes (SWCNTs) and multi-walled carbon nanotubes (MWCNTs). When single graphene sheet is rolled having a diameter between 0.4 and 2 nm and with a length up to 20 cm, then it is called SWCNTs. Whereas, MWCNTs consist of several concentric SWCNTs fitted inside each other having a diameter up to 100 nm. There have been different methods developed for synthesizing CNTs having certain desired properties. Chemical vapor deposition, laser ablation, and arc discharge are the three main methods for the production of CNTs. Due to its vast application range, new discoveries are also under investigation.

In the case of arc discharge and laser ablation technique for a short time burst (μs–ms) high temperature is desired. Although, high-quality carbon nanotubes with superior smoothness and crystallinity are produced through this above said two techniques, however, the disadvantage is that it requires a very high energy source and a large quantity of solid carbon/graphite as a primary source. Therefore, these two techniques have limited their application for large-scale production of CNTs.

The CVD technique provides a better approach for controlling CNT orientation, its size and shape (diameter and length), alignment and yield compared to the high temperature-based arc discharge and laser ablation techniques. Hence, CVD has become the most dominant technique for the synthesis of CNT. As earlier said, in the case of CVD, using the application of electron beam lithography the rate of CNT formation from metal or non-metal sources and its site-specific density can be controlled.

Hydrocarbon precursors are used to form CNTs. This hydrocarbon is formed as CNTs when it flows through a pattern metal catalyst with a continuous flow of gas. Previously, it was reported that substrate iron silicon with ethylene and nickel-glass substrate with acetylene were used to develop CNTs with controlled length. Van der Waals force helped the neighboring CNTs to interact with each other to form dense, rigid bundles of CNT.

Generally, the formation of CNT follows two steps. In the first step, there is a catalyst surface where a meta-stable carbide is formed, which slowly grows an amorphous carbon rod. Then, in the next step, those carbon rods are slowly grown into hollow tubes at the end of the growth phase through graphitization [22]. Carbon monoxide, benzene, xylene, methane, and ethylene are the most commonly used precursor for growing CNTs. Fe, Co, Ni, and transition metals are the most popular metal catalysts used for the preparation of CNTs. Though Mg, Al, Pt, Mn, Pd, Cu, Sn, Mo, Cr, and Au also have been used as metal catalysts in the CVD process, they are not very popular. In Ref. [23], MWCNTs are used for the removal of Cu^{2+}, Pb^{2+}, Zn^{2+}, Cd^{2+} ions, and showed the highest efficiency was achieved at a pH value lower than 2.0.

1.8.3 Graphene Oxide QDs

Over the last decade, graphene oxide quantum dots (GOQDs) have gained a lot of interest in the field of semiconductor QDs and its nanotechnology applications. It is a strange type of nanoparticle from the carbon family, which is made up of single or few-layer graphene oxide (GO) sheets and has lateral dimensions of less than 50 nm.

GOQDs are so popular because they own very unique properties such as hydrophilicity, good photoluminescence, high photostability, low cytotoxicity, and high biocompatibility. GOQDs have been investigated in different branches of applications such as in biosensors, metal sensors, labels for cellular imaging, Photocatalysis and substrate for organic light emitting diodes.

GOQDs have been used very successfully toward metal sensing application as it shows strange reactivity toward metal ions. In the presence of metal ions, GOQDs show instability and forms surface chelation-based metal nanoparticle complexes. As the edges and the basal plane of GOQDs are decorated with carboxyl, carbonyl and hydroxyl groups, the chelation of metal ions on the surface of GOQDs is easy.

Previous investigations have shown the formation of aggregation of GOQDs in the presence of metal ions. This agglomeration and precipitation occur due to the adsorption of metals on the surface of GOQDs and dragging them down toward the bottom. Thus, GOQDs showed much application in meta sensing and removal from water. Compared to other derivatives of carbon, GOQDs showed greater adsorption capabilities toward metals due to their higher edge-to-surface ratio and the intrinsic presence of larger content of functional groups.

Therefore, in the case of optical property change, the presence of certain metal ions (Hg^{2+}, Fe^{3+}, Cu^{2+}, Pb^{2+}) quenches the fluorescence intensity of the GOQDs. The progressive decrement in fluorescence peak intensity of GOQDs is due to the increased rate of non-radiative recombination of excitons present in GOQDs.

1.9 OTHER NANOPARTICLES FOR SENSING AND REMOVAL OF METALS

1.9.1 Metal Oxide Nanoparticles

Metallic and metal oxide nanomaterials have emerged as promising materials for heavy metal ion removal from water and wastewater in recent years. They each have their own sets of advantages, such as faster kinetics, higher adsorption capacity, low cost, and effective nanomaterials for heavy metal removal. Various research groups have conducted numerous scientific studies to demonstrate the role of metallic and metal oxide-based nanoparticles in wastewater decontamination. Because of their instability due to agglomeration in nature, bare nanoparticles are less commonly used as adsorbents. Furthermore, separating bare nanoparticles from wastewater is a time-consuming and difficult process. Therefore, the capping or functionalization of these nanomaterials is required to improve their stability and simplify the separation process. Magnetic and non-magnetic metal oxide nanomaterials have been classified based on their intrinsic magnetic properties. Magnetic nanoparticles (MNPs), such as maghemite (γ-Fe_2O_3), hematite (α-Fe_2O_3), and magnetite (Fe_3O_4) are excellent adsorbing materials for collecting and removing toxic elements from polluted water. Among non-magnetic nanomaterials, TiO_2, ZnO, MgO, CeO, Al_2O_3, CuO, and other non-magnetic nanomaterials have been investigated to remove heavy metals from contaminated water. There are numerous reports in the literature on the use of these magnetic nanoparticles for the removal of a variety of heavy metals in their ionic forms, including chromium, arsenic, lead, cobalt, copper, and nickel. Magnetic nano adsorbents are cost-effective, highly efficient, and using an external magnetic field, they can be easily separated from reaction media. In general, bare MNPs are prone to oxidation and rapidly agglomerate in the aqueous medium, limiting their application. Thus, maintaining their stability by avoiding agglomeration and precipitation is a critical problem. Different moieties can be used to functionalize bare MNPs to overcome these constraints. MNPs with modified surfaces have higher adsorption capabilities and oxidation stability, as well as increased selectivity for a specific metal ion. In addition to electrostatic and Van der Waals interactions, which are responsible for metal ion adsorption on the surface of the adsorbent, functionalization allows for complex formation, chemical binding, and ligand combination. Surface functionalization of MNPs with hydrophilic coatings, such as Polyethylene glycol (PEG), Polyvinyl alcohol (PVA), Oleic acid, and Polyvinyl pyrrolidine (PVP), prevents agglomeration. It increases the homogeneity of MNPs in the solution and results in a higher surface area to volume ratio.

Using dimercaptosuccinic acid (DMSA) anchored Fe_3O_4 magnetic nanorods (MNRs) synthesized through a simple method using Punica Granatum rind extract, a non-toxic waste material, a novel, and environmentally friendly approach to remove toxic heavy metal Pb(II) was developed. The monolayer adsorption capacity of DMSA@Fe_3O_4 MNRs on Pb(II) adsorption was found to be 46.18 mg/g at 301 K [24].

As an inorganic metal oxide, nanostructured MgO is a cheap, abundant, non-toxic, and environmentally friendly nanomaterial with good adsorption capacity for heavy metal ions, organic dyes, fluorides, and phosphates removal from aqueous solution. Furthermore, MgO nanoparticles are used as an antibacterial agent with antibacterial activity against Gram-positive and Gram-negative bacteria as well as bacterial spores. For simultaneous bacterial disinfection and toxic heavy metals removal from aqueous solution, highly active MgO nanoparticles synthesized through sol-gel and calcination processes were used. In comparison to commercial MgO, the synthesized MgO nanoparticles had high efficiencies for both heavy metal ion removal (Cd^{2+} and Pb^{2+}) and E. coli inactivation [25].

Aluminum oxides regulate the composition of soil water, sediment water, and other natural water systems. Aluminum oxide has a wide range of applications as an adsorbent and catalyst due to its high surface area and mechanical strength. Cadmium metal ion (Cd^{2+}) removal from an aqueous solution was investigated using aluminum oxide (Al_2O_3). Batch adsorption kinetic experiments revealed that the initial solution pH, initial metal ion concentration, and adsorbent doses all had a significant impact on Cd^{2+} ion adsorption. The amount of metal ion (Cd^{2+}) adsorption on aluminum oxide (Al_2O_3) increases with the initial metal ion concentration and solution pH but decreases with adsorbent dosages [26]. Tabesh et al. have investigated nanosized γ-Al_2O_3 particles for the removal efficiency of heavy metal ions (lead and cadmium) in the adsorption process. γ-Al_2O_3 nanoparticles were synthesized by a modified sol-gel method with an average crystallite size of about 6–13 nm the as-synthesized alumina nanoparticles were investigated for heavy metal ion removal from contaminated water and the impact of different experimental parameters like pH, adsorbent dosage, and exposure time was also investigated. The sorption of Pb^{2+} and Cd^{2+} onto γ-Al_2O_3 nanoparticles follows Freundlich isotherm, with maximum adsorption capacities of 47.08 mg/g for Pb^{2+} and 17.22 mg/g for Cd^{2+}, respectively [27].

TiO_2 nanoparticles are high-capacity adsorbents for removing arsenic from groundwater. The use of TiO_2 for arsenic (As) adsorptive removal from wastewater, followed by spent adsorbent regeneration and arsenic recovery using NaOH, was investigated. findings show that As(III) does not form an aqueous complex with other metal ions. In the 21 treatment cycles using regenerated TiO_2, an average of 3890 mg/L As (III) at pH 1.4 in the wastewater was lowered 59 µg/L in the effluent with a final pH of 7 in the effluent. Heavy metals such as Cd, Cu, and Pb were also reduced to less than 0.02 mg/L from concentrations of 369, 24, and 5 mg/L, respectively [28].

Nano ZnO, as an environmentally friendly material, can be used in the catalyst industry, gas sensors, and solar cells. Nanostructured ZnO can effectively remove heavy metals from contaminated water. A modified sol-gel technique was used to fabricate Ga-doped ZnO (GZO) nanoparticles with varying Ga contents. The addition of Ga to ZnO increases its electrical conductivity. Cd(II) and Cr(VI) were removed from the aqueous solution using nanopowders. The results showed that incorporating Ga into zinc oxide increases the capacity of nanopowder adsorption [29]. An amended sol-gel process was used to synthesize Al-doped ZnO (ZnO:Al) nanopowders with a content of 0%–5% of Al. ZnO:Al 1% nanoparticles were used to effectively adsorbed 222, 175, 106, and 67 mg/l of Cr (IV), Cd(II), Ni(II), and Co(II), respectively, from aqueous solution, respectively. ZnO:Al 1% nanoparticles can be effectively regenerated and reused for adsorption-desorption cycles, according to the report [30].

1.9.2 Polymeric Nanomaterials

Polymeric nano adsorbents have recently gained popularity in wastewater treatment. Because of their well-defined porous nature, greater gas permeability, minute interfiber pore size, and large specific surface area, and polymer-based nanoparticles have also been used as the core substance of the adsorbents, resulting in improved adsorption performance. These have high skeletal strength, adjustable surface functional moieties, and are easily degradable, making them an excellent candidate for adsorbents.

Chitosan is a non-toxic, environmentally safe, and biocompatible hydrophilic natural biopolymer that can form complexes with a variety of metal ions. The presence of amino groups promotes chelation with metal ions, and chemical modifications of chitosan increase its selectivity and sorption ability. Adsorption of Hg (II) ions from water by chitosan alginate NPs (CANPs) was investigated by Dubey et al. Linking chitosan to various substances improves its sorption ability, resulting in chitosan stabilization in acidic media. Both the tripolyphosphate and calcium ions interact electrostatically with the positive and negative ions of chitosan and alginate, respectively. This study found that CANPs have a remarkable high maximum adsorption potential for Hg (II) ions, with a value of 217.39 mg/g at 30°C [31]. Using a conventional ionic gelation method with sodium tripolyphosphate as a cross-linker, chitosan nanoparticles were synthesized from the extracted chitosan of shrimp shell waste. The chitosan nanoparticles were evaluated as adsorbents of Fe (II) and Mn (II) ions from aqueous solutions using batch equilibrium experiments. The removal efficiency and maximum adsorption capacity for Fe (II) were found to be 99.8%, 116.2 mg/g, and Mn (II) were found to be 95.3%, 74.1 mg/g, respectively [32].

Polymers are being considered as adsorbents of interest, either as a framework into which inorganic nanosized materials can be introduced to improve their chemical, mechanical, thermal, and sorption properties, or as a bed or prototype to synthesize and develop nanoparticles. Surfactant-clay (organo-clay) and chitosan biopolymer were used to fabricate chitosan/nanoclay composites. The batch adsorption process was used to investigate the adsorption of copper (II) and nickel (II) ions from an aqueous solution onto the prepared chitosan/nanoclay composites. For Cu and Ni, the best adsorption capacities were 176 and 144 mg/g, respectively [33]. Poly(ethylene-co-vinyl alcohol) (EVOH) nanofibers with an average diameter of 260 nm were prepared for the removal of Cr(VI) metal ions from an aqueous solution. According to the adsorption kinetics report, the adsorption equilibrium time was less than 100 minutes, and the maximum adsorption capacities were 90.74 mg/g [34].

1.9.3 Nanocomposites

Nanocomposites of various metals, metal oxides, and polymers have also been investigated as nano adsorbents to enhance adsorption and separation capability. A simple precipitation-calcination process was used to synthesize magnetic Fe@MgO nanocomposites, which were used to remove heavy metal ions and dyes from water. The Fe@MgO core-shell nanocomposites, as synthesized, had a wide surface area, a mesoporous structure, and a high magnetic saturation value, which made magnetic separation easier. Fe@MgO nanocomposites had superior adsorption properties, with maximum adsorption capacities of 1476.4 mg/g for Pb(II) and 6947.9 mg/g for methyl orange (MO) respectively [35]. Huai Li et al. have investigated magnetic graphene oxide nanocomposites (Fe_3O_4/SiO_2–GO) with core/shell nanoparticles prepared by a covalent bonding technique. Synthesized Fe_3O_4/SiO_2–GO nanocomposites were evaluated for the removal of chromium from wastewater and showed very fast removal of Cr(III) from the wastewater. Maximum adsorption occurred within 5 minutes and using a permanent magnet, the adsorbent can be separated faster. The adsorption kinetics followed the pseudo second-order model, with better efficiency of Fe_3O_4/SiO_2–GO nanocomposite on the removal of Cr(III) in high pH solution (>3) [36]. An electrospinning method was used to prepare iron oxide aluminum oxide mixed nanocomposite fiber and its output was assessed as heavy metal ion adsorbent. In the analysis of sorption activity of Cu^{2+}, Pb^{2+}, Ni^{2+}, and Hg^{2+} ions with respect to initial levels, contact time, and pH, the batch adsorption experiments were performed. The removal rate was in the order of $Cu^{2+} < Pb^{2+} < Ni^{2+} < Hg^{2+}$ order. Nanocomposite fibers showed maximum removal of 90% of Hg^{2+} ions. Applying Langmuir equations, the maximum sorption potential for Cu^{2+} was found at 4.98 mg/g, Ni^{2+} at 32.36 mg/g, Pb^{2+} at 23.75 mg/g, and Hg^{2+} at 63.69 mg/g. Regeneration studies of mixed nanocomposite adsorbents have also shown that Fe_2O_3–Al_2O_3 mixed nanocomposites have a desorption efficiency of approximately 90% [37].

1.10 MEMBRANE FILTRATION

Because of its simplicity, energy consumption, and manufacturing scalability, membrane filtration technology is gaining popularity and being favored over other technologies. Membranes rely on either pressure-driven or electrical technology to operate. The use of pressure-driven membrane technology to purify water to any desirable level is ideal. Membranes isolate compounds based on the size of the pore and the molecule. It is a reliable, consistent, and automatic wastewater treatment process. The intrinsic trade-off between membrane selectivity and permeability is the problem of membrane technology. Because of the pressure-driven method, this methodology needs a lot of energy. Membrane fouling complicates the process and limits the lifetime of the membranes. The type of membrane content determines the efficiency of the membrane system. The incorporation of functional nanomaterials into membranes improves membrane permeability, fouling resistance, mechanical stability, and thermal stability [38–41].

Nanofiltration (NF) is especially important because the majority of heavy metals are multivalent cations, and its collaborative separation mechanisms involving steric effect and electrostatic repulsion are particularly important. Membrane scaling, on the other hand, remains a significant problem during wastewater treatment. Membrane scaling occurs when highly concentrated minerals like calcium, sulfate, and carbonate reach their solubility limits and accumulate on membrane surfaces as a result of reversible and irreversible interactions. It usually results in increased flow resistance and shorter membrane replacement cycles [42].

1.11 REGENERATION OF NANOMATERIALS

One of the most important aspects of nanoparticle regeneration in water purification is that it controls the economics of water treatment technology. pH-dependent solvents are important for nanoparticle regeneration. They can also be accomplished in the treatment system by using a separation device or immobilizing nanomaterials. Another method for nanomaterials is immobilization. However, existing immobilization methods have not proved to be very reliable. To immobilize nanoparticles without affecting their performance, simple and low-cost methods must be developed. Magnetic separation is an additional method for separating magnetic nanoparticles. The subdue adsorption of heavy metals on Fe_3O_4@SiO_2–NH_2 at lower pH suggested that acid treatment is a promising approach for adsorbent regeneration. The stability of nano adsorbent Fe_3O_4@SiO_2–NH_2 was evaluated by analyzing leached Fe ion before adsorbent regeneration in acidic conditions [43].

Nanoparticles can be regenerated and reused for water treatment, making them a cost-effective material. Thus, the ability of nanoparticles to regenerate themselves can be considered an added benefit for their popularity in wastewater treatment.

1.12 TOXICITY AND ENVIRONMENTAL IMPACT OF NANOMATERIALS

Toxicity is one of the most crucial issues of nanotechnology. Nanotechnology's main concerns are the hazards of nanoparticles and the exposure to harm. These nanomaterials have unique physical, biological, mechanical, and optical properties that make them useful in a variety of applications but also make them harmful to humans and the environment. Humans and the environment can be affected by the biological and chemical effects of reactive nanomaterials. Nanoparticles can allow access to the human body through absorption, ingestion, transdermal delivery, and inhalation, and translocate into different body parts such as the brain, heart, lung, liver, kidneys, bone marrow, and nervous system through the bloodstream. The biological effects of nanoparticles are caused by both physical and chemical modes of action. The formation of reactive oxygen species (ROS), the breakdown and release of toxic ions, the disturbance of cell membrane transport activity, oxidative damage caused by catalysis, and lipid peroxidation are all established chemical processes. The production of ROS induces oxidative stress and inflammation, which causes damage to proteins, membranes, and DNA, ultimately leading to cell death. Because of the size, shape, and surface properties of nanoparticles, physical processes that induce biological effects are observed. Transport pathways, protein conformation-folding, and protein aggregation-fibrillation are all influenced as a result of membrane damage and membrane function disturbance. Toxicity can be determined by the nanomaterials synthesis processes. Non-biodegradable nanoparticles demonstrate their toxicity in the environment, and these nanoparticles eventually agglomerate, accumulate, aggregate, and depose in the environment. Through altering environmental physicochemical parameters such as pH, salinity, and ionic pressure, catalytic nanoparticles interfere with other macromolecules or contaminants. These nanoparticles exert their toxic impact and disrupt the ideal environmental situation. Improper waste management in the industry, such as direct dumping of used nano adsorbents, leakage, and other methods, greatly contribute to pollution, especially in groundwater and soil where nanomaterial can be transferred from one medium to another.

1.13 LIMITATIONS AND FUTURE PROSPECTS

Nanotechnology-based water and wastewater treatment systems have shown considerable promise in laboratory experiments. Any of these innovations are commercialized, while others need extensive research before being scaled up. Their commercialization is challenging; we must resolve numerous technological obstacles, as well as making them cost-effective and sustainable. Until nanotechnology can be used to treat natural and wastewaters on a large level, further research is needed. To verify nanomaterial-enabled sensing, studies should be performed under practical conditions to determine the effectiveness of existing nanotechnology. Only by showing long-term performance in the treatment of water and wastewater, nanotechnology-based innovations can be commercialized. Adoption of emerging technology often strongly depends on cost-effectiveness and possible threats. With a few exceptions, such as nanostructured TiO_2, iron oxide, and polymeric nanoparticles, nanomaterials are currently very expensive. The cost-effectiveness of these nanomaterials can be accomplished by their regeneration and reuse.

1.14 CONCLUSION

Nanotechnology has made significant advances in recent years, opening the way for the development of safe, cost-effective, and environmentally sustainable approaches to environmental remediation. The physicochemical properties of several nanomaterials make them a suitable candidate for treating wastewater. Also at low concentrations, the nanoparticles can eliminate heavy metal ions, with the added advantage of high selectivity and adsorption power. Adsorption is a highly effective and straightforward physicochemical technique for removing heavy metals from wastewater. These methods are more cost-effective, require less time and energy, and produce significantly less waste than typical bulk materials-based methods. According to the literature, nanoparticles synthesized with modification or surface functionalization can be a successful solution for the removal of multiple hazardous materials from wastewater and may offer environmentally sustainable solutions for environmental protection that do not affect the natural environment. However, some precautions must be taken to ensure that nanoparticles do not pose a danger to human health or the atmosphere. As an outcome, it is anticipated that shortly, nanotechnology will be modified, innovated, and adapted to expand the scope of environmental remediation. Nanotechnology is a recognized and important research field because of the possibilities for future advancement, cross-disciplinary scope, and a wide area of cost-effective applications.

REFERENCES

1. L. Monser and N. Adhoum, Modified activated carbon for the removal of copper, zinc, chromium and cyanide from wastewater, *Sep. Purif. Technol.*, vol. 26, no. 2–3, pp. 137–146, Mar. 2002, doi: 10.1016/S1383-5866(01)00155-1.
2. J. W. Haus, H. S. Zhou, I. Honma, and H. Komiyama, Quantum confinement in semiconductor heterostructure nanometer-size particles, *Phys. Rev. B*, vol. 47, no. 3, pp. 1359–1365, Jan. 1993, doi: 10.1103/PhysRevB.47.1359.
3. M. Stanisavljevic, S. Krizkova, M. Vaculovicova, R. Kizek, and V. Adam, Quantum dots-fluorescence resonance energy transfer-based nanosensors and their application, *Biosens. Bioelectron.*, vol. 74, pp. 562–574, 2015, doi: 10.1016/j.bios.2015.06.076.
4. P. Pyykkö, Theoretical chemistry of gold, *Angew. Chemie Int. Ed.*, vol. 43, no. 34, pp. 4412–4456, Aug. 2004, doi: 10.1002/anie.200300624.
5. N. Goswami et al., Protein-directed synthesis of NIR-emitting, tunable HgS quantum dots and their applications in metal-ion sensing, *Small*, vol. 8, no. 20, pp. 3175–3184, Oct. 2012, doi: 10.1002/smll.201200760.
6. S. Chen, S. Chen, Y. Yu, and J. Wang, Inner filter effect-based fluorescent sensing systems: A review Analytica Chimica Acta Inner filter effect-based fluorescent sensing systems: A review, *Anal. Chim. Acta*, vol. 999, no. September, 2019, [Online], doi: 10.1016/j.aca.2017.10.026.

7. Y. Ma et al., A dual-emissive fluorescent sensor fabricated by encapsulating quantum dots and carbon dots into metal-organic frameworks for the ratiometric detection of Cu^{2+} in tap water, *J. Mater. Chem. C*, vol. 5, no. 33, pp. 8566–8571, 2017, doi: 10.1039/c7tc01970j.
8. H. Xiong, B. Wang, W. Wen, X. Zhang, and S. Wang, Fluorometric determination of copper(II) by using 3-aminophenylboronic acid-functionalized CdTe quantum dot probes, *Microchim. Acta*, vol. 186, no. 6, 2019, doi: 10.1007/s00604-019-3515-5.
9. T.-W. Sung and Y.-L. Lo, Highly sensitive and selective sensor based on silica-coated CdSe/ZnS nanoparticles for Cu^{2+} ion detection, *Sensors Actuators B Chem.*, vol. 165, no. 1, pp. 119–125, 2012, doi: 10.1016/j.snb.2012.02.028.
10. M. Liu, H. Zhao, S. Chen, H. Wang, and X. Quan, Photochemical synthesis of highly fluorescent CdTe quantum dots for 'on–off–on' detection of Cu(II) ions, *Inorganica Chim. Acta*, vol. 392, pp. 236–240, 2012, doi: 10.1016/j.ica.2012.01.026.
11. H. Elmizadeh, M. Soleimani, F. Faridbod, and G. R. Bardajee, Ligand-capped CdTe quantum dots as a fluorescent nanosensor for detection of copper ions in environmental water sample, *J. Fluoresc.*, vol. 27, no. 6, pp. 2323–2333, 2017, doi: 10.1007/s10895-017-2174-3.
12. J. Han et al., Fabrication of CdTe nanoparticles-based superparticles for an improved detection of Cu^{2+} and Ag+, *J. Mater. Chem.*, vol. 22, no. 6, pp. 2679–2686, 2012, doi: 10.1039/C1JM14215A.
13. Z. Liu, S. Liu, P. Yin, and Y. He, Fluorescence enhancement of CdTe/CdS quantum dots by coupling of glyphosate and its application for sensitive detection of copper ion, *Anal. Chim. Acta*, vol. 745, pp. 78–84, 2012, doi: 10.1016/j.aca.2012.07.033.
14. R. K. Sajwan, Y. Bagbi, P. Sharma, and P. R. Solanki, L-cysteine and 3-mercaptopropionic acid capped cadmium selenide quantum dots based metal ion probes, *J. Lumin.*, vol. 187, pp. 126–132, 2017, doi: 10.1016/j.jlumin.2017.02.061.
15. Y. Feng et al., Efficient fluorescence energy transfer system between fluorescein isothiocyanate and CdTe quantum dots for the detection of silver ions, *Luminescence*, vol. 31, no. 2, pp. 356–363, 2016, doi: 10.1002/bio.2967.
16. D. Huang, C. Niu, M. Ruan, X. Wang, G. Zeng, and C. Deng, Highly sensitive strategy for Hg^{2+} detection in environmental water samples using long lifetime fluorescence quantum dots and gold nanoparticles, *Environ. Sci. Technol.*, vol. 47, no. 9, pp. 4392–4398, May 2013, doi: 10.1021/es302967n.
17. B. Liu, F. Zeng, G. Wu, and S. Wu, Nanoparticles as scaffolds for FRET-based ratiometric detection of mercury ions in water with QDs as donors, *Analyst*, vol. 137, no. 16, pp. 3717–3724, 2012, doi: 10.1039/C2AN35434A.
18. X. E. Zhao et al., A ratiometric fluorescent nanosensor for the detection of silver ions using graphene quantum dots, *Sensors Actuators, B Chem.*, vol. 253, pp. 239–246, 2017, doi: 10.1016/j.snb.2017.06.086.
19. S. Basiri, A. Mehdinia, and A. Jabbari, Green synthesis of reduced graphene oxide-Ag nanoparticles as a dual-responsive colorimetric platform for detection of dopamine and Cu^{2+}, *Sensors Actuators B Chem.*, vol. 262, pp. 499–507, 2018, doi: 10.1016/j.snb.2018.02.011.
20. T. Liu et al., Carbon quantum dots prepared with polyethyleneimine as both reducing agent and stabilizer for synthesis of Ag/CQDs composite for Hg^{2+} ions detection, *J. Hazard. Mater.*, vol. 322, pp. 430–436, 2017, doi: 10.1016/j.jhazmat.2016.10.034.
21. Y. Zhang et al., Water-soluble, nitrogen-doped fluorescent carbon dots for highly sensitive and selective detection of Hg^{2+} in aqueous solution, *RSC Adv.*, vol. 5, no. 50, pp. 40393–40401, 2015, doi: 10.1039/c5ra04653j.
22. T. Wang and W. Yue, Carbon nanotubes heavy metal detection with stripping voltammetry: A review paper, *Electroanalysis*, vol. 29, no. 10, pp. 2178–2189, Oct. 2017, doi: 10.1002/elan.201700276.
23. S. Kosa, G. Alzhrani, and M. Salam, Removal of heavy metal ions from aqueous solution by multi-walled carbon nanotubes modified with 8-hydroxyquinoline: Kinetic study, *Chem. Eng. J.*, vols. 181–182, pp. 159–168, Mar. 2014, doi: 10.1016/j.cej.2011.11.044.
24. S. Venkateswarlu, B. N. Kumar, B. Prathima, Y. SubbaRao, and N. V. V. Jyothi, A novel green synthesis of Fe_3O_4 magnetic nanorods using Punica Granatum rind extract and its application for removal of Pb(II) from aqueous environment, *Arab. J. Chem.*, vol. 12, no. 4, pp. 588–596, 2019, doi: 10.1016/j.arabjc.2014.09.006.
25. Y. Cai et al., Highly active MgO nanoparticles for simultaneous bacterial inactivation and heavy metal removal from aqueous solution, *Chem. Eng. J.*, vol. 312, pp. 158–166, 2017, doi: 10.1016/j.cej.2016.11.134.
26. T. K. Sen and M. V. Sarzali, Removal of cadmium metal ion (Cd^{2+}) from its aqueous solution by aluminium oxide (Al_2O_3): A kinetic and equilibrium study, *Chem. Eng. J.*, vol. 142, no. 3, pp. 256–262, 2008, doi: 10.1016/j.cej.2007.12.001.

27. S. Tabesh, F. Davar, and M. R. Loghman-Estarki, Preparation of γ-Al_2O_3 nanoparticles using modified sol-gel method and its use for the adsorption of lead and cadmium ions, *J. Alloys Compd.*, vol. 730, pp. 441–449, 2018, doi: 10.1016/j.jallcom.2017.09.246.
28. T. Luo, J. Cui, S. Hu, Y. Huang, and C. Jing, Arsenic removal and recovery from copper smelting wastewater using TiO_2, *Environ. Sci. Technol.*, vol. 44, no. 23, pp. 9094–9098, Dec. 2010, doi: 10.1021/es1024355.
29. I. Ghiloufi, J. El Ghoul, A. Modwi, and L. El Mir, Ga-doped ZnO for adsorption of heavy metals from aqueous solution, *Mater. Sci. Semicond. Process.*, vol. 42, pp. 102–106, 2016, doi: 10.1016/j.mssp.2015.08.047.
30. L. Khezami et al., Effect of aluminum loading on structural and morphological characteristics of ZnO nanoparticles for heavy metal ion elimination, *Environ. Sci. Pollut. Res.*, vol. 27, no. 3, pp. 3086–3099, 2020, doi: 10.1007/s11356-019-07279-0.
31. R. Dubey, J. Bajpai, and A. K. Bajpai, Chitosan-alginate nanoparticles (CANPs) as potential nanosorbent for removal of Hg (II) ions, *Environ. Nanotechnol., Monit. Manag.*, vol. 6, pp. 32–44, 2016, doi: 10.1016/j.enmm.2016.06.008.
32. M. E. A. Ali, M. M. S. Aboelfadl, A. M. Selim, H. F. Khalil, and G. M. Elkady, Chitosan nanoparticles extracted from shrimp shells, application for removal of Fe(II) and Mn(II) from aqueous phases, *Sep. Sci. Technol.*, vol. 53, no. 18, pp. 2870–2881, Dec. 2018, doi: 10.1080/01496395.2018.1489845.
33. U. Malayoglu, Removal of heavy metals by biopolymer (chitosan)/nanoclay composites, *Sep. Sci. Technol.*, vol. 53, no. 17, pp. 2741–2749, Nov. 2018, doi: 10.1080/01496395.2018.1471506.
34. D. Xu, K. Zhu, X. Zheng, and R. Xiao, Poly(ethylene-co-vinyl alcohol) functional nanofiber membranes for the removal of Cr(VI) from water, *Ind. Eng. Chem. Res.*, vol. 54, no. 27, pp. 6836–6844, Jul. 2015, doi: 10.1021/acs.iecr.5b00995.
35. L. Ge et al., Facile fabrication of Fe@MgO magnetic nanocomposites for efficient removal of heavy metal ion and dye from water, *Powder Technol.*, vol. 326, pp. 393–401, 2018, doi: 10.1016/j.powtec.2017.12.003.
36. H. Li, Z. Chi, and J. Li, Covalent bonding synthesis of magnetic graphene oxide nanocomposites for Cr(III) removal, *Desalin. Water Treat.*, vol. 52, no. 10–12, pp. 1937–1946, Mar. 2014, doi: 10.1080/19443994.2013.806224.
37. A. Mahapatra, B. G. Mishra, and G. Hota, Electrospun Fe_2O_3–Al_2O_3 nanocomposite fibers as efficient adsorbent for removal of heavy metal ions from aqueous solution, *J. Hazard. Mater.*, vol. 258–259, pp. 116–123, 2013, doi: 10.1016/j.jhazmat.2013.04.045.
38. K. R. Kunduru, M. Nazarkovsky, S. Farah, R. P. Pawar, A. Basu, and A. J. Domb, 2 – Nanotechnology for water purification: Applications of nanotechnology methods in wastewater treatment, *Water Purif.*, A. M. B. T.-W. P. Grumezescu, Ed. Academic Press, 2017, pp. 33–74.
39. Shah, M. P. *Removal of Emerging Contaminants through Microbial Processes.* Springer. 2021.
40. Shah, M. P. *Advanced Oxidation Processes for Effluent Treatment Plants.* Elsevier. 2020.
41. Shah, M. P. *Microbial Bioremediation and Biodegradation.* Springer. 2020.
42. C.-C. Ye, Q.-F. An, J.-K. Wu, F.-Y. Zhao, P.-Y. Zheng, and N.-X. Wang, Nanofiltration membranes consisting of quaternized polyelectrolyte complex nanoparticles for heavy metal removal, *Chem. Eng. J.*, vol. 359, pp. 994–1005, 2019, doi: 10.1016/j.cej.2018.11.085.
43. J. Wang, S. Zheng, Y. Shao, J. Liu, Z. Xu, and D. Zhu, Amino-functionalized Fe_3O_4@SiO_2 core–shell magnetic nanomaterial as a novel adsorbent for aqueous heavy metals removal, *J. Colloid Interface Sci.*, vol. 349, no. 1, pp. 293–299, 2010, doi: 10.1016/j.jcis.2010.05.010.

2 Emerging Pollutants Removal Using Biochar in Wastewater

A Critical Review

Nagireddi Jagadeesh and Baranidharan Sundaram
National Institute of Technology

CONTENTS

DOI: 10.1201/9781003164982-2

LIST OF ABBREVIATIONS

BET	Brunauer Emmet and Teller
COD	Chemical oxygen demand
EPA	Environmental protection agency
GAC	Granulated activated carbon
GWP	Global warming potential
HAP	Hydroxyapatite
MSGB	Magnetic switchgrass biochar
N	Nitrogen
P	Phosphorous
POPs	Persistent organic pollutants
PPCPs	Pharmaceuticals & personal care products
SGB	Switchgrass biochar
SSA	Specific surface area
WWTPs	Wastewater treatment plants

2.1 INTRODUCTION

The emerging pollutants/contaminants were synthetic or naturally occurring chemicals detected in water and adversely affected human and ecological health. During the last decade, the research for identifying the source, occurrence, and adsorption of emerging pollutants such as pharmaceuticals and personal care products (PPCPs), persistent organic pollutants (POPs), steroid hormones, nanomaterials, cosmetic products, endocrine-disrupting chemicals, and perfluorinated compounds has increased significantly (Visanji et al., 2018). The primary source of emerging pollutants in wastewater is the discharging of contaminated surface water and industrial effluents. The surface water is contaminated when emerging contaminants such as antibiotics, pesticides, hormones, microplastics, and PPCPs are mixed in it (Galindo-Miranda et al., 2019). The addition of emerging pollutants to the wastewater has been growing every year and turns these characteristics of wastewater more toxic than the parent water source (Agüera et al., 2013). United States Environmental Protection Agency (US EPA), along with many research articles, has stated that there are unfamiliar compounds in wastewater and aquatic environments (Deblonde et al., 2011). There was no research until 2005 in developing countries due to the lack of ideas on emerging pollutants and their effects on the ecosystem and human health. The research on emerging pollutants started and rose gradually in an aquatic environment in developing countries since 2005, but there has been a lack of proper technology and research publication on emerging pollutants removal from wastewater (Visanji et al., 2018).

Even though the knowledge and research options have increased in removing emerging pollutants from wastewater, their availability would risk aquatic animals, humans, and animals. This potential risk would lead to decreased male fertility, including congenital disabilities, changes in the reproduction system, and causes cancer sometimes. The monitoring of emerging pollutants in wastewater is difficult due to their trace concentrations varying from nanograms per litre (ng/L) to milligrams per litre (mg/L) and the nonavailability of expensive, sophisticated, and high analytical instruments (Visanji et al., 2018).

Many physical, chemical, and biological methods were investigated to remove emerging pollutants from wastewater (Dhangar & Kumar, 2020). Among these, the reduction of emerging contaminants using the adsorption technique was the most welcoming method. This study summarises the preparation, production, properties, and removal mechanism of emerging pollutants from wastewater using the most economical and highly efficient adsorbent called biochar.

2.2 BIOCHAR AS AN ADSORBENT

Biochar is a reliable and economically cheaper material obtained by carbonisation of biomass (Xiang et al., 2020). The product biochar is obtained by heating biomass in the absence of oxygen. Currently, most of the studies have been choosing biochar as an adsorbent material. It has the following advantages over other adsorbents like activated carbon: (a) biochar has a high specific surface area (SSA), which helps in the removal of pollutants more efficiently; (b) its availability and cost of regeneration, and its microstructure helps in standing biochar apart from the other adsorbents; and (c) it reduces the greenhouse effect and global warming potential (GWP) and enhances the crop yielding (Cha et al., 2016). This study aims to remove emerging pollutants such as organic, inorganic, heavy metals, and N, P, and K from wastewater using biochar as a useful adsorbent material.

2.3 PRODUCTION OF BIOCHAR

The production of biochar is based on the type of raw material used and the cracking temperature of combustion. Different types of raw materials for the production of biochar available are sewage sludge, animal manure, fruit peels, sugarcane bagasse, and agricultural residues. The methods for the production of biochar are pyrolysis (most common), gasification, and hydrothermal carbonisation. The production of biochar demands little oxygen and temperature in the range of 300–800 degrees centigrade.

2.3.1 From Agricultural Waste Biochar

The farm waste feedstock, such as wood waste, dry biomass, nutshells, straw, and husk from farm crops, is used to produce agrarian waste biochar. These farm feedstocks are subjected to the drum pyroliser reactor in which the temperature is maintained in the range of 400°C–600°C. The biochar is produced from peanut shells, cottonwood, and pinewoods with the addition of magnetic chloride hexahydrate ($MgCl_2$ $6H_2O$) to make the magnetic biochar by increasing 10°C per minute till reaching 600°C in the muffle furnace (Thines et al., 2017).

Fruit peels are enormous waste by-products obtained from various sources. This plenty of waste by-products causes an increase in load on landfill, and palm oil mill effluent has hazardous characteristics when it disposes into the environment (Lam et al., 2018; Shah, 2020, 2021). The pyrolysis of different fruit peels, such as banana and orange peels, is dried, cut into 0.5–1 cm size, and then transferred into a muffle furnace. The peel pieces are burned at a temperature gradient of 10°C for 10 minutes to reach the final temperature of 300°C–500°C.

2.3.2 From Algal Biomass

The accumulated waste algal biomass is converted into algal biochar using thermochemical processes such as slow pyrolysis, torrefaction, and hydrothermal carbonisation. Among these thermochemical processes, the yield of biochar produced through the slow pyrolysis process is 28%–31%, and the result is a high percentage of biochar (64%–68%) when the temperature is 500°C–600°C (Srinivasan et al., 2017). The end product from hydrothermal carbonisation is called hydrochar, which has equal properties of biochar and is used as an adsorbent to remove emerging pollutants from wastewater. The torrefaction method is a type of preliminary thermal process for the production of good-yield carbon-rich biochar (Yu et al., 2017).

2.3.3 From Animal Manure Biochar

The animal manure sources are mostly poultry litter, swine solids, and a mixture of swine solids and ryegrass. Among these sources, poultry litter, with more than 25% total solids (TS), is categorised as solid poultry manure, a waste by-product of chicken farming (Mukhtar, 2019). The poultry litter quantity has been increasing significantly to generate a vast amount of waste and leads to waste management issues. The animal manure biochar is prepared from poultry litter in a pyrolysis reactor at 620°C.

Cow manure is used to produce biochar for the removal of emerging pollutants in water and soil. The wet cow manure is dried, and then the pH of both wet and dried manure is measured and noted for reference before being subjected to pyrolysis followed by anaerobic digestion. The manure is converted into biochar through a slow pyrolysis process at a temperature of 380°C–420°C with a supply of little oxygen for keeping the temperature constant (Rehman et al., 2020).

2.3.4 From Sewage Sludge

Sludge is the primary unavoidable product from the WWTPs. Sludge is produced during primary treatment, secondary treatment, anaerobic/aerobic/anoxic digestion, and the dewatering process (Chen et al., 2020; Shah, 2021). Before converting sludge into biochar, the physicochemical properties, such as moisture content, solids percent in the sludge, composition of heavy metals, inorganic salts, pathogen count in the sludge, and nonorganic compounds such as lipids, nucleic acids, etc., are observed. The wet sludge is made into dry powder using various dewatering techniques such as freeze-thaw method, heat treatment, microwave treatment, acid–base treatment, coagulant, and flocculant, and treatment by an advanced oxidation process.

The conversion of raw, dewatered, anaerobic sludge into sludge biochar (0.4–0.73 kg/kg dry sludge) using thermochemical strategies such as conventional pyrolysis (300°C–700°C and <1 hour), microwave-assisted pyrolysis (>700°C and 10–20 minutes), and hydrothermal carbonisation 180°-C–250°C and 1–24 hours) (Tarelho et al., 2020).

2.3.5 From Sugarcane Bagasse

The sugarcane bagasse is obtained as a waste output from the sugar industry. This sugarcane bagasse is burned at 600°C for 2–3 hours and ground the obtained biochar powder to 0.8–2 mm size. The final size of the particles of sugarcane biochar is measured using sieve analysis. The SSA is calculated using Brunauer Emmet and Teller (BET) theory, and the % of porosity of biochar is calculated using the pore volume method (Figure 2.1 and Table 2.1) (Manyuchi et al., 2020).

2.4 FACTORS AFFECTING BIOCHAR PRODUCTION

2.4.1 Temperature

The structure and characteristics of biochar depend on the pyrolysis temperature, and the yield of biochar production is inversely proportional to the increase in the temperature. As the temperature increases, there is an enhancement of SSA, size of the particle, and pH value. The biochar produced at high temperatures (300°C–700°C) possesses an aromatic structure with high carbon composition, which improves properties like SSA and pore structure of the biochar. The advantages of biochar adsorbents, which are prepared at low temperatures (<300°C), have aliphatic structure and excellent adsorption properties in the removal of biological pollutants such as bacteria, *E. coli*, and fungi (Qambrani et al., 2017).

2.4.2 Heating Rate

The heating rate of pyrolysis temperature during the biochar production indirectly affects the output materials such as biochar, bio-fuel/bio-oil, and syngas production. The production of bio-oil dominates when the rate of heating is higher (150°C/S–500°C/S), whereas the production of biochar

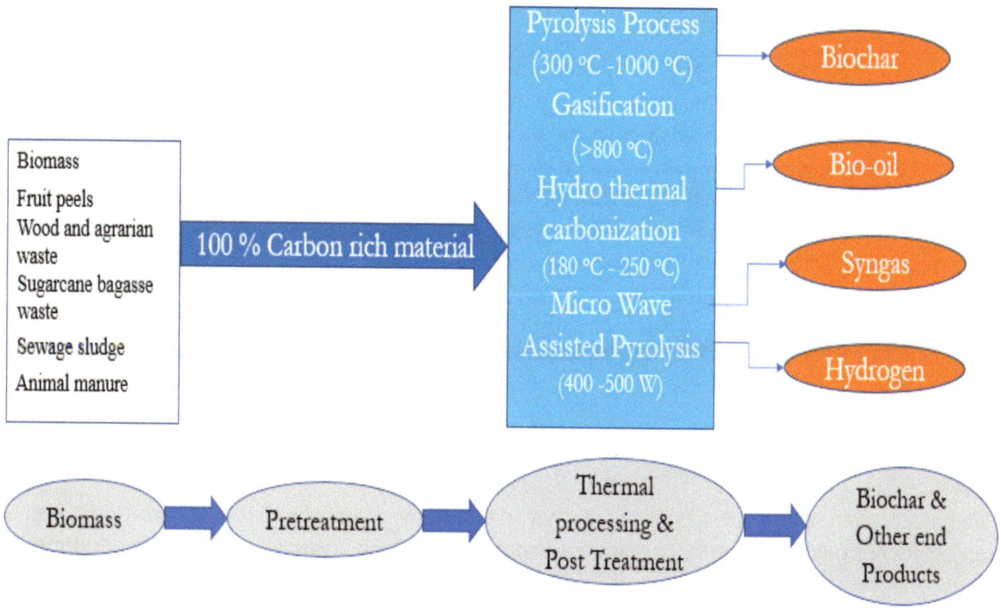

FIGURE 2.1 Biochar production.

TABLE 2.1
Production of Biochar Using Various Techniques

Production Method	Thermal Gradient Variation	Temperature Range (°C)	Time of Combustion	Output or Yield		
				Syngas (%)	Bio-oil (%)	Biochar (%)
Hydro carbonisation	Slow	<350	1–24 hours	15–20	5–10	60–85
Slow pyrolysis	Slow	300–700	1–2 hours	35	30	35
Flash pyrolysis	Flash	700–1000	10–60 minutes	5–15	70–80	15–20
Gasification	Faster	700–1500	10–30 seconds	80–85	0–5	10–15
Fast pyrolysis	Faster	>1000	1–5 seconds	15	70	15

more than bio-oil production when the heating temperature is lower (10°C/S–20°C/S) (Qambrani et al., 2017). The lower heating rates encourage biochar production without major losses from the optimum feedstocks, higher temperatures produce less biochar-adsorbent material and more syngas production with destructed pore structure, and the moderate heating rates with medium pyrolysis temperature produces efficient properties of biochar adsorbent.

2.4.3 Particle Size

Particle size has a more significant influence on the pyrolysis end products. It plays a key role in transferring the heat and mass transfer between the particles during the pyrolysis production process. The particle size of the adsorbent depends on the feedstock material and type of pyrolysis method adopted. The particles with a size of less than 1.8 mm are subjected to flash/fast pyrolysis to get the uniform biochar with high bio-fuel/bio-oil and syngas. The particle size greater than 1.8 mm is subjected to more temperature gradients, giving less gas with equal amounts of biochar and bio-oil/bio-fuel (Qambrani et al., 2017).

2.4.4 Feedstock Composition

Feedstock composition is the key component that greatly affects the biochar adsorbent.

The end product of biochar material depends on the type of feedstock material used for its production. The selection of adsorbent for the removal of emerging pollutants such as organic/inorganic/heavy metals/biological relies on the type of raw material used for its production.

According to Uchimiya et al., different feedstocks produce different end-adsorbent biochar products with various carbon percentages which ultimately turns into producing graphitic structure. According to Hammes et al., the feedstocks used for the production of biochar adsorbents with low hydrogen/carbon (H/C) and oxygen/carbon (O/C) ratios are compared with other adsorbents like soot and charcoal, which has high hydrogen/carbon (H/C) and oxygen/carbon (O/C) ratios but can't be used as effective adsorbents due to their raw material feedstocks used for preparation.

2.4.5 Properties of Biochar

The physicochemical properties of biochar vary with the type of raw material used for the preparation of biochar means of pyrolysis reactor, the particle size of feedstock, and the time of combustion of raw materials for the production of biochar (including the temperature variation). The chemical properties of biochar are mostly dependent on functional groups such as hydroxyl, methyl, carbonyl, carboxyl groups, the high SSA of the particle, the pore volume, and stable molecular composition with better adsorption capacity.

2.4.6 Physical

Biochar with porous structures adsorbs more pollutants due to more SSA of the adsorbent. Among the different production methods of biochar, pyrolysis is the best method because it utilises high temperatures, which helps expand the micropore structure of biochar. The increase in the micropore structure would increase the biochar pores SSA, ultimately producing an effective adsorbent.

2.4.7 Chemical

The biochar chemical properties are elemental composition, pH, cation or anion exchange capacity, functional groups, ash content, and hydrophobicity. Among these chemical properties, biochar's crucial chemical properties for the effective removal of pollutants from the wastewater are cation or anion exchange capacity and hydrophobicity.

2.4.7.1 Cation or Anion Exchange Capacity

Biochar with high cation exchange capacity would adsorb heavy metals in the wastewater, and biochar with high anion exchange capacity would adsorb the phosphates. The adsorption capacity of anions and cations is dependent on the type of feedstock and the type of pyrolysis temperature (Qambrani et al., 2017).

The cation exchange capacity would be more efficient when the feedstock made up of grass at 250–300 degrees centigrade would remove more metals than the biochar made up of oak and pinewood biochar.

2.4.7.2 Hydrophobicity

It is used to determine organic carbon content in the feedstock and enhances the adsorption capacity greatly. It helps in the adsorption of pollutants as well as the harmful pathogens present in the wastewater.

2.5 COMPARISON OF BIOCHAR OVER OTHER ADSORBENTS

2.5.1 Activated Carbon

Activated carbon production consumes more energy and less yield. The production of activated carbon is expensive than biochar, which is almost 15–20 times costlier than biochar production from the pyrolysis process. Both biochar and activated carbon show great capacity towards pollutants removal like polychlorinated biphenyls. Activated carbon removes 4%–10% more polychlorinated biphenyls than biochar (pyrolysis of pellets at 700°C or softwood heating at 450°C).

Activated carbon removes bichlorinated dibenzo-p-dioxins/dibenzofurans more efficiently than biochar. The surface area of activated carbon is more than biochar, and it can adsorb Hg (II) efficiently than biochar. Also, the adsorption rate of pollutants depends on the sorption bond between the contaminant and the adsorbent. The higher sorption capacity of activated carbon would reduce the toxic effect better than biochar on contamination of plants and pathogens.

2.5.2 Bone Char

Bone char is another type of biochar produced through the partial pyrolysis process from the slaughterhouse's waste (animal bone waste). The composition of bone char includes carbon and hydroxyapatite (HAP), which is phosphorus. The usage of bone char as an adsorbent material in the field of wastewater treatment is limited. It adsorbs fluoride ions efficiently from an aqueous drinking water solution (Alkurdi et al., 2019a).

The adsorption of emerging pollutants such as antibiotics from the pharmacy industry, dyes from coloured industries, personnel care products, heavy metals (Pb, Zn, Cd, Co, Cu, Ni, Mg, and Mn), N, P, and biological pollutants from wastewater required biochar prepared from various feedstocks (Alkurdi et al., 2019b). The results of previous studies show that bone char's limitations for the only adsorption of fluorine (effective removal) from drinking water and adsorption of heavy metals from wastewater are partly and less than biochar (Alkurdi et al., 2019b). This is evident that biochar production with various feedstocks adsorbs emerging pollutants, and it has multiple advantages over bone char.

2.6 EMERGING POLLUTANTS

According to UNEP, emerging pollutants are not well-known chemical compounds that aren't monitored regularly in the water and wastewater streams. These emerging pollutants include mainly PPCPs, surfactants, pesticides, paints, and microplastics (turning larger plastic mass into smaller pieces).

2.6.1 Municipal Wastewater

The removal of emerging pollutants from the municipal wastewater treatment plant is carried out using biochar is employed after the secondary treatment of wastewater is known as the biofiltration stage. The combination of biochar with a biofilter is employed after secondary treatment to remove valuable nitrogen and phosphorous from sewage (Cole et al., 2017). The engineered biochar obtained by applying modifications to the physicochemical properties of pristine biochar mixed with aluminium hydroxide ($Al(OH)_3$) has better adsorption capacity towards the removal of phosphorous, which can be reused in fertiliser industries (Zheng et al., 2019).

Municipal wastewater contains ammonium as a pollutant, the activated sludge process cannot remove ammonium from the sewage, and it is being carried out with effluent. The biochar produced with sewage sludge at 450°C through the pyrolysis process has excellent pollutant removal towards ammonium from wastewater. This biochar with high SSA and good functional groups would help remove ammonium from wastewater (Zhang et al., 2019). The combination of biochar produced with sewage sludge/biowaste produced from ASP mixed with biofilters will adsorb total organic carbon, chemical oxygen demand (90%), total nitrogen (64%), total phosphorus (78%), and total suspended solids (89%) from wastewater (Manyuchi et al., 2018).

2.6.2 Industrial Wastewater

The entry of emerging pollutants into wastewater and its treatment are interdependent. The emerging pollutants are entered the environment through industrial wastewater, municipal wastewater, agricultural wastewater, and stormwater (Xiang et al., 2020). The emerging pollutants in industrial wastewater are mostly found as heavy metals, and organic pollutants entered from the smelting process industry, mining, chemical, leather, and battery manufacturing industries (Xiang et al., 2020).

The removal of heavy metals (majorly Cu II and Cd II) from this industrial wastewater is done through the adsorption process using chitosan and nylon 6, which act as effective adsorbents (Prakash et al., 2012). Also, the simultaneous removal of both heavy metals and organic pollutants from industrial wastewater is done by adsorption mechanism using various interfaces like solid/liquid or liquid/liquid or liquid/gas which enhances the properties of both adsorbent and adsorbate for the effective removal of pollutants (Ajiboye et al., 2021). The pollutants from dye-based industries are removed using gliricidia biochar adsorbent prepared from gliricidia septum. This adsorption process is governed by biochar's physicochemical properties such as pH, SSA, and pore volume (Wathukarage et al., 2019).

The palm oil mill effluent (POME) contains huge BOD concentrations, COD and oil, and grease content is removed by adsorption of biochar made of banana and orange peel waste. Determination of COD and BOD of POME is performed initially. A POME sample is mixed with biochar and stirred for 10 minutes and kept the sample in the chiller for 1–3 days. Finally, the BOD, COD, and oil and grease concentrations are analysed. The results are compared with initial BOD, COD, oil and grease concentrations (Lam et al., 2018).

2.6.3 Agricultural Wastewater

The influx of emerging pollutants such as pesticides and toxic metals into agricultural wastewater has increased exponentially due to increased food demand. Sulphur, endosulfan, mancozeb, and phorate are the most common pesticides used in India. Adsorption is the best technique for the removal of pesticides from agricultural wastewater. Among various adsorbents, biochar has better adsorption capacity in removing pesticides from wastewater with cost-benefit and acceptable removal efficiency (Saleh et al., 2020). Atrazine is a herbicidal pest used for maise and sugarcane to prevent the occurrence of broadleaf weeds. The removal of atrazine mixed with agricultural wastewater is done using the adsorption process with biochar made of corn straw. The adsorption of atrazine onto the adsorbent surface using adsorption kinetics isotherms physical properties and their actions such as hydrogen bonding, pi-electron bonding and interactions, and van der Waals force attraction which results in 96%–97% of the atrazine removal from agricultural wastewater (Suo et al., 2019). Pentachlorophenol is also a significant pest found in agrarian wastewater, and it is removed using zero-valent iron magnetic sludge biochar prepared from the paper mill waste (Devi & Saroha, 2014).

Diazinon is highly used as an indoor pest controller to kill insecticides, ants, silverfish, and fleas in residential buildings and gardens during the 1980s. Diazinon is removed from wastewater using biochar made of carbonised coconut shells prepared using a muffle furnace through the pyrolysis process. Coconut shell undergoes a pyrolysis process at a temperature of 700°C using a muffle furnace. The pyrolysed biomass is treated with H_3PO_4 and acid-alkali (NaOH) to get the final adsorbent. A diazinon (1 mg/L) sample of 100 mL is mixed adsorbent and stirred, shaken, and rotated at 150 rpm. The samples are filtered to find the adsorption efficiency using solid-phase extraction (SPE) (Tables 2.2–2.4) (Baharum et al., 2020).

TABLE 2.2
Removal of Heavy Pollutants from Industrial Wastewater

Pollutants Removed	Adsorbent	Feedstock/Raw Material	Adsorption Process	Author
Cd (II) & Cu (II)	Gasifier biochar	Pinewood	Gasifier biochar is mixed with CH_3COOH and NaOH solutions for determining the properties of biochar using FT-IR & scanning electronic microscope.	Burk et al. (2020)
Zn (II) & Cu (II)	Magnetic biochar	Water hyacinth	Magnetic biochar is prepared using chemical coprecipitation of $FeCl_2$ and $FeCl_3$ on water hyacinth feedstock followed by the slow pyrolysis process. The properties are observed using FT-IR & scanning electronic microscope.	Nyamunda et al. (2019)
Cd & sulfamethoxazole	Rice straw biochar	Rice straw	Using adsorption kinetics such as Langmuir isothermal kinetics	Han et al. (2013)
Palm oil mill effluent (POME), which contains BOD, COD & oil content	Peel biochar	Banana & orange fruit peels	Determination of COD & BOD of POME is performed initially. A sample of POME is mixed with biochar and stirred for 10 minutes and kept the sample in the chiller for 1–3 days. Finally, the BOD, COD, and oil and grease concentrations are analysed.	Lam et al. (2018)
Pb^{+2}	Maple wood biochar	Maple wood	The biochar is made of maple wood undergoes pyrolysis at 500°C. Sample of wastewater is mixed with the maple wood biochar under controlled process; the adsorption process carried out through the complexation of oxygen functional groups.	Mohanty et al. (2018)
Ag^{+}	Bamboo wood biochar	Dried bamboo wood	The biochar is made of bamboo wood undergoes pyrolysis at 600°C. Sample of wastewater is mixed with the bamboo wood biochar treated with nitric acid and zero-valent iron. The adsorption is carried out by the electrostatic force of attraction of iron and the outer layer of the pollutant.	Xiong et al. (2017)
As^{+5} & Cr^{+3}	Sewage sludge biochar	Sewage sludge	Sewage sludge biochar prepared through pristine pyrolysis at 300°C. The adsorption process is carried out by mixing the waste sample with biochar through the process of chemisorption	Inyang et al. (2015)

2.7 TREATMENT OF EMERGING POLLUTANTS FROM WASTEWATER USING BIOCHAR AS AN EFFICIENT ADSORBENT

In the current situation, the occurrence of emerging pollutants in wastewater has been increasing due to the uncontrolled disposal of industrial discharges, agricultural runoff into the water streams (Xiang et al., 2020). Wastewater that contains heavy metals, organic pollutants, N, P, K, and other contaminants causes severe damage to both environment and humans (Enaime et al., 2020). The wastewater with these pollutants is disposed into land/river would cause serious health effects to aquatic life, the surrounding environment, and animals and humans. To overcome these emerging pollutants in the wastewater problem, biochar has been chosen as an

TABLE 2.3
Removal of Emerging Pollutants (N & P) from Municipal Wastewater

Pollutants Removed	Adsorbent	Feedstock/Raw Material	Adsorption Process	Author
Phosphorus	Engineered/ modified pristine biochar	Wheat straw	The engineered biochar obtained by applying modifications to the physicochemical properties of pristine biochar mixed with aluminium hydroxide (Al $(OH)_3$).	Zheng et al. (2019)
Ammonium	Sewage sludge biochar	Sewage sludge	The biochar with high SSA and good functional groups would help remove ammonium from wastewater under a controlled chemosorption process.	Zhang et al. (2019)
Total organic carbon	Waste sludge biochar	Sludge derived from ASP	Biochar with Si/O functional groups is oxidised the organic carbon through the process of ozonate catalysis from wastewater.	Chen et al. (2019)
Nutrients recovery (N/ NH_4 & P)	Municipal biochar	Municipal biowaste (organic waste 100–300 kg/m^3)	The biochar produced using organic waste is mixed with the wastewater sample to determine the N, P, COD, BOD, and TSS. The results are compared with the initial wastewater sample results.	Manyuchi et al. (2018)
PO_4^{3-}	Tomato leaves biochar	Tomato leaves	The dried tomato leaves are undergoing a fast pyrolysis process at 600°C. The adsorption of phosphorous is carried out by treating the wastewater sample with Mg enriched biochar through precipitation followed by surface deposition of pollutant	Yao et al. (2012)
NH_4^+	Sugarcane harvest residue biochar	Sugarcane residue	Biochar is made of sugarcane residue through the pyrolysis process at 550°C. The adsorption of nitrogen is carried out by treating the wastewater sample with MgO and pi-pi electrons' electrostatic attraction.	Li et al. (2017)

adsorbent to remove pollutants due to its more significant effect on removal capacity in the liquid phase (Saleh et al., 2020). Biochar can remove 40% of organic pollutants, 45% of heavy metals, and 15% of N, P, and K from the wastewater. The other adsorbents can remove up to only 5% at most (Shen et al., 2020).

2.7.1 Removal of Organic Pollutants

The organic contaminants that pollute wastewater are listed as antibiotics, industrial dyes, agricultural pesticides, oils, phenolic compounds, POPs, and herbicides. The presence of organic contaminants in the wastewater reduces the dissolved oxygen levels and affects aquatic life (Xiang et al., 2020). The role of organic contaminants in the contamination of sewage is mainly due to two factors, such as human-made activities and natural sources. The presence of organic pollutants in wastewater is much harmful than heavy metals and other pollutants (Dai et al., 2019). The adsorption capacity and effective removal of organic pollutants from biochar waste are internally dependent on the electron interaction and attraction, pore volume filling, and cation exchange (Kazemi Shariat Panahi et al., 2020). Effective biochars, such as switchgrass

TABLE 2.4
Removal of Pesticides from Agricultural Wastewater

Pollutants Removed	Adsorbent	Feedstock/Raw Material	Adsorption Process	Author
Atrazine	P-doped biochar	Corn straw	The adsorption of atrazine on to the adsorbent surface due to physical properties actions such as hydrogen bonding, pi- electron bonding and interactions and van der Waals force of attraction	Suo et al. (2019)
Pentachlorophenol	Zero - valent iron magnetic sludge biochar	Papermill waste	The adsorption of pentachlorophenol using zero-valent iron magnetic sludge biochar, the wastewater sample is mixed with the adsorbent under controlled conditions.	Devi and Saroha (2014)
Diazinon	Coconut shell modified biochar	Coconut shells	Coconut shell undergoes pyrolysis process at a temperature 700°C using muffle furnace. The pyrolysed biomass is treated with H_3PO_4 and acid-alkali (NaOH) to get the final adsorbent. A diazinon (1 mg/L) sample of 100 mL is mixed adsorbent and stirred, shaken, and rotated at 150 rpm. The samples are filtered to find the adsorption efficiency using solid-phase extraction (SPE)	Baharum et al. (2020)
Tetracycline	Sawdust biochar	Sawdust	Biochar is prepared through the pyrolysis process at 600°C. The aqueous solution sample is mixed with sawdust biochar by doping with Zn and Fe, the adsorption carried out by bridge enhancement.	Yao et al. (2017)
Metribuzin herbicide	Pristine biochar	Switchgrass	Switchgrass is pyrolysed at 425°C to obtain the biochar. The wastewater sample is mixed with switchgrass biochar, and adsorption has carried out through the magnetisation process with electrostatic action and hydrogen bonds presence.	Essandoh et al. (2017)

biochar (SGB) and magnetic SGB, were used to remove herbicides in the wastewater by employing low soluble pH and variable temperature conditions (Xiang et al., 2020).

Biochar can also remove antibiotics such as fluoroquinolones, macrolides, and sulfonamides from wastewater by electron contributor-receiver interactions with linked attracting groups on the surface of the adsorbent. The biochar prepared with magnetic pine sawdust has greater potential to remove sulfamethoxazole under the pH transforming conditions (pH=4–9) and a little quicker sorption rate from the aqueous solution as an adsorbent (Dai et al., 2019). Engineering biochar prepared from the almond shell at a pyrolysis temperature of 800°C for 45 minutes has an excellent efficiency towards removing pesticides such as dibromo chloropropane from well water (Kazemi Shariat Panahi et al.,

2020). For removing colour/dye from wastewater, biochar produced from hibiscus cannabinus (acid-treated Kenaf) prepared from a slow pyrolysis process (1000°C for 1 hour) adsorbs methylene blue effectively. Oil pollution is due to an increase in China's industrialisation, which demands more significant petroleum quantities. Petroleum production involves exploring, mining, and transportation to the refinery plant. The biochar prepared with maple wood (300°C–700°C) has an adequate adsorption capacity to remove seawater oils. The POPs are mixed with the food chain and enter the human body. These chemicals that enter the body have low degradability and stay longer within the human body to cause various diseases. Biochar prepared with different feedstocks (soybean, cow dung, chicken manure, rice straw, swine grass, and corn stover (pH=7.0°C & 25°C) is applied to remove these POPs from wastewater. Among the different biochars, soybean biochar has the greater removal capacity (Dai et al., 2019).

2.7.2 Removal of Heavy Metals

The adsorption mechanism of heavy metals from an aqueous solution using biochar is dependent on both the physical and chemical properties of the adsorbent. The physical adsorption achieved through the growth of micropores in the pore structure and high SSA of the adsorbent and the interaction between the adsorbent and heavy metals (adsorbate) is weak van der Walls force of attraction (Zhang et al., 2020).

The chemical adsorption depends on the type of redox reaction that takes place between the adsorbent and the metal. As per Harvey et al., the adsorption between the heavy metal and biochar is due to the SSA and polarisability nature of the acid–base character of biochar prepared from various feedstocks. The adsorption rate of biochar depends on the functional groups such as carboxyl, carbonyl, and hydroxide attached to the link. The disassociation of oxygen-containing functional groups, most of the time biochar carries a negative charge on their valency shell, which can comfortably remove the toxic heavy metals named as cadmium (Cd), lead (Pb), aluminium (Al), copper (Cu), nickel (Ni), arsenic (As), mercury (Hg), and manganese (Mn) from wastewater (Qambrani et al., 2017).

The pollutants ad mechanism of biochar on heavy metals shows that the pollutants like Cu(II), Ni(II), Pb(II), Zn(II), Cd(II), and Hg(II) are removed using the interaction of the physical adsorption (SSA) process with carboxyl (R-COOH), and hydroxyl group (O-H) groups, Cr(VI) is reduced to (the oxygen-containing functional group adsorbs Cr (III) and Al(III) at lower pH levels (Enaime et al., 2020).

The other studies focus on removing Hg (II) from mercury-contaminated wastewater sites, using biochar prepared with Brazilian pepper at various combustion temperatures of 300°C–600°C. Among these, the biochar prepared at 600°C with standard pH conditions shows more significant adsorption of Hg(II) with carboxylic and phenolic functional groups (Zhang et al., 2019).

The biochar prepared using sugarcane bagasse got excellent adsorption capacity towards the removal of Pb(II) at a pH of 9.6°C, 25°C, and Ca (II) & Mg(II) at a pH of 5°C & 25°C from the wastewater stream (Anastopoulos et al., 2017).

2.7.3 Removal of Nitrogen and Phosphorous

The occurrence of N and P in wastewater leads to eutrophication. Biochar can remove nitrogen and phosphorous from aqueous solutions with nitrate, phosphate, and ammonium as their usual reactive forms in the tertiary treatment of wastewater (Xiang et al., 2020). The sorption potential of biochar in the removal of N & P depends on mainly three factors: (a) the type of feedstock used for the preparation of biochar, (b) the type of thermochemical process used for its preparation, and (c) the effect of chemical modification (Dai et al., 2020).

Previous experimental studies show that removing nitrogen from wastewater is carried out with a solution prepared by pig house slurry and ammonium nitrogen chlorine stock solution of various concentrations (250–1250 mg/L). On the other hand, the biochar was prepared using the pyrolysis process using softwood chips and rice husk at a temperature of 500°C. The biochar is applied to the slurry for the effective removal of nitrogen, and it was observed that at lower concentrations of 250 mg/L (81%) and 500 mg/L (89%), adsorption of N is higher than concentrations of 1000 mg/L (76%) and 1250 mg/L (80%) (Kizito et al., 2015).

Phosphorous removal from wastewater is achieved by treating an aqueous solution with the biochar produced from calcium-rich biochar. The rate of reduction of the initial concentration of phosphorous in an aqueous solution varies linearly with the calcium-rich biochar (Dai et al., 2020). The adsorption kinetic equations such as Langmuir and Freundlich's isotherms were used to measure the adsorption rate to remove phosphate from an aqueous solution depending on phosphate's chemical reaction ions and Ca (II) ions (Antunes et al., 2018).

2.8 CONCLUSION

Biochar is a functional material prepared under controlled thermal decomposition of organic feedstocks from crops, forestry residues, sewage sludge, algal biomass, and poultry manure. Due to its intrinsic properties, biochar has been widely used as an adsorbent to remove organic pollutants, heavy metals from water and wastewater. The study reviewed the various procedures for producing biochar, properties, factors affecting, and applications in removing emerging pollutants from wastewater.

The basic summary of this study is (a) the importance of biochar as an adsorbent for the removal of pollutants from wastewater; (b) the physicochemical properties (SSA, pore structure, pH, temperature, and functional groups); production (various methods of pyrolysis, gasification, hydro carbonisation, and different feedstocks; maple wood, chicken manure, soybean stock, sugarcane bagasse, swine grass, agricultural, sewage sludge, crop residuals, and fruit peels used as raw material for the production of biochar) and factors affecting (Temperature, heating rate, particle size, and feedstock composition); and the biochar as an adsorbent material, (c) comparison of biochar with other adsorbent materials and its betterment over those adsorbents, (d) Removal of emerging pollutants such as organic pollutants, heavy metals, and N, P, and biological pollutants like bacteria, fungi, and *E. coli*.

Biochar was used as an efficient adsorbent for removing emerging pollutants from wastewater with different adsorbent dosages and at various pH levels, which shows the importance of biochar over other adsorbents. Available functional groups of biochar carried out the removal of heavy metals (Pb, Cd, Hg, Cu, Ni, Mn, Co, As, and Mg) from aqueous solution with standard pH levels and temperature conditions.

REFERENCES

Agüera, A., Martínez Bueno, M. J., & Fernández-Alba, A. R. (2013). New trends in the analytical determination of emerging contaminants and their transformation products in environmental waters. *Environmental Science and Pollution Research*, *20*(6), 3496–3515. https://doi.org/10.1007/s11356-013-1586–0.

Ajiboye, T. O., Oyewo, O. A., & Onwudiwe, D. C. (2021). Simultaneous removal of organics and heavy metals from industrial wastewater: A review. *Chemosphere*, *262*, 128379. https://doi.org/10.1016/j.chemosphere.2020.128379.

Alkurdi, S. S. A., Al-Juboori, R. A., Bundschuh, J., & Hamawand, I. (2019a). Bone char as a green sorbent for removing health threatening fluoride from drinking water. *Environment International*, *127*(March), 704–719. https://doi.org/10.1016/j.envint.2019.03.065.

Alkurdi, S. S. A., Herath, I., Bundschuh, J., Al-Juboori, R. A., Vithanage, M., & Mohan, D. (2019b). Biochar versus bone char for a sustainable inorganic arsenic mitigation in water: What needs to be done in future research? *Environment International*, *127*(February), 52–69. https://doi.org/10.1016/j.envint.2019.03.012.

Anastopoulos, I., Bhatnagar, A., Hameed, B. H., Ok, Y. S., & Omirou, M. (2017). A review on waste-derived adsorbents from sugar industry for pollutant removal in water and wastewater. *Journal of Molecular Liquids*, *240*, 179–188. https://doi.org/10.1016/j.molliq.2017.05.063.

Baharum, N. A., Nasir, H. M., Ishak, M. Y., Isa, N. M., Hassan, M. A., & Aris, A. Z. (2020). Highly efficient removal of diazinon pesticide from aqueous solutions by using coconut shell-modified biochar. *Arabian Journal of Chemistry, 13*(7), 6106–6121. https://doi.org/10.1016/j.arabjc.2020.05.011.

Burk, G. A., Herath, A., Crisler, G. B., Bridges, D., Patel, S., Pittman, C. U., & Mlsna, T. (2020). Cadmium and copper removal from aqueous solutions using chitosan-coated gasifier biochar. *Frontiers in Environmental Science, 8*(November), 1–11. https://doi.org/10.3389/fenvs.2020.541203.

Cha, J. S., Park, S. H., Jung, S. C., Ryu, C., Jeon, J. K., Shin, M. C., & Park, Y. K. (2016). Production and utilisation of biochar: A review. *Journal of Industrial and Engineering Chemistry, 40*, 1–15. https://doi.org/10.1016/j.jiec.2016.06.002.

Chen, Y. di, Wang, R., Duan, X., Wang, S., Ren, N. qi, & Ho, S. H. (2020). Production, properties, and catalytic applications of sludge derived biochar for environmental remediation. *Water Research, 187*, 116390. https://doi.org/10.1016/j.watres.2020.116390.

Chen, C., Yan, X., Xu, Y. Y., Yoza, B. A., Wang, X., Kou, Y., … Li, Q. X. (2019). Activated petroleum waste sludge biochar for efficient catalytic ozonation of refinery wastewater. *Science of the Total Environment, 651*, 2631–2640. https://doi.org/10.1016/j.scitotenv.2018.10.131.

Cole, A. J., Paul, N. A., de Nys, R., & Roberts, D. A. (2017). Good for sewage treatment and good for agriculture: Algal based compost and biochar. *Journal of Environmental Management, 200*, 105–113. https://doi.org/10.1016/j.jenvman.2017.05.082.

Dai, Y., Wang, W., Lu, L., Yan, L., & Yu, D. (2020). Utilisation of biochar for the removal of nitrogen and phosphorus. *Journal of Cleaner Production, 257*. https://doi.org/10.1016/j.jclepro.2020.120573.

Dai, Y., Zhang, N., Xing, C., Cui, Q., & Sun, Q. (2019). The adsorption, regeneration and engineering applications of biochar for removal organic pollutants: A review. *Chemosphere, 223*, 12–27. https://doi.org/10.1016/j.chemosphere.2019.01.161.

Deblonde, T., Cossu-Leguille, C., & Hartemann, P. (2011). Emerging pollutants in wastewater: A review of the literature. *International Journal of Hygiene and Environmental Health, 214*(6), 442–448. https://doi.org/10.1016/j.ijheh.2011.08.002.

Devi, P., & Saroha, A. K. (2014). Synthesis of the magnetic biochar composites for use as an adsorbent for the removal of pentachlorophenol from the effluent. *Bioresource Technology, 169*, 525–531. https://doi.org/10.1016/j.biortech.2014.07.062.

Dhangar, K., & Kumar, M. (2020). Tricks and tracks in removal of emerging contaminants from the wastewater through hybrid treatment systems: A review. *Science of the Total Environment, 738*, 140320. https://doi.org/10.1016/j.scitotenv.2020.140320.

Enaime, G., Baçaoui, A., Yaacoubi, A., & Lübken, M. (2020). Biochar for wastewater treatment-conversion technologies and applications. *Applied Sciences (Switzerland), 10*(10). https://doi.org/10.3390/app10103492.

Essandoh, M., Wolgemuth, D., Pittman, C. U., Mohan, D., & Mlsna, T. (2017). Adsorption of metribuzin from aqueous solution using magnetic and nonmagnetic sustainable low-cost biochar adsorbents. *Environmental Science and Pollution Research, 24*(5), 4577–4590. https://doi.org/10.1007/s11356-016-8188-6.

Galindo-Miranda, J. M., Guízar-González, C., Becerril-Bravo, E. J., Moeller-Chávez, G., León-Becerril, E., & Vallejo-Rodríguez, R. (2019). Occurrence of emerging contaminants in environmental surface waters and their analytical methodology – A review. *Water Science and Technology: Water Supply, 19*(7), 1871–1884. https://doi.org/10.2166/ws.2019.087.

Han, X., Liang, C. F., Li, T. Q., Wang, K., Huang, H. G., & Yang, X. E. (2013). Simultaneous removal of cadmium and sulfamethoxazole from aqueous solution by rice straw biochar. *Journal of Zhejiang University: Science B, 14*(7), 640–649. https://doi.org/10.1631/jzus.B1200353.

Inyang, M., Gao, B., Zimmerman, A., Zhou, Y., & Cao, X. (2015). Sorption and cosorption of lead and sulfapyridine on carbon nanotube-modified biochars. *Environmental Science and Pollution Research, 22*(3), 1868–1876. https://doi.org/10.1007/s11356-014-2740–z.

Kazemi Shariat Panahi, H., Dehhaghi, M., Ok, Y. S., Nizami, A. S., Khoshnevisan, B., Mussatto, S. I., … Lam, S. S. (2020). A comprehensive review of engineered biochar: Production, characteristics, and environmental applications. *Journal of Cleaner Production, 270*, 122462. https://doi.org/10.1016/j.jclepro.2020.122462.

Lam, S. S., Liew, R. K., Cheng, C. K., Rasit, N., Ooi, C. K., Ma, N. L., … Chase, H. A. (2018). Pyrolysis production of fruit peel biochar for potential use in treatment of palm oil mill effluent. *Journal of Environmental Management, 213*, 400–408. https://doi.org/10.1016/j.jenvman.2018.02.092.

Li, R., Wang, J. J., Zhou, B., Zhang, Z., Liu, S., Lei, S., & Xiao, R. (2017). Simultaneous capture removal of phosphate, ammonium and organic substances by MgO impregnated biochar and its potential use in swine wastewater treatment. *Journal of Cleaner Production, 147*, 96–107. https://doi.org/10.1016/j.jclepro.2017.01.069.

Manyuchi, M. M., Mbohwa, C., & Muzenda, E. (2018). Potential to use municipal waste bio char in wastewater treatment for nutrients recovery. *Physics and Chemistry of the Earth, 107*(March), 92–95. https://doi.org/10.1016/j.pce.2018.07.002.

Manyuchi, M. M., Mbohwa, C., & Muzenda, E. (2020). Bio char production from sugar cane bagasse and its application sugarcane processing wastewater treatment. *Proceedings of the International Conference on Industrial Engineering and Operations Management, 0*(March), 2900–2906.

Mohanty, S. K., Valenca, R., Berger, A. W., Yu, I. K. M., Xiong, X., Saunders, T. M., & Tsang, D. C. W. (2018). Plenty of room for carbon on the ground: Potential applications of biochar for stormwater treatment. *Science of the Total Environment, 625*, 1644–1658. https://doi.org/10.1016/j.scitotenv.2018.01.037.

Mukhtar, S. (2019). Poultry production: Manure and wastewater management. *Encyclopedia of Animal Science*, (January 2005), 744–747. https://doi.org/10.1201/9781482276664–219.

Nyamunda, B. C., Chivhanga, T., Guyo, U., & Chigondo, F. (2019). Removal of Zn (II) and Cu (II) ions from industrial wastewaters using magnetic biochar derived from water hyacinth. *Journal of Engineering (United Kingdom), 2019*. https://doi.org/10.1155/2019/5656983.

Prakash, N., Sudha, P. N., & Renganathan, N. G. (2012). Copper and cadmium removal from synthetic industrial wastewater using chitosan and nylon 6. *Environmental Science and Pollution Research, 19*(7), 2930–2941. https://doi.org/10.1007/s11356-012-0801–8.

Qambrani, N. A., Rahman, M. M., Won, S., Shim, S., & Ra, C. (2017). Biochar properties and eco-friendly applications for climate change mitigation, waste management, and wastewater treatment: A review. *Renewable and Sustainable Energy Reviews, 79*(May), 255–273. https://doi.org/10.1016/j.rser.2017.05.057.

Rehman, A., Nawaz, S., Alghamdi, H. A., Alrumman, S., Yan, W., & Nawaz, M. Z. (2020). Effects of manure-based biochar on uptake of nutrients and water holding capacity of different types of soils. *Case Studies in Chemical and Environmental Engineering, 2*(July), 100036. https://doi.org/10.1016/j.cscee.2020.100036.

Saleh, I. A., Zouari, N., & Al-Ghouti, M. A. (2020). Removal of pesticides from water and wastewater: Chemical, physical and biological treatment approaches. *Environmental Technology and Innovation, 19*, 101026. https://doi.org/10.1016/j.eti.2020.101026.

Shah, M. P. (2020). *Microbial Bioremediation and Biodegradation*. Springer.

Shah, M. P. (2021). *Advanced Oxidation Processes for Effluent Treatment Plants*. Elsevier.

Shah, M. P. (2021). *Removal of Emerging Contaminants through Microbial Processes*. Springer.

Shen, M., Song, B., Zhu, Y., Zeng, G., Zhang, Y., Yang, Y., … Yi, H. (2020). Removal of microplastics via drinking water treatment: Current knowledge and future directions. *Chemosphere, 251*, 126612. https://doi.org/10.1016/j.chemosphere.2020.126612.

Srinivasan, M., Uthandi, S., Karthikeyan, S., Palanisamy, M., Mukund, S., Sivakumar, U., & Sivasubramanian, V. (2017). Bio-char production from micro algal biomass of Chlorella vulgaris Bi-char production from micro algal biomass of Chlorella vulgaris Bio-char production from micro algal biomass of Chlorella vulgaris. *Phykos, 47*(1), 99–104. https://www.researchgate.net/publication/316644446.

Suo, F., You, X., Ma, Y., & Li, Y. (2019). Rapid removal of triazine pesticides by P doped biochar and the adsorption mechanism. In *Chemosphere* (Vol. 235). https://doi.org/10.1016/j.chemosphere.2019.06.158.

Tarelho, L. A. C., Hauschild, T., Vilas-Boas, A. C. M., Silva, D. F. R., & Matos, M. A. A. (2020). Biochar from pyrolysis of biological sludge from wastewater treatment. *Energy Reports, 6*(xxxx), 757–763. https://doi.org/10.1016/j.egyr.2019.09.063.

Thines, K. R., Abdullah, E. C., Mubarak, N. M., & Ruthiraan, M. (2017). Synthesis of magnetic biochar from agricultural waste biomass to enhancing route for waste water and polymer application: A review. *Renewable and Sustainable Energy Reviews, 67*, 257–276. https://doi.org/10.1016/j.rser.2016.09.057.

Visanji, Z., Sadr, S. M. K., Johns, M., Savic, D., & Memon, F. A. (2018). Emerging pollutants in developing countries: Optimising sustainable treatment solutions, *3*, 2204–2191. https://doi.org/10.29007/txmw.

Wathukarage, A., Herath, I., Iqbal, M. C. M., & Vithanage, M. (2019). Mechanistic understanding of crystal violet dye sorption by woody biochar: Implications for wastewater treatment. *Environmental Geochemistry and Health, 41*(4), 1647–1661. https://doi.org/10.1007/s10653-017-0013–8.

Xiang, W., Zhang, X., Chen, J., Zou, W., He, F., Hu, X., … Gao, B. (2020). Biochar technology in wastewater treatment: A critical review. *Chemosphere, 252*, 126539. https://doi.org/10.1016/j.chemosphere.2020.126539.

Xiong, X., Yu, I. K. M., Cao, L., Tsang, D. C. W., Zhang, S., & Ok, Y. S. (2017). A review of biochar-based catalysts for chemical synthesis, biofuel production, and pollution control. *Bioresource Technology, 246*, 254–270. https://doi.org/10.1016/j.biortech.2017.06.163.

Yao, Y., Ding, D., Shao, H., Peng, Q., & Huang, Y. (2017). Antibacterial activity and physical properties of fish gelatin-chitosan edible films supplemented with D-limonene. *International Journal of Polymer Science, 2017*. https://doi.org/10.1155/2017/1837171.

Yao, Y., Gao, B., Zhang, M., Inyang, M., & Zimmerman, A. R. (2012). Effect of biochar amendment on sorption and leaching of nitrate, ammonium, and phosphate in a sandy soil. *Chemosphere*, *89*(11), 1467–1471. https://doi.org/10.1016/j.chemosphere.2012.06.002.

Yu, K. L., Lau, B. F., Show, P. L., Ong, H. C., Ling, T. C., Chen, W. H., … Chang, J. S. (2017). Recent developments on algal biochar production and characterisation. *Bioresource Technology*, *246*, 2–11. https://doi.org/10.1016/j.biortech.2017.08.009.

Zhang, A., Li, X., Xing, J., & Xu, G. (2020). Adsorption of potentially toxic elements in water by modified biochar: A review. *Journal of Environmental Chemical Engineering*, *8*(4), 104196. https://doi.org/10.1016/j.jece.2020.104196.

Zhang, C., Zeng, G., Huang, D., Lai, C., Chen, M., Cheng, M., … Wang, R. (2019). Biochar for environmental management: Mitigating greenhouse gas emissions, contaminant treatment, and potential negative impacts. *Chemical Engineering Journal*, *373*(January), 902–922. https://doi.org/10.1016/j.cej.2019.05.139.

Zheng, Y., Wang, B., Wester, A. E., Chen, J., He, F., Chen, H., & Gao, B. (2019). Reclaiming phosphorus from secondary treated municipal wastewater with engineered biochar. *Chemical Engineering Journal*, *362*, 460–468. https://doi.org/10.1016/j.cej.2019.01.036.

3 Microbial Biofilms for Wastewater Treatment

Bandita Dutta
Maulana Abul Kalam Azad University of Technology

Dibyajit Lahiri
University of Engineering & Management

Sougata Ghosh
RK University

Moupriya Nag
University of Engineering & Management

Rina Rani Ray
Maulana Abul Kalam Azad University of Technology

CONTENTS

DOI: 10.1201/9781003164982-3

3.1 INTRODUCTION

Treatment of industrial wastewater is a prerequisite as it has caused a significant problem to the environment throughout the world (Segura et al. 2015). High quantities of hazardous pollutants like heavy metals, organic pollutants, dyes, pesticides, phenols, detergents or any humic substances are the major contaminants of wastewater which are responsible for the toxic effect on living organisms due to the decomposition of the food chain chemically and biologically through the extreme bioaccumulation (Segura et al. 2015, Bahafid et al. 2013). There are some environmental regulations for removing a wide range of pollutants from industrial wastewater. Depending upon the wastewater characteristics various physicochemical treatments such as membrane treatment, adsorption, and coagulation-flocculation have been implemented in order to remove the pollutants from the industrial effluents (Shammas 2005, Leyva-Ramos et al. 2008). Apart from the use of conventional physicochemical treatments, some biological treatment technologies are also used for the treatment of wastewater. Biosorption is one of the treatment methodologies which are based on the accumulation of biological materials or transformation of the pollutants from the contaminated sites (Fomina & Gadd 2014). Another treatment methodology is related to the implementation of living or dead microorganisms in form of suspended or biofilm-immobilized biomass which is a dense population of microbial cells which is fixed to the solid surface or suspended to the carrier (Bahafid et al. 2013, Pan et al. 2014, Davey & O'toole 2000).

This bioremediation technology has a significant capacity to utilize the catabolic pathways in order to allow the degradation of the hazardous contaminants to resist the stressful conditions to the toxic materials present in the wastewater. Microorganisms like bacteria, fungi and algae help in utilization, neutralization, degradation, and mineralization of a large amount of organic and inorganic pollutants from wastewater and contaminated environment (Miranda et al. 2017). Among the various mechanisms of bioremediation, the biofilm-associated remediation technique is considered to be cost-effective and environmental friendly for the purpose of cleaning the environment. The biofilm-forming microbial species are well adapted to strive at extremes of environmental conditions and competes with oxygen and nutrient availability thereby creating tolerance toward pollutants and acting as an effective agent for bioremediation. Indigenous bacterial species forming the biofilm are profoundly observed within the contaminated sites and resist themselves from the variation of environmental conditions (deCarvalho 2018).

A biofilm is a syntrophic association of microbial species that remain embedded within self-synthesized matrix (extracellular polymeric substance or EPS) that comprise vital nutrients providing the sessile communities with its desirable nutrients (Mortensen 2014). It is a consortium of multiple species like fungi, bacteria, archaea, and algae. The synthesis and morphology of EPS vary with the change in environmental conditions (Yin et al. 2019). It also acts as a potent biomarker in understanding the quality of the water in stream with the persistence of heavy metals within them (Pool et al. 2013). The removal of various organic and inorganic pollutants from the wastewater can be achieved by the use of EPS produced by the microbial cells as coagulants (Flemming et al. 2016). The passage of contaminated water through the biofilm results in the degradation of harmful contaminated organic materials by the microbial cells existing within the biofilm. Thus, biofilm acts as an efficient mechanism of bioremediation involving the processes of absorption, immobilization, and degradation of various contaminated environmental pollutants. Thus, this focuses on the applicability of biofilm for the treatment of wastewater. This chapter will also focus on the applicability of various types of the biofilms in remediation of toxic pollutants.

3.2 WASTEWATER CHARACTERISTICS

The characteristics of wastewater are one of the important aspects in determining the treatment process, which can balance between technology and economy. Parameters like biochemical oxygen demand (BOD), chemical oxygen demand (COD), suspended solids (SS), ammonia, nitrogen, pH, heavy metal concentration, and other biological molecules determine the wastewater characteristics. Wastewater characteristics may also vary depending upon the production of toxic pollutants due to industrial activities comprising petroleum, textile, pharmaceutical, and food-processing industries (Lin et al. 2012). Industrial wastewater contains inert organic solid particles which are suspended and colloidal in nature and some pathogenic microorganisms, having organic strength (>1000 mg COD/L), high alkaline and acidic pH and high salinity. The pH variation, high salinity, and presence of high concentrated toxic materials inhibit the biological degradation by the microbial cells as well as the chemical degradation by the flocculation (Lefebvre & Moletta 2006). Wastewater having the BOD/COD ratio of 0.5 or greater can be treated by biological treatment systems that are not only economic but also eco-friendly (Metcalf & Eddy 2003). Depending upon the aerobic and anaerobic conditions the biological degradation process differs. In the aerobic process, microbial cells utilize the dissolved oxygen to convert the organic matter into biomass and CO_2 whereas in the anaerobic process the organic wastes get converted into methane, CO_2 and H_2O in absence of oxygen. Wastewater having the biodegradable COD concentration <1000 mg/L is considered to be treated through the aerobic process, whereas COD concentration over 4000 mg/L can be treated through the anaerobic process (Chan et al. 2009).

3.3 ROLE OF THE MICROORGANISM IN WASTEWATER TREATMENT

The microbial communities have an important role in organic and inorganic pollutants removal processes from the water. Different bacteria and fungi have different mechanisms in degradation of the pollutants from the wastewater. The specific rhizosphere microbial communities are formed through the amelioration and scrapping specific to the plants. Actinobacteria is the most dominant group of the bacteria found in wastewater, whereas Proteobacteria is mainly found in the roots and biofilm samples as depicted in Figure 3.1 (Haichar et al. 2008).

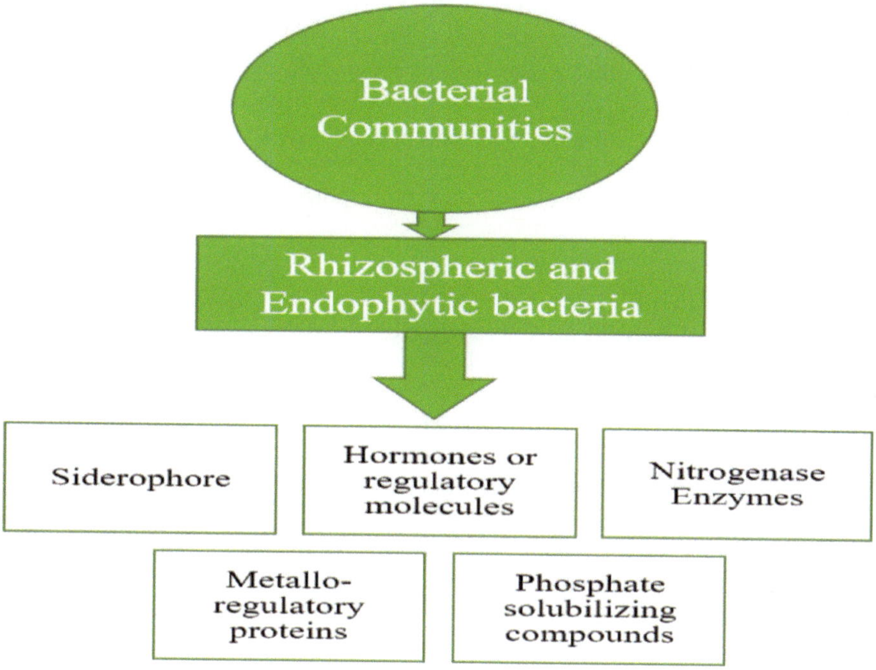

FIGURE 3.1 Bacterial communities and their role.

3.3.1 Role of Endophytes

The metal uptake is enhanced by the microorganisms present in the plants root and the soil that facilitates the breakdown of the complex organic and inorganic compounds into simple molecules. *Rhizobacteria* is one of the endophytes, which stimulates plant growth, biomass production, and uptake of the metals from the soil of the plants (Ryan et al. 2008). The presence of endophytic bacteria in different tissues of the plant increases the ability to take up different biotic and abiotic stresses (Bacon & White 2016). Under the stressed condition, the plant uses different direct and indirect mechanisms to overcome the stress. Siderophore production, antimicrobial metabolites, phosphate-solubilizing compounds, nitrogen-fixing abilities, and phytohormones all come under the direct mechanism, whereas bioremediation and biocontrol come under the indirect mechanism (Berg & Hallmann 2006).

Endophytes also stimulate the degradation of the xenobiotics and nullify the toxic effect of the contaminants like heavy metals and metal ions. For example, *Pseudomonas aeruginosa* increases cadmium tolerance by accumulating and translocating the cadmium into the plants (Shi et al. 2016). Endophytes not only improve xenobiotic remediation but also help in the degradation of pollutants by genetically modified organisms. Toxic metals like iron, nickel, lead, manganese, and chromium are removed from the polluted water with the help of *P. australis.* (Shahid et al. 2019). Endophytic bacteria *T. domigensis* enhance the remediation of textile pollutants. Inoculation of the *Leptochloa fusca* with the consortia of the three endophytes resulted in the increase in the pollutants removal rates while the toxicity level of the polluted water was decreased (Ashraf et al. 2018).

3.3.2 Role of Rhizospheric Bacteria

In degradation of the organic waste, rhizospheric bacteria have an important role, which also translocates the potentially toxic metals (Faußer et al. 2012). This bacterial population qualitatively and quantitively differs from the bacterial population present in the bulk soil, which can also infect the roots and rhizosphere biota. The plant roots attract the bacteria chemotactically by secreting the exudates and metabolites (Ehrenfeld et al. 2005). The plant roots help the microbial colonies to form the biofilm and exchange oxygen, CO_2, nutrients and biochemicals. The root microbial populations emit methane or other gasses and release oxygen to oxidize the iron and ammonia. Different enzymes and organic acids are released by the rhizospheric microbes, which modify the organic wastes and make small molecules in order to provide nutrients to the plant roots to the wetlands (Görres et al. 2013). The sulfide also gets oxidized or reduced by the rhizospheric microbes apart from the regulation of oxygen concentration, redox potential, and release of nitrogen (Husson 2013).

3.3.3 Role of the Fungi

By increasing their bioavailability and transformation fungi has the potential to remediate heavy metals. Fungi like *Allescheriella* sp., *Stachybotrys* sp., *Phlebia* sp., *Pleurotus pulmonarius* and *Botryosphaeria rhodina* are able to bind metals (D'Annibale et al. 2007). With the help of biosorption process, lead-contaminated waste materials can be remediated by fungal species like *Aspergillus parasitica* and *Cephalosporium aphidicola* (Tunali et al. 2006). The fungi *Hymenoscyphus ericae, Neocosmospora vasinfecta*, and *Verticillum terrestre* have resistance to Hg and are able to transform the toxic state of Hg (II) into a non-toxic form (Kelly et al. 2006). *Penicillium, Aspergillus*, and *Rhizopus* are the genera of fungi, which have the efficiency to remove heavy metals from polluted wastewater (Ahalya et al. 2003, Chihpin 1996).

Fungi present in the wetlands are specific in their physical, chemical, and genetic interactions. The association of the root and fungi plays an important role in emission of metal-chelating siderophores, denitrification, and metal detoxification as seen in Table 3.1 (Liu et al. 2015).

TABLE 3.1
Bacteria and Their Role in Wastewater Treatment

Bacteria/Forms	Nature of Bacteria	Role	Ref
Acinetobacter sp.	Perchlorate reducing bacterium	Help in the removal of perchlorate	Bhaskaran et al. (2013)
Denitrifying polyphosphate accumulating microorganisms	-	Helps in the removal of NO_3^-, NH_4^+, -N, TN, TP, and ortho-P.	Zhao et al. (2013)
Bacillus cereus, Acinetobacter, and *Bacillus licheniformis*	Endophytic bacteria	Helps in the reduction of BOD and COD	Ijaz et al. (2015)
Biofilms associated with plant	Natural	Helps in the maintenance of the concentration of Ni and Zn	Ladislas et al. (2015)
Pseudomonas sp., *Acinetobacter* sp. and *Klebsiella* sp.	Endophytic bacteria	Helps in reducing COD and BOD	Li and Guo (2017)
Biofilm	Nitrifying and denitrifying bacteria	Helps in the removal of ammonia, phosphorus and orthophosphorus	Zhang et al. (2016)
Community comprising of Proteobacteria, Firmicutes, Cyanobacteria, Acidobacteria, Planctomycetes, Bacteroidetes, Actinobacteria, and Chlorobi	Periphyton	Helps in the reduction of total nitrogen and total phosphate	Liu et al. (2016)
Dechloromonas, Thiobacillus and *Nitrospira*	Heterotrophic and autotrophic	Helps in reducing the total nitrogen	Gao et al. (2017)
Klebsiella sp., *Acinetobacter junii, Acinetobacter* sp. and *Bacillus subtilis*	Bacterial species responsible for degrading hydrocarbons	Helps in the reduction of BOD and COD	Rehman et al. (2018)
Bacillus cereus, Pseudomonas sp. and *Acinetobacter lwofii*	Bacterial species responsible for the degradation of phenol	Helps in breaking the complex recalcitrants	Saleem et al. (2018)
Bacterial biofilm		Helps in the removal of total nitrogen, total phosphate and total organic carbon (TOC)	Li et al. (2010)
Proteobacteria	Nitrosomonadaceae	Helps in the removal of total nitrogen, total phosphate and total organic carbon (TOC)	Wu et al. (2016)
Nitrifying and denitrifying	Carrying *amoA*, *nirK* and *nirS* genes	Helps in reducing COD	Faulwetter et al. (2011)
Biofilms		Helps in the removal of copper	Tanner and Headley (2011)
Ammonifying bacterial strains	Engineering bacterial strain	Reduction in organic nitrogen amount	Zhao et al. (2011)
Bacillus subtilis, Klebsiella sp., *Acinetobacter junii*, and *Acinetobacter* sp.	Hydrocarbon degrading bacteria	Reduction in BOD and COD	Rehman et al. (2019)

3.4 BIOFILM ASSOCIATED WASTEWATER REMEDIATION

A biofilm is a multi-layered community of sessile cells that form a syntrophic association and remains embedded in hydrated EPS. This EPS helps the cells to colonize upon the living, inert or inanimate surfaces (López et al. 2010). The composition of the matrix contains various nutrients like carbohydrates, proteins, lipids, nucleic acids and other minerals that provide nutrients to the in-dwelling cells. This influences the organisms that are living in the matrix of the biofilm to become virulent, as this encapsulation gives rise to antimicrobial resistance, and is also associated with the phenotypic and genotypic changes within the organisms. This biofilm matrix is the preferred way for the bacteria to live in as it provides the cells with optimal conditions for the exchange of genetic materials involving the process of horizontal gene transfer, and hence biofilm becomes the natural state of its existence (Hall-Stoodley et al. 2004, Shah 2020). The microbial community remains distributed throughout the matrix. At the bottom layers of a biofilm, sessile communities are bound together through a polysaccharide matrix containing organic and inorganic materials. The above layer is amorphous in nature and it is extended toward the surroundings.

3.5 BIOFILM IN WASTEWATER TREATMENT

Since 1893 in England biofilm-based technology is used in wastewater treatment with the utilization of trickling filters (Lohmeyer 1957). Despite having detrimental effects toward the environment, biofilm is considered as an effective agent in biodegradation of the complex pollutants. Compared to the planktonic microbial cells, biofilm-mediated water treatment is efficient and cost-effective, due to their adaptive ability in stressed condition and survival in the harsh environment in presence of different hydrodynamic forces. The complex structures of the EPS help the biofilm to overcome the different environmental challenges (Kreft & Wimpenny 2001, Miqueleto et al. 2010). Biofilm showed its effectiveness in degradation of the xenobiotics due to its beneficial physical and physiological interactions (Das et al. 2012). Different technologies have been widely used in the removal of organic as well as inorganic wastes, contaminants like petroleum, dyes, heavy metals, and pesticides where beneficial aspects of biofilm-like adsorption, immobilization and degradation have been implemented (Mitra & Mukhopadhyay 2016, Quintelas et al. 2010).

Biofilm-based system has several advantages over the suspended growth system in wastewater treatment due to its requirement of low space, operational flexibility, reduced hydraulic retention time, increased biomass residence time, high biomass concentration and ability to degrade the waste materials in slower microbial growth rate, controlled reaction rate and population dynamics (Lazarova & Manem 2000).

Biofilm reactor is mainly classified into two categories depending on the state of the support material, one is fixed-bed reactor and another is mobile beds reactor. In a fixed-bed system, biofilm is usually formed on static material like rocks, plastic profiles, membranes, or sponges. The flow liquid passing through the static material provides the microorganisms with proper nutrients and oxygen. Moving bed system has the biofilm present in the continuous moving media, where high water velocity, mechanical stirring, and high air pressure are maintained. A shear force is applied by altering the stirring intensity and flow velocity and by backwashing to the biofilm reactor to maintain the biofilm thickness (Rodgers & Zhan 2003).

The carrier material for the biofilm is selected based on its size, porosity, density, resistance to erosion and large surface area (m^2/m^3) in order to maintain high biological activity even in a small volume bioreactor (Christensson & Welander 2004).

Bioaugmentation, which is a wastewater treatment using microorganism, is enhanced by using specific bacterial strains. Increased diversity as well as the genetic alteration of the microorganism to the system enhance bioaugmentation. Environment is one of the important aspects inside the bioreactor where the adaptation to the new environment of the microorganisms plays an important

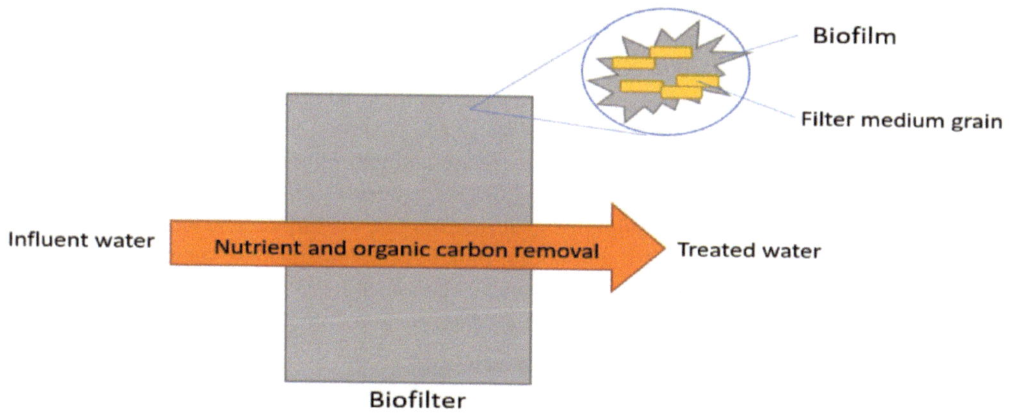

FIGURE 3.2 Biofilm-associated removal of pollutants from wastewater.

TABLE 3.2
Biofilm Bioreactors Associated Wastewater Remediation

Microorganism or Culture	Type of Reactor	Pollutant	Support Material	Reference
Bacterial species in wastewater +*Geobacter metallireducens*	MFC	COD	Graphite	Liu et al. (2004)
Desulfovibrio desulfuricans	MFC	Sulfate	Activated carbon cloth	Zhao et al. (2008)
Arthrobacter viscosus	Suspended biofilm	Cr (III) and Cr(VI)	NaY zeolite	Silva et al. (2008)
Wastewater treatment plant associated secondary sludge	Membrane biofilter reactor	Toluene	Hollow-fiber membrane	Parvatiyar et al. (1996)
Proteobacteria, Bacteroidetes, Nitrospirae, Cyanobacteria, Actinobacteria	Natural ventilation trickling filters (NVTFs)	COD	Sponge Zeolite Ceramsite	Zhang et al. (2016)
Bacillus coagulans	Column lab scale	Cr (VI)	Granular activated carbon (GAC)	Quintelas et al. (2008)
Mixed culture of herbicide degrading bacteria	Biofilm column reactor	MCPP; 2,4-D (herbicides)	GAC	Oh and Tuovinen (1994)
Sphingomonas xenophaga QYY	Laboratory scale (MBR)	Bromoamine acid Ammonia	Sponge Zeolite Ceramsite	Qu et al. (2009)
Rhodococcus sp. and *Pseudomonas* sp.	Laboratory and pilot scale (FBR)	Chlorophenol	Silica-based spherical Celite R-633 (Celite Co.) microcarriers	Puhakka et al. (1995)

role in degradation. Thus the use of biofilm-mediated bioaugmentation protects the microorganism against toxic compounds as shown in Figure 3.2 and Table 3.2 (Stephenson & Stephenson 1992).

3.6 DIFFERENT TYPES OF THE BIOFILM-BASED BIOREACTOR

Wastewater treatment can be classified into two broad categories like attached growth system and dispersed growth system based on the structural configuration.

3.6.1 Dispersed Growth System

The suspended and dispersed form of the biomass, without any support attachment in liquid media, the growth system is considered as dispersed or suspended. Microorganisms present inside the biomass absorb the organic matter and nutrients in the close vicinity of their growth and form the microcolonies. The microcolonies settle as sludge and get removed from the system or resuspended. The dispersed biomass gets removed from the system as sewage. The hydraulic retention time should be greater than the doubling time of microorganisms in order to increase the bacterial cell activity and population size. If the doubling time became greater than the hydraulic retention time, bacterial cells will be easily washed out from the system. This is the main factor to be considered in designing the biological reactor as the reactor volume is directly related to the retention time (Lubberding 1995).

3.6.2 Activated Sludge Technology

Activated sludge system is a multi-chambered reactor unit, where aerobic microorganisms are able to degrade the organic components present in the wastewater and produce high quality of the effluent. Constant oxygen supply is required to maintain the proper aeration inside the tank. Along with aerobic bacteria, some anaerobic and nitrifying bacteria are also present inside the tank helping in carbon oxidation and nitrification to produce carbon dioxide and nitrogen. Small cluster and flocs are produced during the aeration process and after that they are transferred into the secondary clarifier where those flocs settle down. The effluent moves on for further treatment and then the sludge is recycled back to the aeration tank. In industrialized countries this activated sludge technology is most commonly used where the biological wastes are removed by sedimentation (Kim et al. 2010, Shah 2020, 2021).

3.6.3 Extended Aeration System

It is a modified activated sludge process, which provides the biological treatment in the removal of biodegradable organic waste under aerobic condition. The raw sewage directly enters the aerobic digestion chamber, where aerobic bacteria digest the solid substances as the sewage is aerated for 24 hours, indicating complete digestion of the wastes. There is no need for a separate sludge digester and primary settling tank as organic solids are settled down in the aeration tank during long detention time. The main advantage of this system is, its construction, operation, and oxygen transfer efficiency which is high, less yield of sludge, exceptional mixing energy from controlled aeration chain environment. Apart from all these the extended aeration plants without additional unit processes do not achieve denitrification and phosphorus removal from wastewater.

3.6.4 Attached Growth System

In this system the biomass increases and gets attached to a solid support, to create biofilm. The attachment to the solid support is regulated by the media composition, cell–cell interactions, and presence of the polymer molecules upon the surface (Boks et al. 2008). The biofilm grows upon the solid media taking all the nutrients and growth factors from the wastewater. In this system, there is density gradient and velocity gradient which maintain the biomass concentration and flow of the liquid. Therefore, the new substrate is continuously added to the system to increase bacterial cell activity.

Depending upon the technology and depollution strategy, the five compartments are required to design the attached growth system bioreactor.

1. Influent, the wastewater having the pollutant of concern at a particular concentration
2. Containment structure

3. Biofilm carrier or material, which is required for the growth and attachment of the microbial cells
4. Collection system for effluent water
5. Proper aeration and mixing system for agitation and carrier distribution

Keeping those above compartments biofilm-based reactor is classified into membrane biofilm reactors (MBR), trickling filter (TF), moving bed biofilm reactors (MBBR), and fluidized-bed reactors (FBR).

3.6.5 Membrane Biofilm Reactor

It is a pollutant removal system that works in an aqueous solution. Biofilm-based biological degradation of the wastewater is assisted by the physical separation using a membrane unit, which usually replaces the secondary settler. This system provides distinct advantages in the treatment of the industrial and urban wastewater for removing organic as well as inorganic wastes (Di Fabio et al. 2013). This also provides flexibility toward influent variability, high volume loading, good disinfection efficiency, reduced sludge production, and excellent quality of the effluent. Both the submerged or immersed configuration and external or side stream configuration have been implemented in this system. In 1989, the first submerged configuration was implemented where the membrane module was directly immersed inside the reactor. (Yamamoto et al. 1989). The reactor configuration provides negative pressure toward the permeate side, having distinct advantages like low energy consumption, low operating costs and less cleaning procedure. The membrane present outside the bioreactor helps in the recirculation of the mixed liquor, and it can be replaced easily. The driving force through this bioreactor is the high cross-flow velocity (CFV) (Liao et al. 2006). In moving or fixed-bed configuration aerated membranes in the bioreactor provides support to the biofilm growth. Other materials like cord media, RBC media, sponge, plastic media, and granular activated carbon (GAC) can also provide support to the biofilm growth inside the system (Ivanovic & Leiknes 2012). This system has efficiency in removing COD, inorganic nitrogen and degradation of 90% of pollutants. Membrane fouling is one of the major limitations of this system due to colloidal and soluble organic materials like suspended solids or other biopolymers. This fouling causes the clogging of the membrane layer which needs to be resolved in order to maintain the plant and reduce the high energy demand. Maintaining the membrane fouling also reduces energy consumption and optimizes the degradation.

3.6.6 Fluidized-Bed Biofilm Reactor

In this system, the small carrier molecules are kept inside the column to form a bed, which usually keeps the column fluidized and maintains the vertical dynamic flow movement of the wastewater. Though aeration is required for recycling, the air inside the column causes turbulence, which may break the biofilm layer upon the carrier. Particles present in the FBBR, increase the gradient size from top to the bottom of the bioreactor. Support materials like zeolite, silica-based materials, or GAC are used for the fluidized bed. Small materials having diameter 1 mm are used for the pilot scale experiments, where the driving force inside the reactor exceeds gravity (30–50 m/h) to keep the particles separated and suspended (Lewandowski & Boltz 2011). Apart from the mass transfer efficiency, this bioreactor can be used for promising tertiary denitrification. The fluidization of the particles increases the mass transfer and treatment efficiency, where the low degree of bed expansion is recommended to decrease the flow velocity, energy consumption and increase effective biomass concentration. For the treatment of the chlorophenol-contaminated groundwater, *Pseudomonas* sp. and *Rhodococcus* sp. were used in fluidized-bed biofilm bioreactor. The chlorophenol was mineralized with 99.9% efficiently and 2,3,4,6-tetrachlorophenol, 2,4,6-trichlorophenol, and pentachlorophenol were removed with influent and less than 1 hour hydraulic retention time (Puhakka et al. 1995).

3.6.7 Moving Bed Biofilm Reactor

The working principle of MBBR is the combination of the activated sludge process and biofilters. This reactor is mainly used for aerobic, anoxic, or anaerobic processes, where the carrier serves as the supportive material for biofilm growth. In the aerobic process, the agitation helps in mixing up the air inside the tank, whereas in the anaerobic process the carrier is moved by the mechanical mixing (Ødegaard 2006). MBBRs are continuous flow reactor units where 'K1' biofilm carrier is used. High-density polyethylene having density is present within the cylinder, crosses the cylinder inside and 'fins' on the outside (Ødegaard 2006). The moving beds have advantages over the fixed beds in minimizing clogging and utilizing the whole bioreactor volume. The shearing forces that create turbulence inside the reactor help in proper diffusion of the compounds with the biofilm and also lower the thickness of the biofilm. Biomass recirculation is one of the major problems of fixed-bed biofilm reactors which is minimized in this system, only excess biomass is separated (Ivanovic & Leiknes 2012). Moving bed reactor provides a large surface area to the microbial cell for growth and efficient environment for mixing, dispersing the volatile acids from aqueous solution and liberation of the biogas (Karadag et al. 2015).

MBBR also helps with carbon oxidation and nitrification-denitrification in a single or combined system on both a lab scale and a large scale. The only disadvantage of this system is that the carrier should be removed in time intervals to provide benefits to the components (Gilbert et al. 2014).

3.6.8 Trickling Filter

TF is a three-phase biofilm reactor which is in use for over 50 years. TF is a clarification unit composed of five compartments like (a) an influent water distribution system, through which wastewaters are coming into the reactor. The distribution occurs based on fixed-nozzle or rotary distributors; (b) a containment structure; (c) a support media; (d) an underdrain system; and (e) a ventilation system. It is a liquid-solid separation system where suspended solids get separated through circular or rectangular secondary clarifiers. The rotary distributor allows intermittent wastewater application and effective substratum wetting providing an advantage to the influent distribution. Dry pockets present inside the column permit the biofilm to resist inside it during the process of aeration (Lewandowski & Boltz 2011). The temperature gradient should be maintained to provide proper ventilation inside the TF. The carbon oxidation and nitrification both are done within the system and the biofilm upon the filter media uses the oxygen from the air for carbon oxidation (Metcalf & Eddy 2003). Stone and gravel are mainly used inside the TF, which restricts the air within the system and cells get the oxygen to form the microbial biofilm. There is an advantage of using the stone bed TF as the void space is limited and there is high organic load due to the excessive microbial cell growth resulting in clogging (Lewandowski & Boltz 2011).

3.7 BIOLOGICAL PROCESS ASSOCIATED IN THE TREATMENT OF WASTEWATER

3.7.1 Nitrogen Fixation

Microbial wastewater treatment is mainly carried out by microbial species like fungi, protozoa, algae, and bacteria. Those microorganisms utilize the carbon and nitrogen from the wastewater through various metabolic and respiratory pathways hence removing the pollutants. Various Bacteroidetes, Actinobacteria or Alpha-, Beta- and Gammaproteobacteria are mainly found in wastewater treatment systems (Wagner & Loy 2002). Wastewater contains organic pollutants like proteins, carbohydrates, fats, oils, or nutrients like nitrogen and phosphorus, which under aerobic condition are biochemically oxidized by heterotrophic bacteria producing carbon dioxide, water, ammonia and biomass which are utilized by those bacteria as well. Methanogenic archaea partially oxidize organic material under the anaerobic condition to yield carbon dioxide, methane, and biomass (Madigan et al. 2003). The removal of the nitrogen is done by a combination of nitrification-denitrification, oxidation of ammonia to nitrate and reduction of nitrate to nitrogen gas. Nitrifying bacteria are chemo-lithotrophic in nature which are utilizing inorganic nitrogen as electron donor, whereas other ammonia-oxidizing bacteria (AOB) such as *Nitrosomonas*, *Nitrosospira*, and *Nitrosococcus* convert ammonia to nitrite and nitrate oxidizing bacteria (NOB) *Nitrobacter*, *Nitrospira*, *Nitrococcus*, and *Nitrospina* nitrite to nitrate.

Equation 3.1 represents the conversion from ammonia to nitrite

$$15CO_2 + 13NH_4^+ \rightarrow 10NO_2^- + 3C_5H_7NO_2 + 23H^+ + 4H_2O$$

Equation 3.2 represents the conversion from nitrite to nitrate

$$5CO_2 + NH4^+ + 10NO_2^- + 2H_2O \rightarrow 10NO_3^- + C_5H_7NO_2 + H^+$$

The nitrates are reduced to nitrogen gas through the denitrification process, which helps in the removal of nitrogen from the water phase. Denitrifying organisms respire the nitrate or nitrite in absence of the oxygen through the enzymatic reactions coupled with the bacterial inner membrane. The redox potential of oxygen is much higher than the nitrate in the aerobic electron transport system (Henze et al. 2002).

Equation 3.3 represents the aerobic electron transport system where acetate acts as electron donor

$$5CH_3COOH + 8NO_3^- \rightarrow 8HCO_3^- + 2CO_2 + 6H_2O + 4N_2$$

The denitrifying microbial flora comprised of *Pseudomonas*, *Alcaligenes*, *Acinetobacter*, *Paracoccus*, *Methylobacterium*, *Bacillus*, and *Hyphomicrobium* are mainly found in the wastewater treatment plant (Figure 3.3).

3.7.2 Phosphorus Remediation

Phosphorus in the wastewater can be removed through the intracellular accumulation of the polyphosphates which is taken up by the cells for growth. The polyphosphate accumulating organisms (POAs) are also responsible to remediate phosphates from the wastewater. Their uptake is enhanced from 1% to 3% to 10% by the alternating anaerobic and aerobic environments (Bitton 2005, Srivastava & Srivastava 2005).

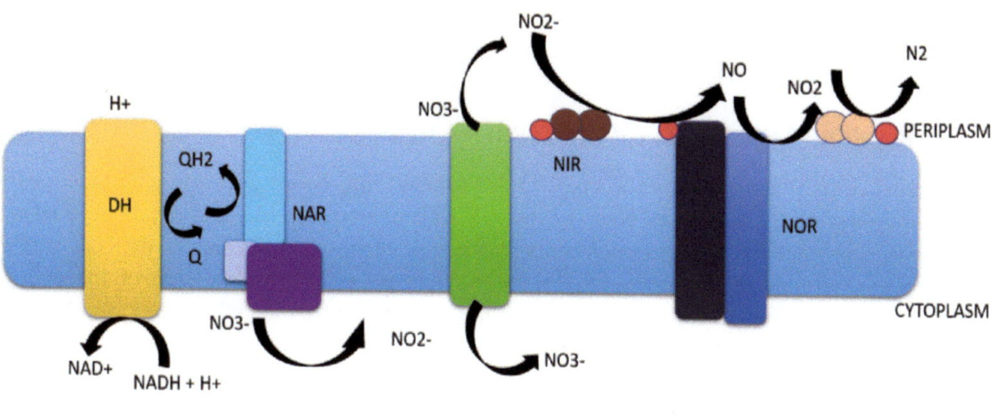

FIGURE 3.3 The enzymatic reactions involved in denitrification in bacteria.

The aerobic step followed by the removal of the biomass helps in the removal of the phosphorus from the wastewater. Some bacteria can accumulate a large amount of phosphorus under aerobic conditions. For example, *Acinetobacter calcoaceticus, Acinetobacter lwoffii*, and *Aeromonas hydrophila* are the group of bacterial species that enhances the amount of phosphorus under aerobic conditions (Srivastava & Srivastava 2005).

3.8 ORGANIC POLLUTANT DEGRADATION

Microbes are well known bio-remediators as they have the capability to break down all organic pollutants by the process of co-metabolism. Rhizospheric bacteria degrade the complex carbon-based compounds into organic carbon and electron acceptor (Stottmeister et al. 2003). The rate of biodegradation is usually regulated by the microbial population, xenobiotics, and the microbes that are heavily influenced by the macrophyte species, possessing the ability to degrade hydrocarbons and aromatic hydrocarbons (Paris et al. 1981).

Biofilm formed in aquatic plants can degrade the organics like phenolics, amines, aliphatic and aldehydes along with organic matter like polychlorinated biphenyls (PCBs) and atrazine. (Simpson 2008).

Methanotrophs are one of the proteobacteria, which utilize methane as carbon source and energy, and by enzymatic reactions can also degrade different harmful organic complexes like chlorinated ethane (Semrau et al. 2010).

3.9 HEAVY METALS REMOVAL

Both the rhizospheric and endophytic bacteria are able to remove the heavy metals, by absorbing metal ions into their cell walls (Mullen et al. 1989).

Microorganisms accumulate heavy metals by specific metal-binding proteins and peptides like phytochelatins and metallothionein. When the toxic metals are exposed to the bacterial cells, the transcription factors of the metal-binding proteins of the cells facilitate the hormone and redox signaling process (Cobbett & Goldsbrough 2002). Cyanobacteria produce proteins that bind to the metals in order to reduce metal toxicity (Huckle et al. 1993). The genetically modified *Ralstoni aeutropha* is able to reduce the harmful Cd (II) by producing metallothionein on the surface of the cell (Valls et al. 2000). *Escherichia coli* also regulate the accumulated Cd toxicity through the production of different proteins and peptides (Mejáre & Bülow 2001). The arsenic (As) and mercury (Hg) get reduced by the microorganism through the production of metallo-regulatory protein. Some *Bacillus* sp. strains have a diverse mechanism that can alleviate the metal stress by an active transport efflux pump (Shin et al. 2012). Methylation by a few endophytic bacteria is able to detoxify the metals. Few Gram-negative bacteria have specific mercury-resistant (*Mer*) operon genes that can degrade the organic mercurials and reduce them into Hg^{+2} (Brown et al. 2003).

3.10 BIOSORPTION AND BIOACCUMULATION

Biosorption of the metal ions by the bacteria is classified into two processes, active and passive. Passive biosorption usually takes place at the cell wall of both the live and dead bacterial cells, through various metabolic processes. Metals ions get attached to the surface of the cell wall with the functional groups (e.g., amine, amide, carbonyl, hydroxyl, sulfonate, etc.) of the cell wall (Vijayaraghavan & Yun 2008). Metal ions which are taken up by the cells are transported to the specific intracellular organelles. Endophytic bacteria like *Microbacterium* sp. and *Pseudomonas fluorescens* on *Brassica napus* have potential in heavy metal Pb bioaccumulation and detoxification (Sheng et al. 2008). *Saccharomyces cerevisiae*, which is commonly known as baker's yeast, is one of the bio-sorbent for the removal of cadmium and zinc, which is attributed to the ion exchange mechanism (Chen & Wang 2007). Similarly, *Cunninghamella elegans* is another efficient sorbent for the heavy metal remediation of textile industry (Tigini et al. 2010).

The EPS of the biofilm, mainly composed of proteins, polysaccharides, nucleic acid, and lipids, binds to the metals in order to reduce their bioavailability (Rajkumar et al. 2013). Biofilm formed by *Azobacter* sp. develop complexes with chromium and cadmium through EPS and reduces the metal uptake by *Triticum aestivum* (Joshi & Juwarkar 2009).

3.11 CONCLUSION

In the field of wastewater treatment technologies, biofilm-based systems have several advantages due to their operational advantages, high resistance to toxic substances and shock loads. Economic issues, sustainability, system performance, and reliability are determining factors for its use at the industrial scale. The development of the biofilm and formation of biomass is influenced by the pH changes, metabolic activities of the cells, which can hinder these technologies through membrane fouling. However, biofilm-associated water treatment is more efficient and environmentally benign alternative to physical and chemical treatment processes.

REFERENCES

Ahalya N., Ramachandra T., Kanamadi R. (2003) Biosorption of heavy metals. *Res. J. Chem. Environ.* 7:71–79.

Ashraf S., Afzal M., Naveed M., Shahid M., Ahmad Zahir Z. (2018) Endophytic bacteria enhance remediation of tannery effluent in constructed wetlands vegetated with *Leptochloa fusca*. *Int. J. Phytorem.* 20:121–128.

Bacon C.W., White J.F. (2016) Functions, mechanisms and regulation of endophytic and epiphytic microbial communities of plants. *Symbiosis.* 68:87–98.

Bahafid W., Joutey N.T., Sayel H. (2013) Chromium adsorption by three yeast strains isolated from sediments in Morocco. *Geomicrobiol J.* 30:422–429.

Berg G., Hallmann J. (2006) Control of plant pathogenic fungi with bacterial endophytes. In *Microbial Root Endophytes*. Berlin/Heidelberg, Germany: Springer, 53–69.

Bhaskaran K., Nadaraja A.V., Tumbath S., Shah L.B., Veetil P.G.P. (2013) Phytoremediation of perchlorate by free floating macrophytes. *J. Hazard. Mater.* 260:901–906.

Bitton G. (2005) *Wastewater Microbiology*. 3rd ed. New Jersey: John Wiley & Sons, Inc.

Boks N.P., Norde W., van der Mei H.C., Busscher H.J. (2008) Forces involved in bacterial adhesion to hydrophilic and hydrophobic surfaces. *Microbiology.* 154(10):3122–3133.

Brown N.L., Stoyanov J.V., Kidd S.P., Hobman J.L. (2003) The MerR family of transcriptional regulators. *FEMS Microbiol. Rev.* 27:145–163.

Chan Y.J., Chong M.F., Law C.L. (2009) A review on anaerobic–aerobic treatment of industrial and municipal wastewater. *Chem. Eng J.* 155:1–18.

Chen C., Wang J.-L. (2007) Characteristics of Zn^{2+} biosorption by Saccharomyces cerevisiae. *Biomed. Environ. Sci. BES.* 20:478–482.

Chihpin H. (1996) Application of Aspergillusoryzae and rhizopusoryzae for Cu (II) removal. *Water Res.* 9:1985–1990.

Christensson M., Welander T. (2004) Treatment of municipal wastewater in a hybrid process using a new suspended carrier with large surface area. *Wat. Sci. Tech.* 49(11–12):207–214.

Cobbett C., Goldsbrough P. (2002) Phytochelatins and metallothioneins: Roles in heavy metal detoxification and homeostasis. *Annu. Rev. Plant Biol.* 53:159–182.

D'Annibale A., Leonardi V., Federici E., Baldi F., Zecchini F., Petruccioli M. (2007) Leaching and microbial treatment of a soil contaminated by sulphide ore ashes and aromatic hydrocarbons. *Appl. Microbiol. Biotechnol.* 74:1135–1144.

Das N., Basak L.V.G., Salam J.A. (2012) Application of biofilms on remediation of pollutants – an overview. *J. Microbiol. Biotechnol. Res.* 2:783–790.

Davey M.E., O'toole G.A. (2000) Microbial biofilms: From ecology to molecular genetics. *Microbiol. Mol. Biol. Rev.* 64:847–867.

deCarvalho C.C. (2018) Marine biofilms: A successful microbial strategy with economic implications. *Front. Mar. Sci.* 5:126.

Di Fabio S., Lampis S., Zanetti L. (2013) Role and characteristics of problematic biofilms within the removal and mobility of trace metals in a pilot-scale membrane bioreactor. *Process Biochem.* 48:1757–1766.

Faußer A.C., Hoppert M., Walther P., Kazda M. (2012) Roots of the wetland plants Typhalatifolia and Phragmitesaustralis are inhabited by methanotrophic bacteria in biofilms. *Flora-Morphol. Distrib. Funct. Ecol. Plants* 207:775–782.

Faulwetter J., Burr M.D., Cunningham A.B., Stewart F.M., Camper A.K., Stein O.R. (2011) Floating treatment wetlands for domestic wastewater treatment. *Water Sci. Technol.* 64:2089–2095.

Flemming H.-C., Wingender J., Szewzyk U., Steinberg P., Rice S.A., Kjelleberg S. (2016) Biofilms: An emergent form of bacterial life. *Nat. Rev. Microbiol.* 14:563.

Fomina M., Gadd G.M. (2014) Biosorption: Current perspectives on concept, definition and application. *Bioresour. Technol.* 160:3–14.

Gao L., Zhou W., Huang J., He S., Yan Y., Zhu W., Wu S., Zhang X. (2017) Nitrogen removal by the enhanced floating treatment wetlands from the secondary e_uent. *Bioresour. Technol.* 234:243–252.

Gilbert E.M., Agrawal S., Karst S.M. (2014) Low temperature partial nitritation/anammox in a moving bed biofilm reactor treating low strength wastewater. *Environ. Sci. Technol.* 48:8784–8792.

Görres C.-M., Conrad R., Petersen S.O. (2013) Effect of soil properties and hydrology on Archaeal community composition in three temperate grasslands on peat. *FEMS Microbiol. Ecol.* 85:227–240.

Haichar F.Z., Marol C., Berge O., Rangel-Castro J.I., Prosser J.I., Balesdent J., Heulin T., Achouak W. (2008) Plant host habitat and root exudates shape soil bacterial community structure. *ISME J.* 2:1221.

Hall-Stoodley L., Costerton J.W., Stoodley P. (2004) Bacterial biofilms: From the natural environment to infectious diseases. *Nat. Rev. Microbiol.* 2(2):95–108.

Henze M., Harremoës P., La Cour Jansen J., Arvin E. (2002) *Wastewater Treatment. Biological and Chemical Processes.* 3rd ed. Förstner U., Murphy R.J., Rulkens W.H. New York: Springer-Verlag.

Huckle J.W., Morby A.P., Turner J.S., Robinson N.J. (1993) Isolation of a prokaryotic metallothionein locus and analysis of transcriptional control by trace metal ions. *Mol. Microbiol.* 7:177–187.

Husson O. (2013) Redox potential (Eh) and pH as drivers of soil/plant/microorganism systems: A transdisciplinaryverview pointing to integrative opportunities for agronomy. *Plant Soil.* 362:389–417.

Ijaz A., Shabir G., Khan Q.M., Afzal M. (2015) Enhanced remediation of sewage e_uent by endophyte-assisted floating treatment wetlands. *Ecol. Eng.* 84:58–66.

Ivanovic I., Leiknes T.O. (2012) The biofilm membrane bioreactor (BF-MBR) – a review. *Desalin Water Treat.* 37:288–295.

Joshi P.M., Juwarkar A.A. (2009) In vivo studies to elucidate the role of extracellular polymeric substances from azotobacter in immobilization of heavy metals. *Environ. Sci. Technol.* 43:5884–5889.

Karadag D., Koroglu O.E., Ozkaya B. (2015) Anaerobic granular reactors for the treatment of dairy wastewater: A review. *Int. J. Dairy Technol.* 68:459–470.

Kelly D.J., Budd K., Lefebvre D.D. (2006) The biotransformation of mercury in pH-stat cultures of microfungi. *Botany.* 84:254–260.

Kim H.S., Gellner J.W., Boltz J.P., Freudenberg R.G., Gunsch C.K., Schuler A.J. (2010) Effects of integrated fixed film activated sludge media on activated sludge settling in biological nutrient removal systems. *Water Res.* 44(5):1553–1561.

Kreft J.U., Wimpenny J.W. (2001) Effect of EPS on biofilm structure and function as revealed by an individual-based model of biofilm growth. *Water Sci. Technol.* 43:135–141.

Ladislas S., Gerente C., Chazarenc F., Brisson J., Andres Y. (2015) Floating treatment wetlands for heavy metal removal in highway stormwater ponds. *Ecol. Eng.* 80:85–91.

Lazarova V., Manem J. (2000) Innovative biofilm treatment technologies for water and wastewater treatment. In *Biofilms II: Process Analysis and Applications*, Bryers J.D., Editor. New York: Wiley-Liss, 159–206.

Lefebvre O., Moletta R. (2006) Treatment of organic pollution in industrial saline wastewater: A literature review. *Water Res.* 40:3671–3682.

Lewandowski Z., Boltz J. (2011) Biofilms in water and wastewater treatment. In *Treatise on Water Science.* Oxford: Academic, 529–570.

Leyva-Ramos R., Jacobo-Azuara A., Diaz-Flores P.E. (2008) Adsorption of chromium(VI) from an aqueous solution on a surfactant-modified zeolite. *Colloids Surf. A Physicochem. Eng. Asp.* 330:35–41.

Li X., Guo R. (2017) Comparison of nitrogen removal in floating treatment wetlands constructed with Phragmitesaustralis and acoruscalamus in a cold temperate zone. *Water Air Soil Pollut.* 228:132.

Li X.-N., Song H.-L., Li W., Lu X.-W., Nishimura O. (2010) An integrated ecological floating-bed employing plant, freshwater clam and biofilm carrier for purification of eutrophic water. *Ecol. Eng.* 36:382–390.

Liao B-Q., Kraemer J.T., Bagley D.M. (2006) Anaerobic membrane bioreactors: Applications and research directions. *Crit. Rev. Environ. Sci. Technol.*

Lin H., Gao W., Meng F. (2012) Membrane bioreactors for industrial wastewater treatment: A critical review. *Crit. Rev. Environ. Sci. Technol.* 42:677–740.

Liu H., Ramnarayanan R., Logan B.E. (2004) Production of electricity during wastewater treatment using a single chamber microbial fuel cell. *Environ. Sci. Technol.* 38:2281–2285.

Liu J., Wang F., Liu W., Tang C., Wu C., Wu Y. (2016) Nutrient removal by up-scaling a hybrid floating treatment bed (HFTB) using plant and periphyton: From laboratory tank to polluted river. *Bioresour. Technol.* 207:142–149.

Liu W.-L., Guan M., Liu S.-Y., Wang J., Chang J., Ge Y., Zhang C.-B. (2015) Fungal denitrification potential in vertical flow microcosm wetlands as impacted by depth stratification and plant species. *Ecol. Eng.* 77:163–171.

Lohmeyer G.T. (1957) Trickling filters and operation tips. *Sewage Ind. Waste.* 29:89–98.

López, D., Vlamakis, H., Kolter, R. (2010) Biofilms. *Cold Spring Harb. Perspect. Biol.* 2(7):a000398.

Lubberding H.J. (1995) Applied anaerobic digestion. In: *International Course on Anaerobic Treatment.* Wageningen, The Netherlands: Wageningen Agricultural University/IHE Delft.

Madigan M.T., Martinko J.M., Parker J. (2003) *Brock Biology of Microorganisms.* 10th ed. New Jersey: Pearson Education, Inc.

Mejáre M., Bülow L. (2001) Metal-binding proteins and peptides in bioremediation and phytoremediation of heavy metals. *Trends Biotechnol.* 19:67–73.

Metcalf W., Eddy C. (2003) *Wastewater Engineering: Treatment and Reuse.* New York: McGraw Hill, 384.

Miqueleto A.P., Dolosic C.C., Pozzi E. (2010) Bioresource technology influence of carbon sources and C/N ratio on EPS production in anaerobic sequencing batch biofilm reactors for wastewater treatment. *Bioresour. Technol.* 101:1324–1330.

Miranda A.F., Ramkumar N., Andriotis C., Höltkemeier T., Yasmin A., Rochfort S. (2017) Applications of microalgal biofilms for wastewater treatment and bioenergy production. *Biotechnol. Biofuel.* 10:120.

Mitra A., Mukhopadhyay S. (2016) Biofilm mediated decontamination of pollutants from the environment. *AIMS Bioeng.* 3:44–59.

Mortensen BK. (2014) Formation and detection of biofilms. *Bactoforce Int.*

Mullen M., Wolf D., Ferris F., Beveridge T., Flemming C., Bailey G. (1989) Bacterial sorption of heavy metals. *Appl. Environ. Microbiol.* 55:3143–3149.

Ødegaard H. (2006) Innovations in wastewater treatment: The moving bed biofilm process. *Water. Sci. Technol.* 53:17–33.

Oh K.H., Tuovinen O.H. (1994) Biodegradation of the phenoxy herbicides MCPP and 2, 4-D in fixedfilm column reactors. *Int. Biodeterior. Biodegrad.* 33:93–99.

Pan X., Liu Z., Chen Z. (2014) Investigation of Cr(VI) reduction and Cr(III) immobilization mechanism by planktonic cells and biofilms of Bacillus subtilis ATCC-6633. *Water Res.* 55:21–29.

Paris D.F., Steen W.C., Baughman G.L., Barnett J.T. (1981) Second-order model to predict microbial degradation of organic compounds in natural waters. *Appl. Environ. Microbiol.* 41:603–609.

Parvatiyar M.G., Govind R., Bishop D.F. (1996) Biodegradation of toluene in a membrane biofilter. *J. Membr. Sci.* 119:17–24.

Pool J.R., Kruse N.A., Vis M.L. (2013) Assessment of mine drainage remediated streams using diatom assemblages and biofilm enzyme activities. *Hydrobiologia.* 709:101–116.

Puhakka J.A., Melin E.S., Järvinen K.T. (1995) Fluidized-bed biofilms for chlorophenol mineralization. *Water Sci. Technol.* 31:227–235.

Qu Y.Y., Zhou J.T., Wang J. (2009) Population dynamics in bioaugmented membrane bioreactor for treatment of bromoamine acid wastewater. *Bioresour. Technol.* 100:244–248.

Quintelas C., Fernandes B., Castro J. (2008) Biosorption of Cr(VI) by a Bacillus coagulans biofilm supported on granular activated carbon (GAC). *Chem. Eng. J.* 136:195–203.

Quintelas C., Silva B., Figueiredo H. (2010) Removal of organic compounds by a biofilm supported on GAC: Modelling of batch and column data. *Biodegradation.* 21:379–392.

Rajkumar M., Prasad M.N.V., Swaminathan S., Freitas H. (2013) Climate change driven plant–metal–microbe interactions. *Environ. Int.* 53:74–86.

Rehman K., Imran A., Amin I., Afzal M. (2018) Inoculation with bacteria in floating treatment wetlands positively modulates the phytoremediation of oil field wastewater. *J. Hazard. Mater.* 349:242–251.

Rehman K., Imran A., Amin I., Afzal M. (2019) Enhancement of oil field-produced wastewater remediation by bacterially-augmented floating treatment wetlands. *Chemosphere.* 217:576–583.

Rodgers M., Zhan X.M. (2003) Moving-medium biofilm reactors. *Rev. Environ. Sci. Biotechnol.* 2(2):213–224.

Ryan R.P., Germaine K., Franks A., Ryan D.J., Dowling D.N. (2008) Bacterial endophytes: Recent developments and applications. *FEMS Microbiol. Lett.* 278:1–9.

Saleem H., Rehman K., Arslan M., Afzal M. (2018) Enhanced degradation of phenol in floating treatment wetlands by plant-bacterial synergism. *Int. J. Phytorem.* 20:692–698.

Segura Y., Martínez F., Melero J.A. (2015) Zero valent iron (ZVI) mediated Fenton degradation of industrial wastewater: Treatment performance and characterization of final composites. *Chem. Eng. J.* 269:298–305.

Semrau J.D., DiSpirito A.A., Yoon S. (2010) Methanotrophs and copper. *FEMS Microbiol. Rev.* 34:496–531.

Shah M.P. (2020) *Microbial Bioremediation and Biodegradation.* Springer.

Shah M.P. (2021) *Advanced Oxidation Processes for Effluent Treatment Plants.* Elsevier.

Shah M.P. (2021) *Removal of Emerging Contaminants through Microbial Processes.* Springer.

Shahid M.J., Tahseen R., Siddique M., Ali S., Iqbal S., Afzal M. (2019) Remediation of polluted river water by floating treatment wetlands. *Water Supply.* 19:967–977.

Shammas N.K. (2005) Coagulation and flocculation. In: *Physicochemical Treatment Processes*, Wang L.K., Hung Y-T., Shammas N.K., Editors. Totowa: Humana Press, 103–139.

Sheng X.-F., Xia J.-J., Jiang C.-Y., He L.-Y., Qian M. (2008) Characterization of heavy metal-resistant endophytic bacteria from rape (Brassica napus) roots and their potential in promoting the growth and lead accumulation of rape. *Environ. Pollut.* 156:1164–1170.

Shi P., Zhu K., Zhang Y., Chai T. (2016) Growth and cadmium accumulation of Solanumnigrum L. seedling were enhanced by heavy metal-tolerant strains of Pseudomonas aeruginosa. *Water Air Soil Pollut.* 227:459.

Shin M.-N., Shim J., You Y., Myung H., Bang K.-S., Cho M., Kamala-Kannan S., Oh B.-T. (2012) Characterization of lead resistant endophytic Bacillus sp. MN3–4 and its potential for promoting lead accumulation in metal hyperaccumulatorAlnus firma. *J. Hazard. Mater.* 199:314–320.

Silva B., Figueiredo H., Quintelas C. (2008) Zeolites as supports for the biorecovery of hexavalent and trivalent chromium. *Microporous Mesoporous Mater.* 116:555–560.

Simpson D.R. (2008) Biofilm processes in biologically active carbon water purification. *Water Res.* 42:2839–2848.

Srivastava S., Srivastava A.K. (2005) Studies on phosphate uptake by Acinetobactercalcoaceticus under aerobic conditions. *Enzyme Microb. Technol.* 36(2–3):362–368.

Stephenson D., Stephenson T. (1992) Bioaugmentation for anhancing biological wastewater treatment. *Biotechnol. Adv.* 10(4):549–559.

Stottmeister U., Wießner A., Kuschk P., Kappelmeyer U., Kästner M., Bederski O., Müller R., Moormann H. (2003) Effects of plants and microorganisms in constructed wetlands for wastewater treatment. *Biotechnol. Adv.* 22:93–117.

Tanner C.C., Headley T.R. (2011) Components of floating emergent macrophyte treatment wetlands influencing removal of stormwater pollutants. *Ecol. Eng.* 37:474–486.

Tigini V., Prigione V., Giansanti P., Mangiavillano A., Pannocchia A., Varese G.C. (2010) Fungal biosorption, an innovative treatment for the decolourisation and detoxification of textile effluents. *Water.* 2:550–565.

Tunali S., Akar T., Özcan A.S., Kiran I., Özcan A. (2006) Equilibrium and kinetics of biosorption of lead (II) from aqueous solutions by Cephalosporiumaphidicola. *Sep. Purif. Technol.* 47:105–112.

Valls M., Atrian S., de Lorenzo V., Fernández L.A. (2000) Engineering a mouse metallothionein on the cell surface of Ralstoniaeutropha CH34 for immobilization of heavy metals in soil. *Nat. Biotechnol.* 18:661.

Vijayaraghavan K., Yun Y.-S. (2008) Bacterial biosorbents and biosorption. *Biotechnol. Adv.* 26:266–291.

Wagner M., Loy A. (2002) Bacterial community composition and function in sewage treatment systems. *Curr. Opin. Biotechnol.* 13(3):218–227.

Wu Q., Hu Y., Li S., Peng S., Zhao H. (2016) Microbial mechanisms of using enhanced ecological floating beds for eutrophic water improvement. *Bioresour. Technol.* 211:451–456.

Yamamoto K., Hiasa M., Mahmood T. (1989) Direct solid-liquid separation using hollow fiber membrane in an activated sludge aeration tank. *Water Sci. Technol.* 21:43–54.

Yin W., Wang Y., Liu L., He J. (2019) Biofilms: The microbial "protective clothing" in extreme environments. *Int. J. Mol. Sci.* 20:3423.

Zhang L., Zhao J., Cui N., Dai Y., Kong L., Wu J., Cheng S. (2016) Enhancing the water purification efficiency of a floating treatment wetland using a biofilm carrier. *Environ. Sci. Pollut. Res.* 23:7437–7443.

Zhang X., Li J., Yu Y. (2016) Biofilm characteristics in natural ventilation trickling filters(NVTFs) for municipal wastewater treatment: Comparison of three kinds of biofilm carriers. *Biochem. Eng. J.* 106:87–96.

Zhao F., Rahunen N., Varcoe J.R. (2008) Activated carbon cloth as anode for sulfate removal in a microbial fuel cell. *Environ. Sci. Technol.* 42:4971–4976.

Zhao F., Zhang S., Ding Z., Aziz R., Rafiq M.T., Li H., He Z. (2013) Enhanced purification of eutrophic water by microbe-inoculated stereo floating beds. *Pol. J. Environ. Stud.* 22:957–964.

Zhao T., Fan P., Yao L., Yan G., Li D., Zhang W. (2011) Ammonifying bacteria in plant floating island of constructed wetland for strengthening decomposition of organic nitrogen. *Trans. Chin. Soc. Agric. Eng.* 27:223–226.

4 Emerging Technologies and Their Advancements Toward Wastewater Treatment from Various Industries

Deepti, Piyal Mondal, and Mihir Kumar Purkait
Indian Institute of Technology

CONTENTS

4.1 INTRODUCTION

Today, over two billion people worldwide experience elevated water stress. The dynamic relationship between water demand and global growth is compounded by climate change, urbanization, industrialization, and bioenergy requirements. Wastewater, the main waste source in the municipal, petrochemical, medicinal, food, fiber, forestry, polymer, and paper sectors, produces high amounts of chemical pollutants, and organic and inorganic compound salts. Unfortunately, most of the freshwater sources in the world are polluted; the majority of wastewater is dumped immediately into the natural system in developed countries without prior or sufficient treatment [1]. The majority of amenities that pollute water, with particulars of the chemicals they generate must be controlled by environmental emission management initiatives. Water-intensive industries are the largest contributors to industrial wastewater. There have been processing changes in the industry sector to minimize wastewater generation. However, this often changes the quality of the wastewater in such a

DOI: 10.1201/9781003164982-4

way that it includes more refractive contaminants than the previous process. It is important to adapt the process improvements not only to minimize the generation of wastewater but also to decrease the concentration of toxic contaminants in wastewater to make it more suitable for treatment. Water shortages and discharges also raise concerns about seeking alternate resources that can help to achieve a balance in demand and effectively conserve water. Wastewater treatment and reuse are some of the promising possibilities. Wastewater treatment and reuse is an environmentally friendly technology because it protects the atmosphere, resulting in fewer emissions being released; uses energy more efficiently; facilitates recycling of unused goods, and treats residual waste in a more technically appropriate manner. Wastewater treatment systems usually make use of physical, chemical, biological, and combination of technologies [2]. Operations including screening, filtration, centrifugation, sedimentation, coagulation, and flotation are mainly used. Biological treatment, which may be aerobic or anaerobic in nature, is often used for secondary treatment. More often distillation, crystallization, evaporation, solvent extraction, oxidation, precipitation, reverse osmosis, electrolysis, and electrodialysis are used in tertiary treatment. The choice of appropriate technology depends on the characteristics and types of water pollutants and the expected application of treated water. While wastewater can be treated by different methods, the quest for safe water and a clean environment has resulted in strict discharge limits being imposed by different environmental protection agencies. On the other hand, there are still factors that affect the quality of wastewater that have varying environmental effects, where effective wastewater management incorporated with primary, secondary, and specialized treatment methods tends to be more feasible [3].

Several types of industries are booming because of the ever-changing requirements and necessities and, as a consequence, different contaminants are loaded into wastewater that entails the need for emerging technologies. Emerging technologies are technical inventions that reflect progressive advancements in a sector for competitive advantage. Emerging technologies are pivotal to tackle ever-changing wastewater loads and there is still a nexus between water and energy. Besides, water reuse has an enormous capacity to supplement already-stressed portfolios of water supplies. Technologies such as advanced oxidation processes (AOPs), ion exchange, membrane-mediated separations, precipitation based on solvents, advanced biological algae treatment using bacteria, and mixed fungi are emerging today, promising safe and clean treated water [4].

With the identification of more and more pollutants, the exponential increase of population and commercial activity, and the declining supply of water supplies, the usage of conventional water and wastewater management systems are constantly being challenged. Traces of toxic compounds, heavy metals, and other toxins are found in industrial wastewater. To achieve some of the several unique targets, which comprise the elimination of (a) suspended solids (b), organic matter (c) heavy metals (d) dissolved solids, and (e) hazardous substances, various emerging treatment methods are used. These solutions offer options for safer public health and environmental safety [5]. This chapter provides a summary of various industries with their production process, along with classification, sources, and the occurrence of wastewater generated. Emerging industrial wastewater treatment systems from different industries are briefly discussed, stressing the theory, its performance quality, benefits, and drawbacks with the potential scope of the techniques.

4.2 DIFFERENT INDUSTRIAL WASTEWATER AND ITS CHARACTERISTICS

4.2.1 Textile Industry

The textile industry is one of the dynamic manufacturing sectors since it is a decentralized and heterogeneous market controlled by small and medium-sized enterprises. The classification of the textile industry is complicated due to the large range of methods, substrates, usage of equipment and components, and the method of finishing. Varied types of fabrics or yarns, means of cloth processing, and processes like preparation, printing, dyeing, coating, and chemical or mechanical finishing are interrelated in the production of the finished fabric [6]. Processes in the textile industry include

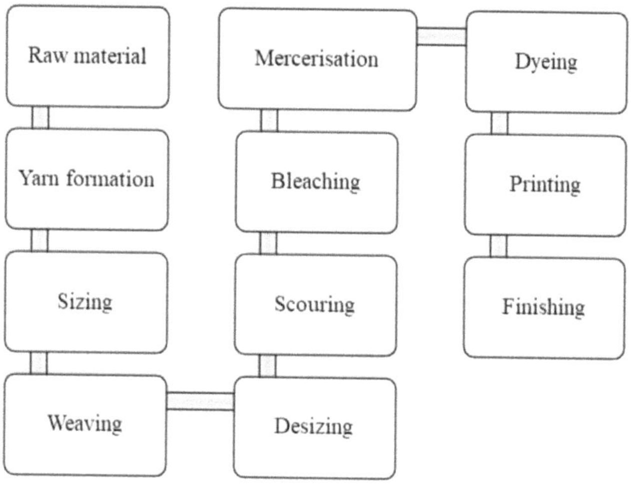

FIGURE 4.1 Steps involved in the processing of fabric.

those that turn natural such as cotton, wool, silk, etc., and synthetic such as viscose, polyester, and acrylic fibers into fabrics and other products. Under this industrial sector, the primary main activities can be identified as follows: sizing -de-sizing, bleaching, mercerization, dyeing, printing, and final finishing. The flow diagram of the steps involved in fabric processing is shown in Figure 4.1.

A majority of textile manufacturing processes are chemical wet processing operations required for the proper preparation, purification, coloring, or finishing of the fabric. This results in the generation of wastewater, the contamination load of which comes not only from the removal of raw material impurities but also from the residual chemical reagents used for manufacturing. These effluents contain different contaminants identified by high emission indicator values, for example, biochemical oxygen demand (BOD), chemical oxygen demand (COD), suspended solids (SS), color, turbidity, and toxicity [7]. The characteristics of wastewater from each phase are shown in Table 4.1. The severe contamination of surface and groundwater due to the treatment of textile wastewater is a foremost environmental concern.

Various methods are adopted for wastewater treatment including membrane filtration, coagulation, precipitation, AOPs, and many others to reach the stringent requirements either for the reuse or for the discharge of the wastewater to the streams [8].

4.2.2 Tannery Industry Wastewater

The tanning method attempts to turn skin into safe and imputrescible leather goods. Rawhide tanning is achieved either by the method of vegetable tanning, which creates dense heavy-duty leather or by the process of chrome tanning, which provides light, thin leather. To produce finished leather, there are four main processes needed: beam house service, tanyard processes, retaining, and finishing [9]. Figure 4.2 shows the flow chart of the unit operation in leather manufacturing.

On average, 30–35 m^3 of wastewater is generated per ton of rawhide. However, depending on the raw material, the final product, and the manufacturing methods, wastewater production ranges in a large range. Organic pollutants arise either from skins or are added during operations.

Tannery effluent has a disagreeable odor and dark brown color with high pollutant parameters such as COD, BOD, total suspended solids (TSS), total dissolved solids (TDS), sulfate, phenol, nitrate, phosphate and a mixture of toxic organic and heavy metal pollutants. Harmful chemical compounds such as chromium (Cr), vegetable tannins, syntans, phenolic acids, pesticides, sulfonated oils are currently used to convert rawhide into leather [10]. These hazardous chemical composites

TABLE 4.1
Characteristics of Wastewater from Different Processes of Textile Industry

Textile Industry Process	Parameters	Composition
Sizing and de-sizing	Ammonia, waxes, starch, enzymes	High BOD, COD, dissolved solids
Scouring	Oils, waxes, fats, surfactants, pectin, solvents, enzymes, disinfectants and insecticides residues	TDS, TSS, high pH, COD
Bleaching	Hydrogen peroxide, organic stabilizer, sodium silicate, chlorine, acids	High pH, SS, alkalinity, TS
Mercerisation	Sodium hydroxide	High pH, TDS, low BOD
Dyeing	Color, metals, salts, auxiliaries, formaldehyde, sulfide, surfactants	High color, high heavy metal content, BOD, TDS, low SS
Printing	Solvents, urea, color, metals	High color, BOD, SS, oil and grease, high pH
Finishing	Waxes, resins, spent solvents, softeners, chlorinated compounds	Low BOD, SS

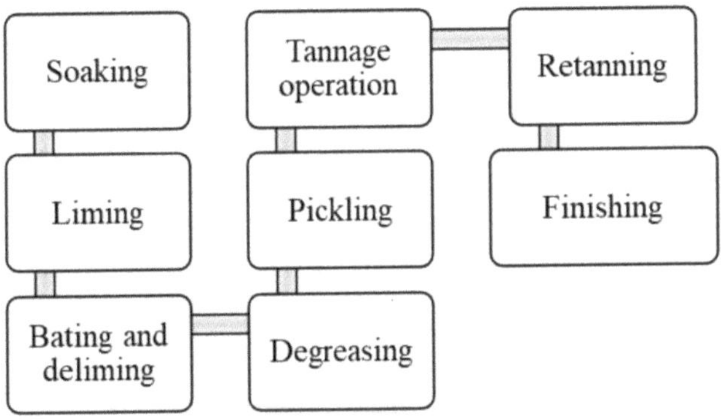

FIGURE 4.2 Unit process flowchart in leather production.

are not completely removed by the rawhide and thus discharged, which causes significant environmental and serious toxic hazards. The characteristics of tannery wastewater from each process are given in Table 4.2.

Several Physico-chemical treatment techniques have also been developed over time to enhance the efficiency of treated wastewater, namely, coagulation, flocculation, sedimentation, adsorption, ozonation, photocatalysis, microfiltration, etc. However, due to secondary contamination concerns, and high operating and treatment expenditures, these treatment systems are of restricted reach and are therefore not environmentally and economically appropriate for wastewater treatment. Emerging approaches to effluent treatment are also available for the treatment of tannery effluents, such as membrane bioreactors (MBRs), anaerobic ammonium oxidation (ANAMMOX), membrane filtration technologies, and advanced oxidation technologies namely electrochemical treatment, Fenton, etc. [11].

4.2.3 Paper and Pulp Industry Wastewater

The process of manufacturing pulp paper consists of four basic steps: preparation of raw materials, pulping, bleaching, and paper-making. Different pulping methods are used depending on the quality of the raw material, viz. (a) mechanical pulping, (b) semi-chemical pulping, (c) sulphite pulping, (d)

TABLE 4.2
Characteristics of Wastewater from Different Processes of Tannery Industry

Tannery Industry Process	Parameters	Composition
Soaking	Salts	High chlorides, sulfides, dissolved solids
Liming	Lime, alkaline reducing agents	High pH, BOD
Bating and deliming	Ammonium salts, enzymes	High pH, BOD, ammonia, TSS
Pickling	Salts, acids	Low pH, High COD, ammonia, TSS, TDS, chloride, sulfide
Tanning operation	Solvents, oil, salts, acids	High color, oil, chromium, salts
Retanning	Waxes, resins, spent solvents, softeners, chlorinated compounds	High BOD, COD, chlorides, TDS, TSS

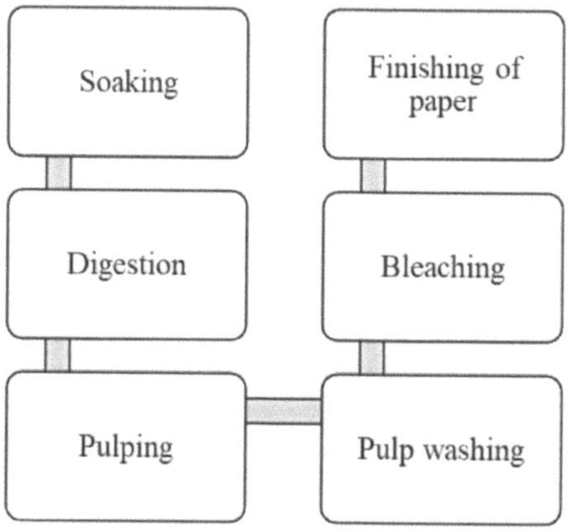

FIGURE 4.3 Schematic flow diagram of a paper manufacturing process.

sulfate pulping, (e) chemical-mechanical pulping. A schematic flow diagram of a paper manufacturing process is shown in Figure 4.3.

Most of the toxic contaminants released from the paper and pulp industry are primarily generated by various stages of the process of paper production and wastewater production. Such industries emit waste containing compounds that are complex organic and inorganic. Around 55%–60% of lingo-cellulosic waste, consisting of various organic and inorganic contaminants that are released into the atmosphere, is discharged from raw materials during the process of wood digestion and pulp bleaching [12]. In general, wastewater is alkaline in nature, has high suspended solids, TDS, low BOD, lignins, dioxins, furans, stilbenes, chlorides, phenols, and compounds of sulfur. The effluent from the process of pulping is rich in lignin and lignin degradation materials, called 'black liquor' [13]. The characteristics of wastewater from each process of the paper and pulp industry are provided in Table 4.3. Various methods have been introduced to decrease the organic and inorganic waste from pulp and paper wastewater, including AOPs, ozonation, gasification, aerobic granulation, microbic fuel cells, and many more [14,15].

4.2.4 Steel Industry Wastewater

After oil and gas, the steel industry is the second-largest industry in the world. Steel is manufactured through two key routes, integrated and the scrap route. Although the only raw material is steel scrap in the scrap route, minerals and fossils are prepared in the integrated route to be suitable for charging them in the blast furnace [16]. Figure 4.4 refers to a schematic diagram showing the main

TABLE 4.3
Characteristics of Wastewater from Main Processes of Paper and Pulp Industry

Paper and Pulp Industry Process	Parameters	Composition
Soaking	Organic matter, bark particles, dirt, pigments, and fibers	High BOD, COD, TDS, color
Digestion	Salts, organic matter	High SS, chlorides, sulfides, TDS, COD, BOD
Pulping	Salts, organic matter	High pH, high heavy metal content
Pulp washing	Solvents, organic matter	High SS, chlorides, sulfides, TDS, COD, BOD, sodium, heavy metals
Bleaching	Sodium bisulfite, sodium or zinc hydrosulfite, calcium or sodium hypochlorite	Oil, lignin, dissolved carbohydrates
Finishing of paper	Cellulose, hemicellulose, lignin, resins, fatty acids, and other phenolic	High pH, phenol, chlorophenol, lignin, BOD, COD, TDS

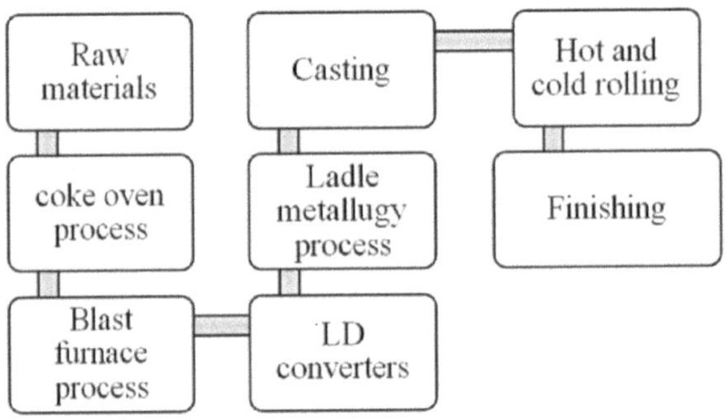

FIGURE 4.4 Main stages of steel production through the integrated route.

stages of the integrated route adopted in the steel industry. Metallurgical coke gets converted from coal in the coke oven plant and iron ore is transformed in the sinter plant. Pig iron which is produced in the blast furnace is processed in the basic oxygen furnace (BOF) to decrease the carbon content and impurities. The obtained liquid steel passes through the ladle for further treatment. The solidification into semi-finished products of liquid steel is carried out by ingot casting followed by hot and cold rolling [17].

Water, particularly for integrated routes, is primarily used for cooling systems, and technical applications, such as wet gas cleaning and sanitary applications. Different techniques, such as mining, pig iron smelting in a blast furnace (BF), Linz and Donawitz (LD) converters and rolling mills generates wastewater. Highly concentrated chlorides, calcium bicarbonate, fluoride, sulfates, cyanide, TSS, volatile suspended solids, TDS, cyanide, thiocyanates, traces of phenol, iron, magnesium oxide, oil, and grease are the constituents present in wastewater produced from all the processes [18]. Several treatment approaches have been adopted, such as coagulation-flocculation, precipitation, advanced oxidation, electrochemical procedures, membrane technology, and a combination of one or more of these processes (Table 4.4) [19].

4.2.5 Dairy Industry Wastewater

In the dairy sector, raw milk is processed into pasteurized and sour milk, yogurt, cheese, butter and cream products, ice cream, whey powder, condensed milk, and various forms of desserts [20].

TABLE 4.4
Characteristics of Wastewater from Main Processes of Integrated Iron and Steel Industry

Iron and Steel Industry Process	Parameters	Composition
Coke oven plant	Solvents, oil	High BOD, COD, ammonia, phenol, cyanide
Blast furnace process	Iron, metal ores	High SS, chlorides, sulfates, TDS
LD converters	Metals, salts	High pH, high heavy metal content
Rolling	Acid, oil, grease	High content of oil and grease, chromium, nickel, zinc, copper, tin

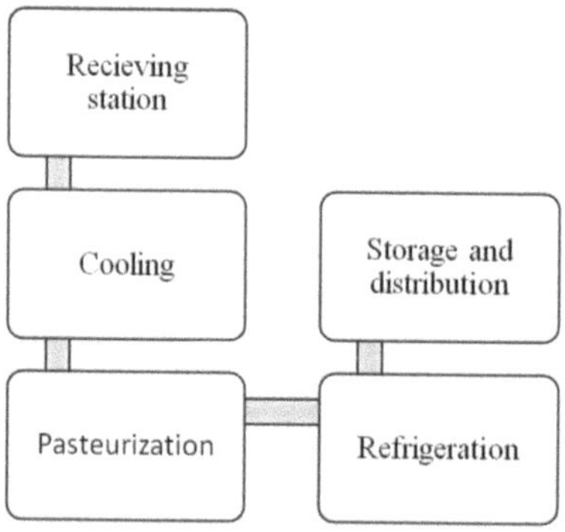

FIGURE 4.5 Schematic flow diagram of a milk dairy process.

Pasteurization, sterilization, isolation, filtration, and homogenization are a set of unit operations. The main unit processes of the dairy industry are shown in Figure 4.5.

The industry produces an immense amount of wastewater, roughly 0.2–10L of waste per liter of raw milk. The generation and characteristics of wastewater are dependent on many parameters; the size of the plant, type of treatment adopted, the quality and methods used for cleaning the plant, the working practices, the approaches used for disposing of effluent, the type and cost of water sources [21]. Lactose, lipids, mineral salts, soluble proteins, and detergents are the main constituents of milk wastewater, along with low concentrations of some of the heavy metals. Dairy effluent characteristics include temperature, pH, BOD, COD, dissolved solids, suspended solids, color, oil, and grease [22].

Wastewater is generally white in color and faintly alkaline in nature. Due to the fermentation of milk sugar into lactic acid, it becomes very acidic with time. While the pH of ice cream wastewater is slightly acidic and decreases in whey wastewater [23]. The organic content of wastewater, in the case of COD and BOD_5, varies across a wide range depending on the method of processing. Physicochemical treatment methods including electrochemical, coagulation-flocculation, adsorption, and membrane treatment have been used to remove suspended, colloidal, and dissolved contaminants from dairy wastewater [24,25]. The anaerobic treatment has been successfully extended to the treatment of highly contaminated wastewater which has the benefits of lower nutrient requirements and less wastewater production with sludge. Since the dairy wastewater industry is high in volume and highly contaminated, granule and biofilm-based reactors are also used to extract organic matter [26,27].

4.2.6 Pharmaceutical Industry Wastewater

Pharmaceutical production methods can be classified into five groups, that is; fermentation, extraction, chemical synthesis, formulation, and packaging [28]. Pharmaceutical systems are usually characterized by batch and continuous processes. In each case, the process consists of the main (upstream) and secondary (downstream) parts: the first involves the processing of the Active Pharmaceutical Component (API or Drug Material, DS), while the latter focuses on combining the API with excipients to create the final branded formulation (Drug Product, DP) [29]. The pharmaceutical industry has a special position on this list of exceptionally water-dependent industries. Most of the processes in the pharmaceutical sector need very pure water which meets strict quality standards. In the pharmaceutical industry, the effluents are typical and must be discarded to preserve the requirements of the pharmaceuticals such as consistency and safety. It is important to categorize the materials and their composition before treatment as the effluents they create include carcinogenic, mutagenic, teratogenic, and other significant adverse reactions [30]. It consists mainly of organic compounds that are more toxic than their inorganic counterparts. As pharmaceutical firms manufacture a range of drugs and cosmetics, their wastewater sources are not standardized. It includes compounds from active biomass and antibiotics to polyaromatic hydrocarbons and phenols. These have a gene-altering, endocrine-disrupting effect on marine organisms and are toxic to another biota. They pose an indirect danger to human health as a result of the fragmentation process. Moreover, the health risk of the release of antibiotic residues into natural habitats is a growing issue in recent years of pharmaceutical wastewater treatment and discharge, as common biological treatment has reduced antibiotic removal efficiency due to its antimicrobial action. The most distinctive characteristic of the pharmaceutical industry's wastewater system is that it includes not only APIs but also various intermediates, solvents, catalysts, and other raw materials required to synthesize and produce a specific drug in its desirable dosage form. And these chemicals are very toxic to the water in the catchment area that is eventually disposed of [31].

Several mitigation methods have been implemented, such as ozone, electrocoagulation, and catalytic oxidation. Bioremediation, among others, has arisen as a promising treatment method. The bioremediation method depends either on the activity of fungal, bacterial, or plant species (phytoremediation) or on the use of aerobic, anaerobic, or membrane bioreactors to handle and sustain the steady state of contaminants in the water ecosystem [32].

4.3 OVERVIEW ON EMERGING TECHNOLOGIES FOR WASTEWATER TREATMENT

4.3.1 Membrane-Based Technology

Technology based on membranes has emerged as a promising alternative. The system is in a fast-growth stage and is believed to be the leading wastewater treatment technology. Study on customized membrane material development closely promotes the advances. The application of the process of nanofiltration (NF) and reverse osmosis (RO) for the removal of organic compounds from wastewater with a low molecular range has been widely reported. Microfiltration (MF) and ultrafiltration (UF) membranes function and resemble a fine screen or sieve using a surface removal mechanism. Particles larger than the largest pore size are rejected by the surface of the membrane and remain on the side of the feed or concentrate, while particles smaller than the largest pore size can pass through the membrane to the side of the filtrate [33]. Membranes are now commonly used in the food/beverage industry; in desalination procedures; in the medical industry, for the purification of body fluids, antibiotics, etc.; and in industrial effluent separation [34]. The several benefits associated with membrane processes have been due to this rise in the use of membranes in industry, such as good product quality, higher energy performance than traditional separation methods, quick operation, no chemical modification of the materials, quick scale-up, and better suitability for the environment [35]. Microfiltration (MF) and ultrafiltration (UF) membranes are used along with bioreactors, which are called MBRs, in the handling of wastewater. In MBR systems, two types of membranes are of concern [36,37]. These are polymeric and ceramic Barredo-Damas et al. [38] evaluated the efficiency of tubular ceramic ultrafiltration membranes under varying operational conditions, treating combined effluents from the textile industry. Parameters like chemical oxygen demand (>70%), color (96%), and turbidity (>93%), were reduced substantially [38].

Rahimi et al. [39] the synthesized nanocomposite ultrafiltration membrane which was applied in a bioreactor was used to treat milk processing wastewater (MPW). Around 92%–99% of COD removal efficiency was achieved in the experiments [39]. Chakrabortty et al. [40] suggested a semi-pilot-based nanofiltration method for the continuous recovery of clean water from pharmaceutical effluent. Rejection of more than 98% of contaminants (COD, API, chloride, sulfate) was achieved [40].

Petrinic et al. [41] studied the technological and economic feasibility of incorporating an ultrafiltration process as a pre-treatment to remove dissolved and colloidal pollutants and to remove membrane fouling before the RO process. The results show that the pollutants from the effluent, such as metal elements, organic and inorganic compounds, are removed between 91.3% and 99.8% by ultrafiltration and RO treatment. Pollutants such as nitrogen, suspended solids (SS), ammonium, nitrogen, COD and BOD, sulfate, and nickel were completely removed and the permeate concentrations were below the detection limits, therefore the quality of both the processes met the reuse standards [41].

4.3.2 Advanced Oxidation Process

Advanced oxidation processes (AOPs) are a group of chemical treatment procedures capable of removing organic wastewater materials by oxidation reactions with a strong, non-selective hydroxyl radical that oxidizes conventionally resistant organic contaminants and can also increase wastewater biodegradability. AOP includes two steps of the oxidation process. The initial phase of the oxidation process is the formation of strong oxidants (hydroxyl radicals) and the second phase is the reaction of these oxidants with organic contaminants. A major mechanism is the production of highly reactive free radicals and a large number of methods are categorized under the wide concept of AOP. It is capable of removing non-degradable organic components and preventing the removal of residual deposits. Theoretically, AOPs may completely convert organic compounds to carbon dioxide and water according to Eq. 4.1 [42].

$$R - H + OH \rightarrow H_2O + R \tag{4.1}$$

AOPs, viz., Fenton process, photolysis, ozonation, wet air oxidation, photocatalysis, anodic oxidation, and sonolysis unselectively attack different types of contaminants and degrade them by the generation of hydroxyl radicals. Each AOP has its hydroxyl radical generation mechanisms [43].

Photocatalytic treatment of textile & dyeing industry effluent was investigated by Bahadur and Bhargava [8] which included heterogeneous photocatalysis (HP) involving nano-TiO_2/UV as a secondary treatment. The total reduction achieved on biological oxygen demand (BOD) and COD was 95% and 91%, respectively [8].

Korpe et al. [11] analyzed the various parameters concerning the chemical oxidation process of tannery wastewater such as pH, ultrasound irradiation time, and hydrogen peroxide dosage. Employing hydrodynamic cavitation at pH 3 and 2 mL/L of H_2O_2 dosage, the suspended TOC removal efficiency was found to be 87% [11].

4.3.3 Nanoparticles and Nanocomposites for Treatment

Nanomaterials are typically described as materials that have at least one dimension less than 100 nm. At such a size, materials often show special physical or chemical properties over their bulky counterparts [44]. Taking account of these size-dependent effects, the existing method of treatment of water and wastewater could be significantly enhanced by incorporating nanomaterials into the system. Numerous studies have shown that nanomaterials have a broad range of capabilities and potential in water and wastewater treatment, especially in the areas of catalytic oxidation, membrane process, adsorption, disinfection, and sensing [45–48]. One exciting approach to the forward-looking application of nanomaterials is the production of nanocomposite materials that support both hosts and impregnated functional nanoparticles. Hosts such as polymers, biopolymers, activated carbons, membranes, or minerals could promote the dispersion and stabilization of the nanoparticles being charged [49]. In recent years, several researchers have been working on nanocomposites for industrial wastewater treatment. Mondal and Purkait [50] synthesized a novel green iron-aluminum nanocomposite for the removal of fluoride from water. A total adsorption capacity of 42.95 mg/g was obtained for an adsorbent dose of 0.25 g/L [50]. Similarly, Sontakke and Purkait [51], developed graphene oxide nanoscrolls (GONS) at low frequency (20 kHz) ultrasonication with tunable measurements. GONS obtained may be used in various engineering applications such as adsorption and membrane filtration for wastewater treatment [51]. Verma and Balomajumder [52] prepared nickel ferrite nanocomposites (NFNCs), which were then used to eliminate Cr(VI) from the electroplating industry. The adsorption method showed that hexavalent chromium (91.6%) was removed in the first hour. NFNCs have been able to remove copper, zinc, and nickel from electroplating effluents, indicating that NFNCs have a high ability to meet environmental standards [52].

4.3.4 Solventing-out Process

One of the earliest concepts of separation is precipitation. While there are several kinds of processes of separation, a new focus has recently been paid to precipitation. Usually, the precipitation action is achieved either by adding a precipitation agent to the original solution or by changing conditions like temperature and pH to create a new (solid) phase from the parent phase. In the precipitation phase, choosing an appropriate precipitation agent or regulating the conditions are the main factors. One of the first separation processes to remove an organic component from an aqueous solution was using salt. The decline in the solubility of an organic component after the addition of salt in aqueous solutions is called "salting-out". While reduction, of the solubility of salt in aqueous solutions, an organic solvent is added, and the process is known as solventing-out. Inorganic salts were precipitated out using organic solvents from aqueous solutions either in industrial applications or methods of analysis.

Several organic solvents have been identified for possible use in precipitation processes [53]. Such organic solvents are isopropylamine, diisopropylamine, propylamine, ethylamine, diethylamine, and dimethylamine. Precipitation relies on the miscibility of the organic solvent in water and its ability to create a solid hydrogen bond with water that affects the hydration of salts. The amount of salt precipitated from aqueous-saline solution depends on the solubility of the salt in the organic solvent; the lower the solubility, the greater the precipitation [19,54]. Deepti et al. [19] studied the separation of chlorides and sulfates from the nanofiltration rejected stream of steel industry using miscible organic solvents such as diisopropylamine (DIIPA), isopropylamine (IPA), and ethylamine (EA). The removal efficiency of chloride and sulfate are 77.50% and 99.82% respectively. It was also observed that there is a huge amount of reduction in TDS and other heavy metals [19]. Similarly Bader [55] investigated the precipitation of chloride and sulfate in several cation forms like potassium, calcium, sodium, magnesium, strontium, and barium from aqueous streams using miscible organic solvents such as isopropylamine and ethylamine as precipitation solvents [55].

4.3.5 Sequential Anaerobic and Aerobic Bio-treatment Process

In wastewater treatment, biological treatment tends to be a promising technology for those wastewater containing biodegradable constituents with a BOD/COD ratio of 0.5 or greater. It also has the benefits of reduced maintenance costs without secondary contamination compared to other wastewater treatment systems [56]. It is possible to use aerobic and anaerobic processes; the former includes the use of free or dissolved oxygen by microorganisms (aerobes) to turn organic waste into biomass and CO_2, while the latter involves the breakdown of complex organic waste into methane, CO_2, and H_2O in the absence of oxygen by three simple steps (hydrolysis, acidogenesis including acetogenesis and methanogenesis). To attain a high treatment efficiency, aerobic biological methods are widely used in the treatment of organic wastewater, while in anaerobic treatment, major progress has been made in the anaerobic method of treatment based on the principle of resource recovery and utilization, while at the same time meeting the goal of emission control [57–59].

The benefits of the aerobic and anaerobic processes are excellent resource recycling capacity, high total quality of treatment, less sludge disposal, and low energy consumption. Biodegradation of chlorinated aromatic hydrocarbons including anaerobic dechlorinations and aerobic ring cleavage, concurrent removal of nitrogen including aerobic nitrification, and anaerobic denitrification have also been shown to perform well with the anaerobic-aerobic system processes.

The aerobic-anaerobic process has been commonly used in the management of urban and industrial wastewater for many years. Efficient organic removal has been reported using such two-stage treatment methods for food solid waste leaching, textile wastewater, pulp, and paper effluent, etc.

Shi et al. [60] investigated a sequential anaerobic-aerobic procedure for the treatment of high salinity pharmaceutical wastewater. Combination of up-flow anaerobic sludge blanket (UASB)+membrane bioreactor (MBR) and UASB+ batch sequencing reactor (SBR) systems achieved outstanding organic removal performance with COD removal of 94.7% and 91.8% respectively [60].

Similarly, Mazhar et al. [61] treated paper and pulp industry effluent with three biological treatments namely; aerobic, anaerobic, and sequential. Substantial reductions in COD, BOD, total suspended (TSS), dissolved solids (TDS) and turbidity with 81%, 71%, 65%, 60%, 68% respectively were reported during sequential treatment [61].

Shoukat et al. [62] investigated the anaerobic-aerobic biological treatment technology for textile effluent based on evaluation and optimization systems. The proposed anaerobic-aerobic system was very effective in the removal efficiencies of COD (99.5%), TKN (99.3%), and dyes (78.4%). The anaerobic sequential batch reactor (SBR) played a key role in improving treatment efficiency in addition to the production of sustainable energy sources of biogas which added value to the hybrid system [62].

4.3.6 Hybrid Technologies

The hybrid process is defined as a combination of two or more treatment approaches, two or more of which are as follows: physical, biological, and chemical unit processes. There are essentially four forms of hybrid systems available: (a) Physical–biological, (b) physical–chemical, (c) chemical–biological, and (d) physical–chemical–biological hybrid system. The physical–biological hybrid system should be used where contaminants comprise high suspended solids, oil, and grease, organic and inorganic elements. The physical–chemical hybrid method is usually adopted for wastewater rich in suspended solids, oil, grease, metals, and ions [63].

A hybrid technique; flocculation accompanied by microfiltration for the treatment of steel-generated cold roll mill (CRM) wastewater was investigated by Deepti et al. [17]. The method adopted showed a substantial reduction in all the parameters. The concentration of chromium reduced from 2.26 to 0.035 mg/L and the iron concentration decreased from 5.7 to 0.51 mg/L [17]. Similarly, Das et al. [64] studied the elimination of cyanide, COD, biological oxygen demand (BOD), and chloride through a combination procedure involving ozonation and electrocoagulation from biological oxidation-treated (BOT) effluent from the steel industry. At optimal conditions, the removal efficiencies of cyanide, COD, BOD, and chloride ions in the hybrid process were 99.8%, 94.7%, 95%, and 46.5% respectively [64]. Changotra et al. [65] concentrated on the treatment of organic pharmaceutical wastewater of low and high strength obtained from the effluent treatment system of the pharmaceutical manufacturing unit. The hybrid method included the treatment of coagulation, biological treatment, and gamma irradiation which were used independently and sequentially to assess the optimal treatment series for the successful treatment of pharmaceutical wastewater. The sequential series of three methods – coagulation, gamma irradiation, and biological treatment – resulted in the elimination of low- and high-strength organics respectively by 92.7% and 90.2% of COD [65].

Bener et al. [66] investigated the application of a hybrid method to pre-treated textile wastewater. The hybrid process includes electrocoagulation, adsorption, and Fenton-like oxidation processes. At the end of the hybrid process, 87% TOC, 49% COD, 96% turbidity, roughly 90% color, and 95%–97% suspended solid removal were achieved [66].

4.4 SUSTAINABILITY OF EMERGING TECHNOLOGIES

In an era where there is increasing concern about the local and global effects of our existing environmental management policies and the need to mitigate sanitation, disease, and poverty issues, there is a growing necessity to implement more environmentally liable, apposite wastewater treatment technologies that balance environmental, economic and social sustainability efficiency. Well, the concept of sustainability is not defined, what is evident is that it aims for the preservation of economic well-being, the conservation of the environment and the prudent utilization of natural resources, and fair social change, recognizing the just interests of all citizens, societies and the environment. It also acknowledges that human and industrial processes need to be structured to ensure that the utilization of natural resources and cycles by humans does not lead to a reduction in the quality of life, either due to reductions in potential economic prospects or to detrimental effects on socioeconomic circumstances, human well-being, and the climate. Various measurement methods such as energy analysis, economic analysis, and life cycle analysis (LCA) may be used to assess the sustainability of wastewater treatment facilities [67]. Sustainability research has focused mainly on low energy needs, circular economy and waste control, and enhanced efficiency of treatment techniques. Taking all of these into account, several works are reported. Sandoval and Salazar [68] investigated a work on the application of electrochemical technologies of industrial wastewater considering the 3Rs as a priority, for removing organic contaminants and some heavy metals. They have discussed that the reuse, recycling, and recovery of water, energy, and nutrients is very important to minimize environmental impact and to enhance the sustainability of the process, to close the water cycle in the slaughtering sector [68].

Foteinis et al. [69] investigated the environmental sustainability of a semi-industrial autonomous solar compound parabolic collector (CPC) plant which is based on a solar photo-Fenton process assisted by ferrioxalate. It aims to treat wastewater from a pharmaceutical firm. Employing the LCA approach, environmental sustainability was determined. At the semi-industrial level, the solar photo-Fenton process was found to be a sustainable technology for treating wastewater containing micropollutants [69].

4.5 FUTURE RESEARCH PERSPECTIVES

To explore the uniqueness of new technology, more study and progress are needed. In the discovery of new modifications and developments, understanding the fundamental properties of each industrial wastewater plays an important role. Besides, advances can lead to an economically viable process that can be scaled up and, relative to conventional methods, eco-friendly in nature. Future experiments should include using more local resources to make it commercially feasible, and for better biocompatible applications, research should rely more on aspects of stability. The study should concentrate on designing sustainable solutions to improve the performance of the disposal of pollutants with minimal eco-toxicological consequences and should focus more on resolving and monitoring the risk management, fate, and toxicological effects of chemical agents used in industrial wastewater treatment. Sludge production, which can be organic or inorganic in nature, is the result of the disposal of wastewater by one or more of the above-discussed means of technologies. Due to their properties, such as viscosity, the presence of toxins in a concentrated form, some of which may be harmful, hazardous, or difficult to dewater and dispose of, sludges are a strange concern. When planning the wastewater treatment plant, this fact requires careful attention. When the solid residues, liquid effluents, and gaseous pollutants are treated appropriately and securely disposed of, the treatment is said to be complete. Further research can also be aimed at investigating the most influencing factors in regulating sludge production and finding a full treatment technology.

4.6 SUMMARY

With the discovery of more and more pollutants, the exponential increase of population and commercial activity, and the declining supply of water supplies, the usage of conventional water and wastewater management systems are constantly being challenged. Emerging treatments such as membrane filtration, advanced oxidation processes (AOPs) and UV irradiation, solvent-based precipitation, and nanocomposites for treatments are highly promising alternative solutions for industrial wastewater treatment for better safety of the environment and public health as addressed in this chapter. The emphasis was on their fundamental principles, basic implementations, and technological advances. It can be inferred that the uses of these innovations will be improved on an exponential scale, along with rising knowledge and advancements in the manufacturing industry. In summary, different technological processes were identified, which will be useful in future research as a resource to evaluate and use for more developments to tackle industrial wastewater and hazardous pollutants with the lowest chance of having a toxicological impact on the environment. This chapter might be useful to the readers for acquiring in-depth knowledge on different industrial processes and type of wastewater generated along with suitable technologies for the treatment.

REFERENCES

1. N.G. Wun Jern, Industrial wastewater treatment (2006). doi:10.1142/P405.
2. F.W. Steer, Yden family: Arms and alliances, *Notes Queries.* 179 (1940) 317. doi:10.1093/nq/179.18.317-a.
3. T.L. Talarico, I.A. Casas, T.C. Chung, W.J. Dobrogosz, Production and isolation of reuterin, a growth inhibitor produced by Lactobacillus reuteri, *Antimicrob. Agents Chemother.* 32 (1988) 1854–1858. doi:10.1128/AAC.32.12.1854.
4. A.I. Shah, M.U. Din Dar, R.A. Bhat, J.P. Singh, K. Singh, S.A. Bhat, Prospectives and challenges of wastewater treatment technologies to combat contaminants of emerging concerns, *Ecol. Eng.* 152 (2020) 105882. doi:10.1016/j.ecoleng.2020.105882.
5. J.A.H. Melián, Sustainable waste water treatment systems (2018–2019), *Sustain.* 12 (2020) 7–11. doi:10.3390/su12051940.
6. A. Hasanbeigi, L. Price, A review of energy use and energy efficiency technologies for the textile industry, *Renew. Sustain. Energy Rev.* 16 (2012) 3648–3665. doi:10.1016/j.rser.2012.03.029.
7. D.A. Yaseen, M. Scholz, *Textile Dye Wastewater Characteristics and Constituents of Synthetic Effluents: A Critical Review*, Springer: Berlin, Heidelberg, 2019. doi:10.1007/s13762-018-2130-z.
8. N. Bahadur, N. Bhargava, Novel pilot scale photocatalytic treatment of textile & dyeing industry wastewater to achieve process water quality and enabling zero liquid discharge, *J. Water Process Eng.* 32 (2019) 100934. doi:10.1016/j.jwpe.2019.100934.
9. G. Lofrano, S. Meriç, G.E. Zengin, D. Orhon, Chemical and biological treatment technologies for leather tannery chemicals and wastewaters: A review, *Sci. Total Environ.* 461–462 (2013) 265–281. doi:10.1016/j.scitotenv.2013.05.004.
10. G. Saxena, D. Purchase, S.I. Mulla, R.N. Bharagava, Degradation and detoxification of leather tannery effluent by a newly developed bacterial consortium GS-TE1310 for environmental safety, *J. Water Process Eng.* 38 (2020) 101592. doi:10.1016/j.jwpe.2020.101592.
11. S. Korpe, B. Bethi, S.H. Sonawane, K.V. Jayakumar, Tannery wastewater treatment by cavitation combined with advanced oxidation process (AOP), *Ultrason. Sonochem.* 59 (2019) 104723. doi:10.1016/j.ultsonch.2019.104723.
12. G.K. Gupta, H. Liu, P. Shukla, Pulp and paper industry–based pollutants, their health hazards and environmental risks, *Curr. Opin. Environ. Sci. Heal.* 12 (2019) 48–56. doi:10.1016/j.coesh.2019.09.010.
13. R. Toczyłowska-Mamińska, Limits and perspectives of pulp and paper industry wastewater treatment – A review, *Renew. Sustain. Energy Rev.* 78 (2017) 764–772. doi:10.1016/j.rser.2017.05.021.

14. D.T. Pio, L.A.C. Tarelho, P.C.R. Pinto, Gasification-based biorefinery integration in the pulp and paper industry: A critical review, *Renew. Sustain. Energy Rev.* 133 (2020) 110210. doi:10.1016/j.rser.2020.110210.
15. U. Tezel, E. Guven, T.H. Erguder, G.N. Demirer, Sequential (anaerobic/aerobic) biological treatment of Dalaman SEKA Pulp and Paper Industry effluent, *Waste Manag.* 21 (2001) 717–724. doi:10.1016/S0956-053X(01)00013-7.
16. V. Colla, I. Matino, T.A. Branca, B. Fornai, L. Romaniello, F. Rosito, Efficient use of water resources in the steel industry, *Water (Switzerland).* 9 (2017) 1–15. doi:10.3390/w9110874.
17. Deepti, A. Sinha, P. Biswas, S. Sarkar, U. Bora, M.K. Purkait, Utilization of LD slag from steel industry for the preparation of MF membrane, *J. Environ. Manage.* 259 (2020) 110060. doi:10.1016/j.jenvman.2019.110060.
18. R. Mukherjee, M. Mondal, A. Sinha, S. Sarkar, S. De, Application of nanofiltration membrane for treatment of chloride rich steel plant effluent, *J. Environ. Chem. Eng.* 4 (2016) 1–9. doi:10.1016/j.jece.2015.10.038.
19. Deepti, A. Sinha, P. Biswas, S. Sarkar, U. Bora, M.K. Purkait, Separation of chloride and sulphate ions from nanofiltration rejected wastewater of steel industry, *J. Water Process Eng.* 33 (2020) 101108. doi:10.1016/j.jwpe.2019.101108.
20. A. Patwardhan, Industrial wastewater treatment (2008). doi:10.1016/0735-1097(93)90602-W.
21. B. Sarkar, P.P. Chakrabarti, A. Vijaykumar, V. Kale, Wastewater treatment in dairy industries – Possibility of reuse, *Desalination.* 195 (2006) 141–152. doi:10.1016/j.desal.2005.11.015.
22. R.V. Kumar, L. Goswami, K. Pakshirajan, G. Pugazhenthi, Dairy wastewater treatment using a novel low cost tubular ceramic membrane and membrane fouling mechanism using pore blocking models, *J. Water Process Eng.* 13 (2016) 168–175. doi:10.1016/j.jwpe.2016.08.012.
23. B. Farizoglu, S. Uzuner, The investigation of dairy industry wastewater treatment in a biological high performance membrane system, *Biochem. Eng. J.* 57 (2011) 46–54. doi:10.1016/j.bej.2011.08.007.
24. R. Kothari, V. Kumar, V.V. Pathak, V.V. Tyagi, Sequential hydrogen and methane production with simultaneous treatment of dairy industry wastewater: Bioenergy profit approach, *Int. J. Hydrogen Energy.* 42 (2017) 4870–4879. doi:10.1016/j.ijhydene.2016.11.163.
25. J.J. Murcia, M. Hernández-Laverde, H. Rojas, E. Muñoz, J.A. Navío, M.C. Hidalgo, Study of the effectiveness of the flocculation-photocatalysis in the treatment of wastewater coming from dairy industries, *J. Photochem. Photobiol. A Chem.* 358 (2018) 256–264. doi:10.1016/j.jphotochem.2018.03.034.
26. J. Akansha, P.V. Nidheesh, A. Gopinath, K.V. Anupama, M. Suresh Kumar, Treatment of dairy industry wastewater by combined aerated electrocoagulation and phytoremediation process, *Chemosphere.* 253 (2020) 126652. doi:10.1016/j.chemosphere.2020.126652.
27. M.A. Sandoval, R. Salazar, Electrochemical treatment of slaughterhouse and dairy wastewater: Toward making a sustainable process, *Curr. Opin. Electrochem.* 26 (2021) 100662. doi:10.1016/j.coelec.2020.100662.
28. X. Shi, K.Y. Leong, H.Y. Ng, Anaerobic treatment of pharmaceutical wastewater: A critical review, *Bioresour. Technol.* 245 (2017) 1238–1244. doi:10.1016/j.biortech.2017.08.150.
29. D.I. Gerogiorgis, H.G. Jolliffe, Continuous pharmaceutical process engineering and economics: Investigating technical efficiency, environmental impact and economic viability, *Chim. Oggi/Chem. Today.* 33 (2015) 29–32.
30. A. Shah, M. Shah, Characterisation and bioremediation of wastewater: A review exploring bioremediation as a sustainable technique for pharmaceutical wastewater, *Groundw. Sustain. Dev.* 11 (2020) 100383. doi:10.1016/j.gsd.2020.100383.
31. Y. Jiang, X. Shi, H.Y. Ng, Aerobic granular sludge systems for treating hypersaline pharmaceutical wastewater: Start-up, long-term performances and metabolic function, *J. Hazard. Mater.* 412 (2021) 125229. doi:10.1016/j.jhazmat.2021.125229.
32. G. Başaran Dindaş, Y. Çalişkan, E.E. Çelebi, M. Tekbaş, N. Bektaş, H.C. Yatmaz, Treatment of pharmaceutical wastewater by combination of electrocoagulation, electro-fenton and photocatalytic oxidation processes, *J. Environ. Chem. Eng.* 8 (2020). doi:10.1016/j.jece.2020.103777.
33. X. Sun, C. Wang, Y. Li, W. Wang, J. Wei, Treatment of phenolic wastewater by combined UF and NF/RO processes, *Desalination.* 355 (2015) 68–74. doi:10.1016/j.desal.2014.10.018.
34. M.K. Purkait, R. Singh, Membrane technology in separation science (2018). doi:10.1201/9781315229263.
35. N.A. Yaranal, S. Kumari, S. Narayanasamy, S. Subbiah, An analysis of the effects of pressure-assisted osmotic backwashing on the high recovery reverse osmosis system, *J. Water Supply Res. Technol.* (2019) 1–21. doi:10.2166/aqua.2019.089.

36. K. Dhangar, M. Kumar, Tricks and tracks in removal of emerging contaminants from the wastewater through hybrid treatment systems: A review, *Sci. Total Environ.* 738 (2020) 140320. doi:10.1016/j.scitotenv.2020.140320.
37. S. Samsami, M. Mohamadi, M.H. Sarrafzadeh, E.R. Rene, M. Firoozbahr, Recent advances in the treatment of dye-containing wastewater from textile industries: Overview and perspectives, *Process Saf. Environ. Prot.* 143 (2020) 138–163. doi:10.1016/j.psep.2020.05.034.
38. S. Barredo-Damas, M.I. Alcaina-Miranda, M.I. Iborra-Clar, J.A. Mendoza-Roca, Application of tubular ceramic ultrafiltration membranes for the treatment of integrated textile wastewaters, *Chem. Eng. J.* 192 (2012) 211–218. doi:10.1016/j.cej.2012.03.079.
39. Z. Rahimi, A.A. Zinatizadeh, S. Zinadini, Milk processing wastewater treatment in a bioreactor followed by an antifouling O-carboxymethyl chitosan modified Fe3O4/PVDF ultrafiltration membrane, *J. Ind. Eng. Chem.* 38 (2016) 103–112. doi:10.1016/j.jiec.2016.04.011.
40. S. Chakrabortty, J. Nayak, P. Pal, R. Kumar, P. Chakraborty, Separation of COD, sulphate and chloride from pharmaceutical wastewater using membrane integrated system: Transport modeling towards scale-up, *J. Environ. Chem. Eng.* 8 (2020) 104275. doi:10.1016/j.jece.2020.104275.
41. I. Petrinic, J. Korenak, D. Povodnik, C. Hélix-Nielsen, A feasibility study of ultrafiltration/reverse osmosis (UF/RO)-based wastewater treatment and reuse in the metal finishing industry, *J. Clean. Prod.* 101 (2015) 292–300. doi:10.1016/j.jclepro.2015.04.022.
42. C.V. Rekhate, J.K. Srivastava, Recent advances in ozone-based advanced oxidation processes for treatment of wastewater – A review, *Chem. Eng. J. Adv.* 3 (2020) 100031. doi:10.1016/j.ceja.2020.100031.
43. A. Tufail, W.E. Price, M. Mohseni, B.K. Pramanik, F.I. Hai, A critical review of advanced oxidation processes for emerging trace organic contaminant degradation: Mechanisms, factors, degradation products, and effluent toxicity, *J. Water Process Eng.* (2020) 101778. doi:10.1016/j.jwpe.2020.101778.
44. M. Anjum, R. Miandad, M. Waqas, F. Gehany, M.A. Barakat, Remediation of wastewater using various nano-materials, *Arab. J. Chem.* 12 (2019) 4897–4919. doi:10.1016/j.arabjc.2016.10.004.
45. M. Abd Elkodous, G.S. El-Sayyad, M.I.A. Abdel Maksoud, R. Kumar, K. Maegawa, G. Kawamura, W.K. Tan, A. Matsuda, Nanocomposite matrix conjugated with carbon nanomaterials for photocatalytic wastewater treatment, *J. Hazard. Mater.* (2020) 124657. doi:10.1016/j.jhazmat.2020.124657.
46. K.M. Aguilar-Pérez, J.I. Avilés-Castrillo, G. Ruiz-Pulido, Nano-sorbent materials for pharmaceutical-based wastewater effluents – An overview, *Case Stud. Chem. Environ. Eng.* 2 (2020) 100028. doi:10.1016/j.cscee.2020.100028.
47. M. Ghosh, M. Mondal, S. Mandal, A. Roy, S. Chakrabarty, G. Chakrabarti, S.K. Pradhan, Enhanced photocatalytic and antibacterial activities of mechanosynthesized TiO_2–Ag nanocomposite in wastewater treatment, *J. Mol. Struct.* 1211 (2020) 128076. doi:10.1016/j.molstruc.2020.128076.
48. S. Manikandan, N. Karmegam, R. Subbaiya, G. Karthiga Devi, R. Arulvel, B. Ravindran, M. Kumar Awasthi, Emerging nano-structured innovative materials as adsorbents in wastewater treatment, *Bioresour. Technol.* 320 (2021) 124394. doi:10.1016/j.biortech.2020.124394.
49. Y. Zhang, B. Wu, H. Xu, H. Liu, M. Wang, Y. He, B. Pan, Nanomaterials-enabled water and wastewater treatment, *NanoImpact.* 3–4 (2016) 22–39. doi:10.1016/j.impact.2016.09.004.
50. P. Mondal, M.K. Purkait, Preparation and characterization of novel green synthesized iron–aluminum nanocomposite and studying its efficiency in fluoride removal, *Chemosphere.* 235 (2019) 391–402. doi:10.1016/j.chemosphere.2019.06.189.
51. A.D. Sontakke, M.K. Purkait, Fabrication of ultrasound-mediated tunable graphene oxide nanoscrolls, *Ultrason. Sonochem.* 63 (2020) 104976. doi:10.1016/j.ultsonch.2020.104976.
52. B. Verma, C. Balomajumder, Synthesis of magnetic nickel ferrites nanocomposites: An advanced remediation of electroplating wastewater, *J. Taiwan Inst. Chem. Eng.* 112 (2020) 106–115. doi:10.1016/j.jtice.2020.07.006.
53. R. Cited, M.S. Bader, (12) United States Patent, 1 (2012).
54. M.S.H. Bader, Separation of chloride and sulfate ions in univalent and divalent cation forms from aqueous streams, *J. Hazard. Mater.* 73 (2000) 269–283. doi:10.1016/S0304-3894(99)00186-7.
55. M.S.H. Bader, Thermodynamics of ions precipitation in mixed-solvent mixtures, *J. Hazard. Mater.* 69 (1999) 319–334. doi:10.1016/S0304-3894(99)00129-6.
56. Y.J. Chan, M.F. Chong, C.L. Law, D.G. Hassell, A review on anaerobic-aerobic treatment of industrial and municipal wastewater, *Chem. Eng. J.* 155 (2009) 1–18. doi:10.1016/j.cej.2009.06.041.
57. H. Liu, C. Yang, W. Pu, J. Zhang, Removal of nitrogen from wastewater for reusing to boiler feedwater by an anaerobic/aerobic/membrane bioreactor, *Chem. Eng. J.* 140 (2008) 122–129. doi:10.1016/j.cej.2007.09.048.

58. N. Supaka, K. Juntongjin, S. Damronglerd, M.L. Delia, P. Strehaiano, Microbial decolorization of reactive azo dyes in a sequential anaerobic-aerobic system, *Chem. Eng. J.* 99 (2004) 169–176. doi:10.1016/j.cej.2003.09.010.
59. Y.T. Ahn, S.T. Kang, S.R. Chae, C.Y. Lee, B.U. Bae, H.S. Shin, Simultaneous high-strength organic and nitrogen removal with combined anaerobic upflow bed filter and aerobic membrane bioreactor, *Desalination.* 202 (2007) 114–121. doi:10.1016/j.desal.2005.12.046.
60. X. Shi, O. Lefebvre, K.K. Ng, H.Y. Ng, Sequential anaerobic-aerobic treatment of pharmaceutical wastewater with high salinity, *Bioresour. Technol.* 153 (2014) 79–86. doi:10.1016/j.biortech.2013.11.045.
61. S. Mazhar, A. Ditta, L. Bulgariu, I. Ahmad, M. Ahmed, A.A. Nadiri, Sequential treatment of paper and pulp industrial wastewater: Prediction of water quality parameters by Mamdani Fuzzy Logic model and phytotoxicity assessment, *Chemosphere.* 227 (2019) 256–268. doi:10.1016/j.chemosphere.2019.04.022.
62. R. Shoukat, S.J. Khan, Y. Jamal, Hybrid anaerobic-aerobic biological treatment for real textile wastewater, *J. Water Process Eng.* 29 (2019) 100804. doi:10.1016/j.jwpe.2019.100804.
63. P.F. Tee, M.O. Abdullah, I.A.W. Tan, N.K.A. Rashid, M.A.M. Amin, C. Nolasco-Hipolito, K. Bujang, Review on hybrid energy systems for wastewater treatment and bio-energy production, *Renew. Sustain. Energy Rev.* 54 (2016) 235–246. doi:10.1016/j.rser.2015.10.011.
64. P.P. Das, A. Anweshan, P. Mondal, A. Sinha, P. Biswas, S. Sarkar, M.K. Purkait, Integrated ozonation assisted electrocoagulation process for the removal of cyanide from steel industry wastewater, *Chemosphere.* 263 (2021) 128370. doi:10.1016/j.chemosphere.2020.128370.
65. R. Changotra, H. Rajput, J. Paul Guin, L. Varshney, A. Dhir, Hybrid coagulation, gamma irradiation and biological treatment of real pharmaceutical wastewater, *Chem. Eng. J.* 370 (2019) 595–605. doi:10.1016/j.cej.2019.03.256.
66. S. Bener, S. Atalay, G. Ersöz, The hybrid process with eco-friendly materials for the treatment of the real textile industry wastewater, *Ecol. Eng.* 148 (2020). doi:10.1016/j.ecoleng.2020.105789.
67. H.E. Muga, J.R. Mihelcic, Sustainability of wastewater treatment technologies, *J. Environ. Manage.* 88 (2008) 437–447. doi:10.1016/j.jenvman.2007.03.008.
68. M.A. Sandoval, R. Salazar, Electrochemical treatment of slaughterhouse and dairy wastewater: Toward making a sustainable process, *Curr. Opin. Electrochem.* 26 (2021) 100662. doi:10.1016/j.coelec.2020.100662.
69. S. Foteinis, J.M. Monteagudo, A. Durán, E. Chatzisymeon, Environmental sustainability of the solar photo-Fenton process for wastewater treatment and pharmaceuticals mineralization at semi-industrial scale, *Sci. Total Environ.* 612 (2018) 605–612. doi:10.1016/j.scitotenv.2017.08.277.

5 Nanobiotechnology in Wastewater Treatment

Gangar Tarun, Satyam, Dalui Sushovan, and Patra Sanjukta
Indian Institute of Technology

CONTENTS

5.1 INTRODUCTION

The problem of water pollution is on a continuous rise and is expected to be at its peak when the global population reaches 9 billion by 2050 (Sharma and Sharma, 2013). Water pollution has led to serious environmental threats. All domains of life, ranging from microscopic forms to large animals, are suffering at some point from this problem. The major cause of such water pollution is anthropogenic sources. With vast industrialization and globalization, the biosphere of the earth has been loaded with all of the harmful synthesized chemicals, which persist in the environment for a long time. Such effluents cause deterioration of the environment as well as the existence of life on earth. Developing nations are the leaders in their contribution to the pollution of existing water bodies. There have been a number of cases and studies where it has been found that industrial and domestic wastewater is directly discharged into the water bodies or is discharged only after primary treatment. As per the reports by the Government of India, in 2015, the total sewage generation in

DOI: 10.1201/9781003164982-5

India is 62,000 MLD (million liters per day), of which 522 working municipal sewage treatment plants (STPs) can treat only 23,277 MLD. This data only points toward the inefficiency of the system and non-seriousness toward the growing environmental concerns. Similar is the case in other developing nations, especially in the Middle East, where is scenario is much worse (Rafique et al., 2019). The average number of global deaths because of water-borne diseases is 10–20 million (Leonard et al., 2003). The point of major concern is that the majority of water bodies around the globe are at an alarming concentration of toxic pollutants as per WHO and various EPAs (Leonard et al., 2003).

The requirement for potable water is increasing and has reached a point where natural resources can no longer support the need for clean potable water. Water is one of the necessities for human existence, which necessitates the scientific community to search for alternate sources of potable water. There is an urgent need for the development of sustainable and integrated water management solution that will enable water security. There is a requirement for the whole new system of policies that will keep a check on the domestic as well as industrial discharge. Reuse and recycling of wastewater could be one approach that various industries can follow to reduce the burden of wastewater treatment plants. Advanced treatment processes along with conventional techniques can be used and effectively applied to remove biodegradable organic matter, dissolved compounds, colloidal substances, heavy metals, suspended solids, nitrogen and phosphorus compounds, and biomass.

The use of conventional methods for the treatment of wastewater now has failed to meet the needs of the existing situation of wastewater. The physical separation, chemical processing, and biological interference, which involve coagulation-flocculation activated carbon adsorption, ozonation, and advanced oxidation processes, membrane processes, membrane bioreaction, attached growth treatment processes, etc. alone, cannot treat the new-age effluents which have compounded to a much greater extent. This requires intervention with a new age technology that can treat these effluents and will also supplement traditional methods.

Nanotechnology, an emerging scientific field also known as the "knowledge-based economy," is getting much attention of the scientific community all around the globe. The main reasons behind this attention are plausibility, security, selectivity, and flexibility of the techniques used in nanotechnology. All these properties of nanomaterials, when unified with the advantages of biomolecules in living cells along with conventional treatment methods, can revolutionize the whole system of water and wastewater treatment (Bhuyan et al., 2015). Nanotechnology operates at a nanoscale (of order 10^{-9} to 10^{-7}m). Nanoparticles have characteristic properties entirely different from their macromolecular counterparts. They differ in magnetic, optical, and electrical properties. This may be attributed to their unique surface area: volume ratio (Amin et al., 2014). In other words, nanoparticles are the particles whose at least one structural component is less than 100 nm. Nanomaterials have been fabricated in a variety of form-factors such as nanotubes, nanofilms, nanowires, quantum dots, colloids, etc. For wastewater treatment, various eco-friendly and cost-effective nanomaterials have been developed which have unique abilities for decontamination of domestic and industrial effluents as well as other contaminated sources of water. Various reputed published literature cite nanotechnology as one of the most advanced processes for the treatment of wastewater.

Based on their application in the wastewater treatment process, nanomaterials can be categorized as follows:

1. Nano-adsorbents: These are the materials with high adsorption capacities on their surface. Materials used for the development of nano-adsorbents (or nanosorbents) include silica, metal oxides, activated carbon, clay, and modified composites.
2. Nanocatalysts: These include materials such as metal oxides and semiconductors. These constitute electrocatalysts, Fenton-based catalysts. They help in improving the chemical oxidation of organic pollutants.

3. Nanomembranes: Membrane filters are used here with enhanced pressure application. Nanofiltration is used for the treatment/filtration of wastewater from the industrial sector because of its small pore size, low cost, user-friendliness, and cost-effectiveness.

5.2 STRATEGIES FOR WASTEWATER TREATMENT

5.2.1 Nanosorption for Removal of Pollutants from Polluted Water

5.2.1.1 Nanosorbents

Nanosorbents are inorganic or organic nanoscale particles that can absorb other substances. Most of their environmental applications are in wastewater treatment and drinking water processing, with others focusing on air pollution or groundwater contamination. Their sorption potential is vastly improved compared to traditional materials because nanoparticles have such a large surface area. Furthermore, the bottom-up approach in nanotechnology allows for customized nanosorbent design to suit individual needs. Adsorption is the most preferred method used for wastewater purification. In the field of nanotechnology, the technique of adsorption finds wide application in the field of wastewater management. Activated carbon is one of the most well-known examples of a traditional sorbent for environmental applications.

Certain metal oxides such as Fe_2O_3, TiO_2, and Al_2O_3 are very economical to use with high adsorption capacities. One of the most interesting properties of nanosorbents is their high porosity and active surface area, which are effective in adsorbing and removing toxic chemicals and contaminants from the test environment. Also, there is no residual side reaction to this technique as nanosorbents are completely free of any payload. Most of the time, adsorption is applied for cleaning industrial effluents free of metallic pollutants, such as Cr, Hg, Pb, Co, Cd, Zn, Cu, Ni, As, etc. Metal nanoparticles and carbon-based nanotubes are some nanomaterials used for adsorption-based wastewater management.

Multiple reactive nanostructured materials can be engineered into nanosorbents (e.g., metal-impregnated activated carbon; palladium-gold nanocomposites). Researchers have developed a nanostructured sorbent made up of branched macromolecules with physically and/or chemically tunable sites assembled into spherical microparticles. Researchers presented a sensitive method for determining formaldehyde based on modified alumina nanoparticles for its preconcentration and followed by UV–vis spectroscopy. The chemical modification of alumina nanoparticles was achieved by immobilizing 2,4-dinitrophenylhydrazine (DNPH) using sodium dodecyl sulfate as a surfactant. The formaldehyde that was retained on the modified adsorbent was then desorbed, and its catalytic effect on the thionine oxidation by bromated ion was determined. Alumina nanoparticles were chemically modified by immobilization of 2,4-dinitrophenylhydrazine (DNPH) via sodium dodecyl sulfate as a surfactant. In practical applications, this innovative design approach offered superior advantages for the nanosorbent. The researchers grafted functional groups onto the surface of Al_2O_3 nanoparticles, which are an inorganic substrate. The sorption capacity of these functionalized nanoparticles for formaldehyde from real samples was extremely high. Considering that amine groups can adsorb formaldehyde from samples, it was extracted using amine grafted γ-Al_2O_3 nanoparticles. Researchers used the ionic gelation process to make chitosan nanoparticles. This nano sorbent was used to extract eosin Y (a model anionic dye) using a batch adsorption technique. The carbodiimide activation process was used to make the Fe_3O_4 nanoparticles–carboxymethylated chitosan conjugate (anionic magnetic nano adsorbent). This conjugate was investigated for its ability to dissolve acid dyes like orange G and acid green. Various nanomaterials used as adsorbents with their respective adsorption capacities and rate constants are listed in Table 5.1. The table highlights the use of various nanomaterials as adsorbents for the effective removal of contaminants such as dyes, pesticides, etc.

Nanoclays, which are nanometer-sized sorbents, is another form of sorbent. Montmorillonite is commonly found in thick earth platelets. As a primary component, platelets can be modified with a

TABLE 5.1
Adsorption of Contaminants by Using Various Adsorbents

Sr. No.	Contaminant	Nanomaterial Used	Adsorption Rate (mg/g)	Rate Constants (h^{-1})	Reference
1	Reactive red M-2BE	Activated carbon	260.7	1.503	Machado et al. (2011)
		Carbon nanotubes	335.7	2.860	2
2	Methylene blue	Graphene oxide/Fe_3O_4 hybrid composite	167.2	–	Xie et al. (2012)
		Reduced graphene oxide/TiO_2	467.6	3.127	Wang et al. (2016)
		Polyvinyl alcohol	123.3	4.854	Agarwal et al. (2016)
3	Bromothymol blue	Oxidized multiwalled carbon nanotubes	55	0.042	Ghaedi et al. (2012)
		Polyvinyl alcohol	276.7	4.266	Agarwal et al. (2016)
4	Maxillon blue	Multiwalled carbon nanotubes	260.7	–	Alkaim et al. (2015)
5	Reactive blue 198	MgO	123.5	5.4	Moussavi and Mahmoudi (2009)
6	Rhodamine B	Polyacrylamide/$Ni_{0.02}Zn_{0.98}O$ nanocomposite	–	8.88	Kumar et al. (2014)
7	Cr(VI)	Single walled carbon nanotubes	1.26	–	Dehghani et al. (2015)
		Multiwalled carbon nanotubes	2.35	0.42	Dehghani et al. (2015)
		Ethylenediamine functionalized nano Fe_3O_4	136.98	47.172	Zhao et al. (2010)
		Polypyrrole/Fe_3O_4 magnetic nanocomposite	169.4	11.28	Bhaumik et al. (2011)
8	Pb(II)	Diethylenetriamine modified multiwalled carbon nanotubes	58.26	–	Vuković et al. (2011)
		Graphene oxide nanosheets	842	–	Zhao et al. (2011)
		TiO_2 nanotubes/carbon nanotubes	192.588	–	Doong and Chiang (2008)
		MnO_2/carbon nanotubes	78.74	0.816	Wang et al. (2007)
9	Cd(II)	Diethyltriamine-modified multiwalled carbon nanotubes	31.45	–	Vuković et al. (2011)
		Graphene	106.3	–	Zhao et al. (2011)
		TiO_2	16.69	2.244	Visa et al. (2009)
		Al_2O_3/multiwalled carbon nanotubes	27.21	5.764	Liang et al. (2015)
		Ni@C composite nanostructure	6.43	–	Ni et al. (2010)
10	Chlorpyrifos	Reduced graphene oxide nanosheets	1200	–	Maliyekkal et al. (2013)
11	Endosulfan	Reduced graphene oxide nanosheets	1100	–	Maliyekkal et al. (2013)
12	Malathion	Reduced graphene oxide nanosheets	800	–	Maliyekkal et al. (2013)

(*Continued*)

TABLE 5.1 (*Continued*)
Adsorption of Contaminants by Using Various Adsorbents

Sr. No.	Contaminant	Nanomaterial Used	Adsorption Rate (mg/g)	Rate Constants (h^{-1})	Reference
13	Nitrofurazone	Powder activated carbon	50.8	0.1129	Wu and Xiong (2016)
		Multiwalled carbon nanotubes	59.9	0.2082	Wu and Xiong (2016)
14	Trichloroethylene	Al_2O_3/multiwalled carbon nanotubes	19.84	1.1048	Liang et al. (2015)

variety of organic compounds to improve their properties (Organoclays). The adsorption capability of the investigated organoclay was significantly higher than that of the unmodified clay, according to studies. Within an hour, they were able to adsorb hydrocarbons up to ten times their own weight. Different nanoclays have also been studied as dye and phosphorus sorbents. Organoclays were found to have exceptionally high contaminant affinity in both experiments, suggesting that nanoclays may be superior sorbents for dye and phosphorus removal from wastewater.

5.2.1.2 Zeolite

Zeolites are microporous crystalline structures composed of aluminosilicate material with pore size, which ranges from nanometer to sub-nanometer range. This material offers several advantages, such as resistance to inflation in water and suspension formation, which makes it a suitable material for the fabrication of nanomembranes. Also, zeolite possesses good chemical and thermal conductivity because of the inert nature of aluminosilicate crystals which makes it very suitable for use in reverse osmosis and nanofilters. Zeolite-based membranes find wide application in the removal of pollutants for the production of clean water and for the separation of hydrocarbon gases.

5.2.2 Nanocatalysts for Oxidation of Organic Pollutants

Nanocatalysis is a rapidly growing field in which nanomaterials are used as catalysts in a variety of homogeneous and heterogeneous catalysis applications. Heterogeneous catalysts are typically very complex, and their performance is influenced by a variety of factors such as electronic properties, scale, morphology, and the association of active sites with supports. It's also difficult to separate their individual contributions to achieve a molecular-level understanding. Nanoparticles of metals, semiconductors, oxides, and other compounds have been commonly used for important chemical reactions in heterogeneous catalysis, which is one of the oldest industrial activities of nanoscience.

Many industries are currently using nanostructured materials for their catalytic processes. Many challenges such as the transition from the fossil to renewable energy sources and sustainable agriculture can be faced by the industries, which can be tackled by these nanocatalysts.

The structural, quantum scale and electronic effects of nanopore confinement result in catalytic nano-effects. In correlation curves between the catalytic behavior and the dissociative binding energy of the reactants, these phenomena induce a transition toward metals with higher binding energy. During the preparation of nanocatalysts and when they are used under such conditions (e.g., during a chemical reaction), significant structural variations in size and morphology occur. Catalysts with high activity and selectivity and with high stability vary based on temperature, pH, solvent, and other factors have always been preferred (Prinsen and Lugue, 2019).

Nano-sized catalysts are extremely active, making them ideal for use in environmental applications such as water treatment. Two metal systems are being investigated for dehalogenation. Pd catalysts (with or without carrier) can support hydro-dehalogenation in the water at room temperature when hydrogen is used as the reducing agent. The range of halogenated hydrocarbons that can be treated is quite extensive. To avoid Pd leaching into the water, Pd polymers are coated with a polymer membrane. Then AOX-reduction process is used to treat the industrial wastewater containing halogenated contaminants (such as halogenated pharmaceuticals or herbicides).

A significant number of unconsumed dyes from the textile and printing industry are poured into the sea on a regular basis. The presence of dyes and pigments in water causes significant environmental harm. Malachite green must be illustrated by the various dyes used in producing goods. This pigment has been used as a dye for linen, jute, leather, fur, cotton, and paper, as well as a food coloring additive. But the conventional methods have some limitations. Incineration can generate harmful volatiles; biological treatment necessitates lengthy treatment times and undesirable odors; ozonation has a low half-life; ozone stabilization is impaired by the presence of salts, pH, and temperature; and adsorption results in phase transference of the contaminant, rather than oxidation and the production of sludge. Nb_2O_5 was used as a catalyst which was used to degrade indigo carmine dye, and it degraded almost all of the dye. Because of its easy recovery and recycling, Nb_2O_5 has a great advantage than any other photocatalysts (TiO_2 and ZnO) (Figure 5.1) (Prado et al., 2008).

The catalytic activities can be tuned by selectively hosting NPs in confined spaces. The improvement of the catalytic activity of sub-nanometer titania clusters can be done and the redox activity of metal clusters within carbon nanotubes can be tuned (as opposed to those assisted on the outer wall of CNTs). In the epoxidation of propylene by H_2O_2, the enclosed titania has a much higher activity than the titania particles connected to the outer walls of the CNTs (the outside titania) (Zhang et al., 2011b).

Another example is that using hydride as a reducing agent; copper can catalyze dehalogenation. Compared to Pd catalysts, copper is significantly less prone to poisoning, making it a welcome alternative. On the other hand, copper has a broader range of halogenated hydrocarbons that can be treated than Pd, and the two metals' abilities are complementary.

In both homogeneous and heterogeneous catalysis, the field of nanocatalysis (the use of nanoparticles to catalyze reactions) has seen exponential development in the last decade. Since nanoparticles have a higher surface-to-volume ratio than bulk materials, they are appealing as catalysts. Catalysts are used to speed up and fuel thousands of various chemical reactions on a regular basis, and they are the foundation of the multibillion-dollar chemical industry as well as essential environmental

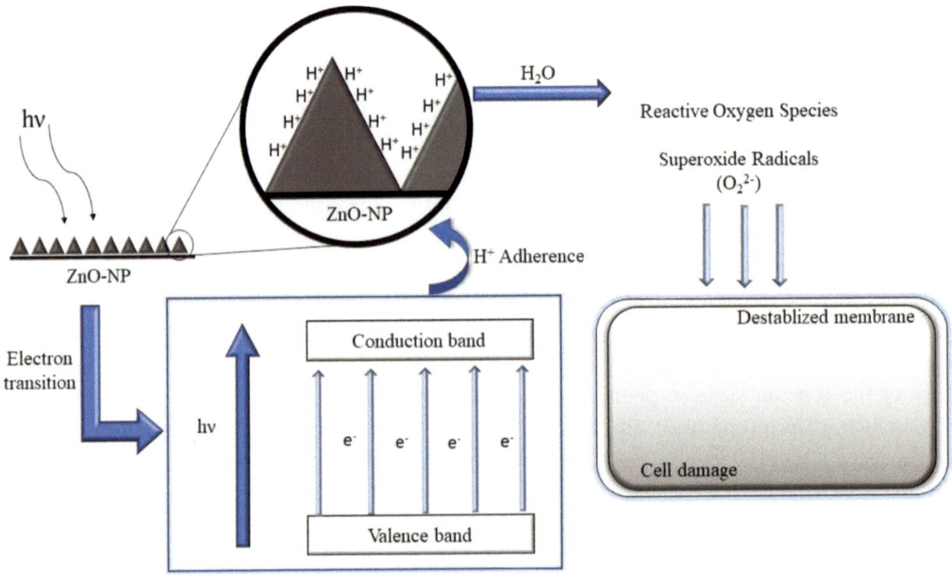

FIGURE 5.1 Photocatalytic activity of ZnO nanoparticles.

protection technologies. The discovery of new catalysts is predicted to be greatly influenced by research in nanotechnology and nanoscience.

5.2.3 Nanomembranes for Filtration of Dissolved Contaminants

Membrane filtration has now become a common practice in the modern era as most urban household use reverse osmosis (RO) system to purify water and make it drinkable. Membranes provide a physical barrier for various pollutants. Similarly, at industrial scale, nanotechnology is being used when nanomaterials are used to fabricate nanomembranes or nanofilters which work not only as physical barriers but also perform chemical separation of pollutants from wastewater.

Nanomembranes have several advantages over conventional membrane filters:

- Ease of operation
- Low capital input
- Low space requirement
- High efficiency

In addition to above, nanomembranes can be functionalized via certain modifications through chemical or biological means so as to solve a particular purpose.

5.2.3.1 Nanofilteration

Nanofilteration is a high-pressure membrane separation technique. It employs high pressure for treatment but as compared to RO where the working pressure range is (20–100 atm) the working pressure is very low (7–14 bar). The useful properties such as charge-based repulsion and high rate of permeation make it a technique of choice in comparison to conventional RO systems. The basic working mechanism behind nanofiltration is simple diffusion and also it occurs due to the size exclusion property of the membrane. The fixed surface charge of nanofiltration membranes helps in purification via selective contaminant binding (Bowen and Mukhtar, 1996). This attributes to lower energy consumption. For wastewater circulation in the system and driving pressure, centrifugal pumps are often used. The working range of a nanofiltration module is 0.9–5.5 m in length and 100–300 mm in diameter (Tchobanoglous et al., 2014; Shah, 2020). Such modules depending upon choice can be either arranged vertically or horizontally. The pore size of nanofilters usually lies in the range of 1–5 nm where all the solutes are effectively removed but due to their surface charge property, some monovalent ions are allowed to pass through. The size exclusion and electrostatic interaction filtration operate for uncharged and ionic species respectively (Choi et al., 2002; Schäfer et al., 2004; Verliefde et al., 2008). This procedure of wastewater treatment meets the quality check standard for most of the industrial grade reuse water. The after-processing of water such as disinfection is not highly desired as nanofiltration removes all organic as well as organic matter.

5.2.3.2 Nanofibers

Nanofibers are the very first class of membrane nanofilters. They possess certain properties such as large surface area to volume ratio, interconnectivity, and high porosity. Electrospinning method is used for the fabrication of nanofibers which makes it highly modifiable as per application. Nanofiber filters have the capacity to filter out even the micro-sized particles.

Composite nanofibers modified with ultrafine cellulose can filter out bacteria and viruses (Saravanan et al., 2016). Union of effective adsorbents with nanofibers can be used to treat water free of toxic heavy metals for e.g. nanofibers fabricated via electrospinning poly (ethylene oxide) chitosan and Fe^{3+} followed by ammonia vapor crosslinking is able to effectively remove trace arsenate from water. Here Fe^{3+} amalgamated with chitosan has high affinity for arsenic salts. Further, such composites can be functionalized via the incorporation of functional groups such as C-O, N-H, and O-H to increase the adsorption capacity. Their ease of regeneration via NaOH makes it a very reliable choice for wastewater treatment (Min et al., 2016; Shah, 2020, 2021).

Another example of a modified nanomembrane includes single-step deposition of polymer nanofibers onto stainless steel which will render the membrane superhydrophobic-superleophilic and hence can be used used to separate mineral oil, grease, and gasoline to from wastewater. The polymer used in this case was polystyrene because of its low free energy and non-wetting property which enhances oil-water separation (Lee et al., 2013).

Similarly, a biopolymer poly(l-lactic acid) can be used as a scaffold and surface modification can be done via β-cyclodextrins and polydopamine (PDA). Here β-cyclodextrins help to improve host-guest inclusions complexes with organic compounds and PDA helps to make surface rougher and imparts negative charge to the surface layer. These modifications impart properties such as high under water oleophobicity and superhydrophilicity. Such nanofiber membranes can be used to target toluene emulsions and methylene blue (positive organic dyes), O/w emulsions (Kang et al., 2018).

5.2.3.3 Biologically Inspired Membrane (Mixed Matrix Membrane)

With the immense possibilities represented by living organisms, various scientific groups have explored applications of biological entities. For the application purpose, biologically synthesized nanoparticles are incorporated into the membranes. Such membranes have the ability to replace conventional polymeric membrane filters as they possess high selectivity and permeability. A perfect blend of nanoparticles with mono- or polymeric membranes results in the formation of mixed matrix membranes also referred to as nanoparticle-entrapped membranes (Pendergast and Hoek, 2011).

Aquaporin-1 is a membrane channel protein known to cause water transport across membrane at the rate of $5*10^9$ molecules/s (Verkman et al., 2017). The utilization of this protein in polymeric membranes has been sought because of its high selectivity and water permeability (Zhao et al., 2012). The method used for the synthesis of aquaporin-1 embedded membrane (biomimetic membrane) is interfacial polymerization. The membrane produced because of this technique has high interfacial area, good mechanical stability, high water permeability rates and good NaCl rejection. This makes such membrane a good choice for desalination applications. Another example includes biomimetic membrane produced by using Aquaporin-Z with polydopamine layer on polysulfone substrate through amidation reaction (Ding et al., 2015).

5.2.3.4 Carbon Nanomaterials

Carbon nanomaterials are one of the most preferred materials for membrane development. Ease of preparation, good rejection ability and high mechanical robustness are some important features which make carbon NM as most desired materials. Hollow carbon nanotubes-based nanomembranes have high solvent permeability and high contaminant rejection rate. CNTs also operate on a similar mechanism as MMM, as their smaller diameter (1–10 nm) allows only water molecules to pass through, rejecting all the contaminants. The noticeable difference is that the fast permeation of water through CNT is extremely fast. Various studies have reported the increased rejection rates of organic dyes and long-chain hydrocarbons along with efficient bacterial (*E. coli*) and viral (poliovirus) elimination (Karan et al., 2012; Srivastava et al., 2004). Graphene sheets have also been used in the nanofiltration in the shape of two-dimensional graphene sheets. Graphene sheets have better thermal, mechanical and chemical stability than carbon nanotubes. Graphene sheet nanofiltration is actively applied in the removal of nanoparticles and organic dyes. The thickness of graphene membrane for nanofiltration is 22–53 nm on a microporous substrate. One of the major drawbacks is that graphene-mediated nanofiltration cannot be applied to ionic species.

5.2.3.5 Metal Oxides

Metal oxides are the most economical alternative for the filtration because of ease of separation, regeneration and low cost. Metal oxide nanoparticles possess certain properties such as high surface area to volume ratio, have more adsorption sites, and they are much easier to reuse by a mere shift in the pH of solution. The superparamagnetic property of some metal-oxide nanoparticles allows for better adsorption capacity than activated carbon (Corsi et al., 2018). The photocatalytic activity of most of the metal oxides helps to degrade different types of organic and inorganic dyes. This property makes the membrane reactive rather than being just a physical filter (Chan et al., 2011).

The fabrication of metal-oxide nanoparticles is brought about by complexation between molecules of oxygen and metal dissolved in solution (Ilakoon, 2014). Strong adsorption capacity and faster kinetics of metal-oxide nanoparticles can be attributed to their large number of surface reactive sites, short intraparticle distance, and high surface area which is not reduced even after compression (Sa and Premlatha, 2016). Titanium dioxide (TiO_2) nanofilters fabricated to form a nanowire mesh are proven to remove 100% humic acid and 90% total organic carbon (TOC) from wastewater (Zhang et al., 2008). Also, various combinations of one type of metal oxides with others may result in enhanced stability and remediation capacity. For example, the amalgamation of TiO_2 with γ-alumina showed enhanced retention of organic dyes (Romanos et al., 2012). In another report, the stability of nanofilters has been improved by the incorporation of zirconium dioxide over a wide pH range.

5.3 ZEROVALENT METAL NANOPARTICLES

These are elemental forms of metals bearing net zero charge on the surface. They are usually used as oxidizing agents (Tsarev et al., 2016). They possess magnetic, electronic, catalytic, mechanical and optical properties which are quite different from their ionic counterparts (Prasad et al., 2017). The most common forms of zerovalent metal ions along with their mechanism of action are represented in Figure 5.2.

5.3.1 Silver Nanoparticles

Ag-NP works directly by producing ROS species and interfering with the disruption of essential cellular machinery which renders the cell non-functional (Prasad et al., 2018). The dissolution of silver nanoparticles and its interaction with the thiol groups of amino acids in enzymes interrupt their functioning (Aziz et al., 2019).

5.3.2 Zinc Nanoparticles

Zerovalent ZnNPs wastewater treatment is based on dehalogenation (Zheng et al., 2011). The reduction potential of ZnNP is far greater than other nanoparticles. This is the reason zinc nanoparticles are used extensively in wastewater treatment (Fu et al., 2014).

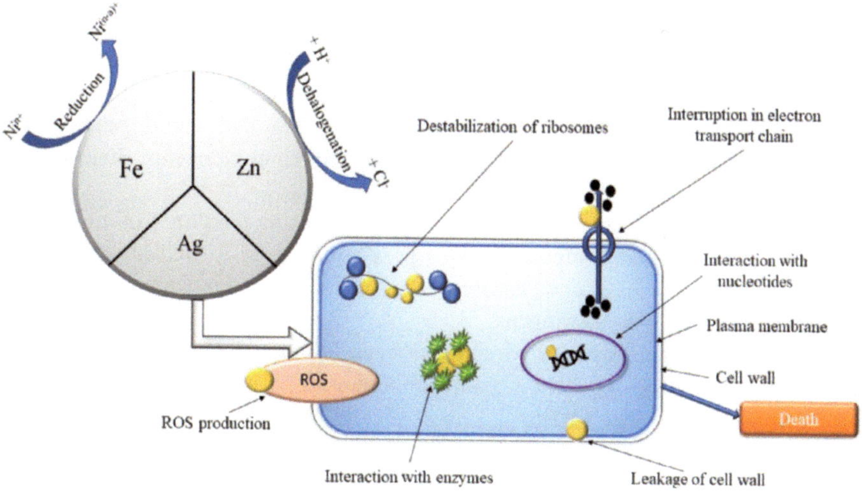

FIGURE 5.2 Zerovalent metal nanoparticles involved in wastewater treatment.

5.3.3 Iron Nanoparticles

Under anaerobic conditions, iron NP is oxidized to form ferrous ions and dihydrogen which are potential reductants for various pollutants in wastewater treatment plants. The basic oxidation of Fe(II) to Fe(III) facilitates the removal of organic as well as inorganic contaminants (Klačanová et al., 2013).

5.4 ANTIMICROBIAL NANOMATERIALS FOR WASTEWATER DISINFECTION

Nanotechnology is quickly progressing and has now proved to be more efficient than standard water treatment systems, allowing for the safe use of alternative water sources. Metallic nanoparticles have unusual antimicrobial properties, making them ideal for use in medical and medicinal applications, as well as wastewater treatment (Bankier et al., 2019). Antimicrobial active nanoparticle-based materials have recently attracted a lot of attention. Nano-scaled materials, such as fibers, plastics, and metals coated with nano-silver, as well as nanocomponents based on titanium dioxide, magnesium oxide, copper and copper oxide, zinc oxide, cadmium selenide/telluride, chitosan, and carbon nanotubes, are being used for biocidal or bacteriostatic properties (Karwowska, 2017).

Antimicrobial additives are classified into four categories based on Silver Ion, Copper, Zinc, and Organic technologies:

- Antimicrobial Silver Ion Additives can be used in a wide variety of fabrics and applications, including paints, coatings, textiles, polymers, and other kinds of products.
- Antimicrobial additives containing zinc are a well-known antibacterial and antifungal compound.
- Copper antimicrobial additives: can be used to provide antimicrobial safety in preservative and hygienic applications, with oils, coatings, and polymers being the best candidates.
- phenolic biocides and quaternary ammonium compounds are examples of organic antimicrobial additives.

Salts, oxides, complexes, and elemental nanoparticles are examples of antimicrobially active nanomaterials. Their efficacy derives from their chemical toxicity, limited scale, and distinctive shape, which helps them to destroy cell membranes (Suresh et al., 2013).

Nanoparticles work by interrupting some of the most important cellular functions rendering the cell non-functional or sometimes lyse the cell. This is done via direct interference in the electron transport chain and as a result hamper membrane stability. Another way through which nanoparticles cause major damage to the cells is by oxidation of cellular components, toxic secondary products formation, or by producing reactive oxygen species.

One of the growing concerns in present times is the increasing resistance of microbial pathogens toward existing antimicrobial compounds. This presses the need to find alternative methods, of which the use of nanoparticles as antimicrobial agents is most popular, which will help disinfection of wastewater free of microbial pathogens.

The most promising nanoparticles which have been found to be effective against a variety of pathogens are chitosan, nano-Ag carbon-based nanomaterial (graphene, fullerenes), and photocatalytic TiO_2.

The above nanoparticles have found promising applications in reducing the microbial load from wastewater but the extent of application depends upon the nature of nanoparticles (Table 5.2 and Figure 5.3).

TABLE 5.2
Mechanisms Involved in Treatment of Wastewater Using Various Nanomaterials

Sr. No.	Mechanism of Operation	Nanomaterials	Application	Reference
1.	ROS production	Fullerenes and derivatives	Photocatalytic reactors, functional groups on carbon cage, solar disinfection	Spesia et al. (2008), Grinholc et al. (2015)
		TiO_2	Reactive membranes, biofouling resistant surfaces, water treatment systems for organic contaminants	Hebeish et al. (2013), Pratap Reddy et al. (2007)
2.	Membrane damage	MgO and Ce_2O_4	Surface coating	Makhluf et al. (2005), Huang et al. (2005), Dos Santos et al. (2014), Maqbool et al. (2016)
		Nano-chitosan	Enzymes and other biological molecules, flocculants in wastewater treatment	Higazy et al. (2010), Rabea et al. (2003)
		Ag	Potable water filters, membranes	Chou et al. (2005), Rai et al. (2009)
3.	Oxidative stress	CNT	Carbon hollow fibers, packed bed filters	Kang et al. (2007), Brady-Estevez et al. (2008)
		Graphene and graphite membranes	Antibiofouling membranes	Zhu et al. (2017), Liu et al. (2011)
4.	Cell wall disruption	Nano-zerovalent Fe	Permeable reactive barrier for groundwater remediation, surface coating in the adsorbent	Kim et al. (2010, 2011)

5.4.1 Mechanisms of Disinfection

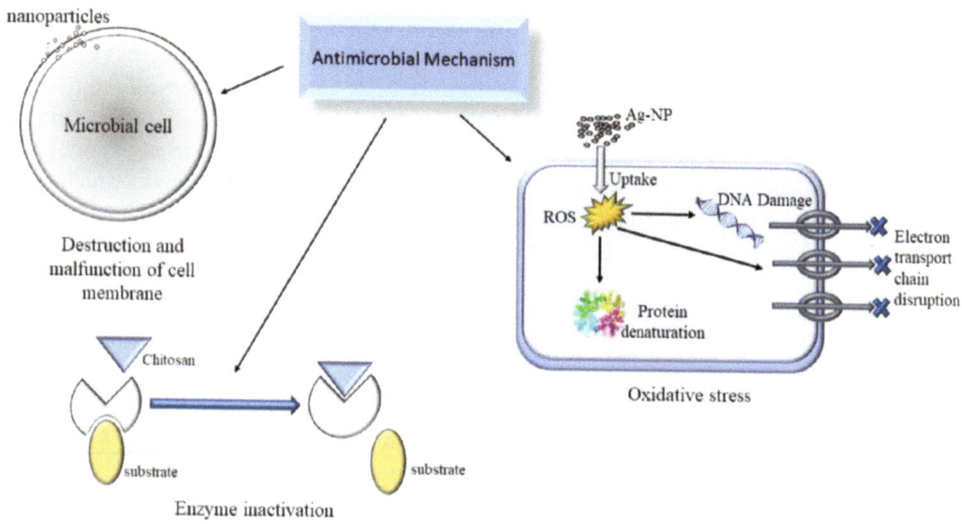

FIGURE 5.3 Antimicrobial mechanism of nanoparticles.

5.4.1.1 Oxidative Stress

Oxidative stress conferred by nanoparticles is a key contributor in the destabilization of cell membrane which results in cell damage (Cheloni et al., 2016). Reactive oxygen species not only is involved in the cell membrane disintegration but also is involved in the elevated level of gene expression of oxidative proteins, an important mechanism in bacterial apoptosis. Also, ROS plays an active role in blocking the expression or inhibition of periplasmic enzyme activity, these enzymes play an active role in maintaining bacterial cell morphology.

Interaction of bacterial cell membrane with Al_2O_3 nanoparticles triggers the loss of membrane integrity and this is the result of internal oxidative stress (Ansari et al., 2015). Similarly, nano-Ag ions are used as catalytic centers for oxygen activation in air and water with an aim of ultimate production of hydroxyl radical and reactive oxygen ions (Shrivastava et al., 2007; Yang et al., 2009).

Ultrasonic treatment of nanoparticles and cells facilitates the dissociation of nanoparticles and helps in their penetration into the bacterial cell membrane. During this process, the metal ions released from the surface inhibit bacterial proliferation. Other reasons for such inhibition may be the increased transport of nutrients, oxygen and waste by ultrasound treatment (Seil and Webster, 2012). Carbon hollow fibers and packed bed filters are the most prominent examples which are used for the treatment purpose (Brady-Estevez et al., 2008; Liu et al., 2011).

5.4.1.2 Dissolved Metal Ions

Metal oxides absorbed directly through the membrane interact with functional groups such as mercapto (-SH), carboxyl (-COOH), and amino (-NH). Additionally, they play an active role in altering enzyme activity, changing cell membrane structure, and affecting the normal physiology of cell. All these changes ultimately cause cell death. However, various studies have emphasized the fact that metal oxides lose their antibacterial property inside lipid vesicles as they have little effect on the pH of cell. On the other hand, superparamagnetic iron oxide directly interferes in the transfer of transmembrane electrons.

5.5 LIMITATIONS OF NANOMATERIALS IN WASTEWATER TREATMENT

The technology of wastewater treatment via nanomaterials involves working at a nanometer scale which makes immediate implementation of the use of nanotechnology difficult. As most of the methods and techniques described above are applied or tested in-vitro and laboratory set conditions. And it is a well-known fact that nanoparticles are toxic entities, to assess all the possibilities of their implementation keeping in mind the safety of use and the impact of their application, proper and elaborate studies need to be done. Although virtually all can be toxic in high enough doses, the more relevant question is: how toxic are nanomaterials in the concentrations at which they may be used? The type of base material, scale, form, and coatings can all influence the toxic effects of nanomaterials. How nanoparticles are used in the workplace, how they are divided between various media, their versatility in both media, and their stability all influence the setting. To evaluate nanoparticle hazards, such specific knowledge about their behavior and toxicity is needed; however, a practical evaluation cannot be made solely based on this information; however, some data on the expected concentration of nanoparticles in environmental systems is required, and there is no reliable research on such concentrations to date (Reijnders, 2006; Taghavi et al., 2013; Ray et al., 2009). Because of the larger surface area, nanoparticles can cause more harm to the human body and environment. As a result, national and international attention has been drawn to the potential danger that nanoparticles pose to society. It is becoming more difficult to detect toxic nanoparticles in wastewater that may contaminate the environment as existing nanoscale materials become smaller (Zhang et al., 2011a). Nanoparticles can interact with the environment in various ways: they can be attached to a carrier and carried underground by bio-uptake, toxins, or organic compounds. Because of the possibility of aggregation, traditional transportation to sensitive areas where the nanoparticles will break up into colloidal nanoparticles would be possible. Inhalation of carbon nanotubes into the lungs has unknown detrimental effects on the human body. Once it enters the lungs, it proved to be more toxic than carbon black and quartz.

The technique of overall production and running cost of wastewater treatment plant involving nanotechnology is not cost-effective because of very sophisticated methodologies of production. However, their reuse requires extensive research and in-depth studies to use this technology at up-scaled level. The effect of nanoengineered materials for water and wastewater technology on the aqueous system is critical to their commercialization. Several studies need to be conducted to identify health effects, including toxicity tests, life cycle analysis, technology evaluation, and mechanisms and dispersal of nanoparticles in water bodies (Baun et al., 2008; Lu et al., 2016).

5.6 CONCLUSION

There is an urgent demand for clean drinking water due to its depleting natural sources. Conventional treatment methods still being used have somehow failed on various grounds when actual statistics are taken into consideration. This calls for immediate attention toward developing newer, sustainable, and safer technologies that can provide a consistent supply of potable water and hence match with contemporary demands. Nanotechnology-based advanced methods and materials have a great promising approach toward the treatment of wastewater. Nanomaterials such as nano-adsorbents, nanocatalysts, and nanomembranes offer great potential for removing various organic, inorganic, and heavy metal contaminants with some materials efficient up to 80%–90%. Certain metal-oxide nanosorbents such as Fe_3O_4, TiO_2, Al_2O_3, and zeolite are effectively applied to make industrial pollutants free of heavy metal contamination and for the removal of gaseous hydrocarbons. Pd, Nb_2O_5, and ZnO nanocatalysts offer a great advantage for the removal of harmful dyes and pigments from industrial wastewater. These also play an important role in AOP and hence the removal of organic load from the system. New-age composite nanofibers modified with biological entities (ultrafine cellulose) can filter bacterial and even viral contaminants. The composite fiber production has helped in the production of nanofilters with specific separation abilities such as toluene, methylene blue, and O/w emulsion removal. On the other hand, zerovalent metal ions work by the production of ROS species (AgNP), dehalogenation (ZnNP), and reduction (FeNP) with the ultimate goal of reducing microbial load. Bio-based nanomaterials offer promising results for the selective removal of pollutants for e.g., the use of aquaporin-1 selective membrane for the desalination of water. Still, there is a need to conduct future studies to check the robustness and applicability of nanomaterials discussed for their up-scaled use. Also, to assess its long-term effects on the environment and humans, various statistical analyses and scale model studies need to be done. However, there needs to be considerable development in this field which majorly involves overcoming limitations discussed in the previous section for its full-scale operation and commercialization.

ACKNOWLEDGMENT

Tarun Gangar, Satyam, Sushovan Dalui and Sanjukta Patra acknowledge the Department of Biosciences and Bioengineering, Indian Institute of Technology, Guwahati for providing necessary infrastructure support. Further, Tarun Gangar, Satyam, and Sushovan Dalui also acknowledge the Indo-EU Horizon 2020 Project (BT/IN/EU-WR/60/SP/2018) funded by the Department of Biotechnology, Ministry of Education, India.

REFERENCES

Agarwal S, Sadegh H, Monajjemi M, Hamdi AS, Ali GA, Memar AO, Shahryari-Ghoshekandi R, Tyagi I, Gupta VK (2016) Efficient removal of toxic bromothymol blue and methylene blue from wastewater by polyvinyl alcohol. *J Mol Liq* 218:191–197.

Alkaim AF, Sadik Z, Mahdi DK, Alshrefi SM, Al-Sammarraie AM, Alamgir FM, Aljeboree AM (2015) Preparation, structure and adsorption properties of synthesized multiwall carbon nanotubes for highly effective removal of maxilon blue dye. *Korean J Chem Eng* 32(12):2456–2462.

Amin MT, Alazba AA, Manzoor U (2014) A review of removal of pollyuatnts from water/wastewater using different types of nanomaterials. *Adv Mat Sci Eng*:1–24.

Ansari MA, Khan HM, Alzohairy MA, Jalal M, Ali SG, Pal R, Musarrat J (2015) Green synthesis of Al_2O_3 nanoparticles and their bactericidal potential against clinical isolates of multi-drug resistant *Pseudomonas aeruginosa. World J Microbiol Biotechnol* 31(1):153–164.

Aziz N, Faraz M, Sherwani MA, Fatma T, Prasad R (2019) Illuminating the anticancerous efficacy of a new fungal chassis for silver nanoparticle synthesis. *Front Chem* 7:65. doi: 10.3389/fchem.2019.00065.

Bankier C, Matharu RK, Cheong YK, Ren GG, Cloutman-Green E, Ciri, L (2019) Synergistic antibacterial effects of metallic nanoparticle combinations. *Sci Rep* 9(1):1–8.

Baun A, Sørensen SN, Rasmussen RF, Hartmann NB, Koch CB (2008) Toxicity and bioaccumulation of xenobiotic organic compounds in the presence of aqueous suspensions of nano-C60. *Aquat Toxicol* 86:379–387.

Bhaumik M, Maity A, Srinivasu VV, Onyango MS (2011) Enhanced removal of Cr (VI) from aqueous solution using polypyrrole/Fe_3O_4 magnetic nanocomposite. *J Hazard Mater* 190:381–390.

Bhuyan T, Mishra K, Khanooja M, Prasad R, Varma A (2015) Bosynthesis of ZNO nanoparticles from *Azadirachta indica* for antibacterial and photocatalytic applications. *Mater Sci Semicond Process* 32:55–61.

Bowen WR, Mukhtar H (1996) Characterization and prediction of separation performance of nanofiltration membranes. *J Membr Sci* 112(2):263–274.

Brady-Estévez AS, Kang S, Elimelech M (2008) A single-walled carbon-nanotube filter for removal of viral and bacterial pathogens. *Small* 4:481–484. doi: 10.1002/smll.200700863.

Chan SHS, Yeong Wu T, Juan JC, Teh CY (2011) Recent developments of metal oxide semiconductors as photocatalysts in advanced oxidation processes (AOPs) for treatment of dye wastewater. *J Chem Technol Biotechnol* 86(9):1130–1158.

Cheloni G, Marti E, Slaveykova VI (2016) Interactive effects of copper oxide nanoparticles and light to green alga *Chlamydomonas reinhardtii. Aquat Toxicol* 170:120–128.

Choi JH, Dockko S, Fukushi K, Yamahoto K (2002) A novel applicatuion of a submerged nanofiltration membrane bioreactor (NF MBR) for wastewater treatment. *Desalination* 146(1–3):413–420.

Chou WL, Yu DG, Yang MC (2005) The preparation and characterization of silver-loading cellulose acetate hollow fiber membrane for water treatment. *Polym Adv Technol* 16:600–607.

Corsi I, Winther Nielse M, Seth R, Punta C, Della Torre C, Libralato G, Lofrano G, Sabatini L, Aiello M, Fiordi L, Ciluzzi F (2018) Ecofriendly nanotechnologies and nanomaterials for environmental application: key issues and consensus recommendations for sustainable and ecosafe nanoremediation. *Ecotoxicol Environ Saf* 154:237–244.

Dehghani MH, Taher MM, Bajpai AK, Heibati B, Tyagi I, Asif M, Agarwal S, Gupta VK (2015) Removal of noxious Cr(VI) ions using single-walled carbon nanotubes and multi-walled carbon nanotubes. *Chem Eng J* 279:344–352.

Ding W, Cai J, Yu Z, Wang Q, Xu Z, Wang Z, Gao C (2015) Fabrication of an aquaporin-based forward osmosis membrane through covalent bonding of a lipid bilayer to a microporous support. *J Mat Chem A* 3:20118–20126. doi: 10.1039/C5TA05751E.

Doong RA, Chiang LF (2008) Coupled removal of organic compounds and heavy metals by titanate/carbon nanotube composites. *Wat Sci Tech* 58:1985–1992.

Dos Santos CCL, Farias IAP, Dos Reis Albuquerque ADJ, E Silva PM, Da Costa One DF, Sampaio FC (2014) Antimicrobial activity of nano cerium oxide (IV) (Ceo_2) against *Streptococcus mutans*, IN: *BMC Proceedings*. BioMed Central, London, UK. 48.

Fu F, Dionysiou DD, Liu H (2014) The use of zero-valent iron for groundwater remediation and wastewater treatment: a review. *J Hazard Mater* 267:194–205.

Ghaedi M, Khajehsharifi H, Yadkuri AH, Roosta M, Asghari A (2012) Oxidized multiwalled carbon nanotubes as efficient adsorbent for bromothymol blue. *Toxicol Environ Chem* 94:873–883.

Grinholc M, Nakonieczna J, Fila G, Taraszkiewicz A, Kawiak A, Szewczyk G, Sarna T, Lilge L, Bielawski KP (2015) Antimicrobial photodynamic therapy with fulleropyrrolidine: photoinactivation mechanism of *Staphylococcus aureus*, in vitro and in vivo studies. *Appl Microbiol Biotechnol* 99:4031–4043. doi: 10.1007/s00253-015-6539-8.

Hebeish AA, Abdelhady MM, Youssef AM (2013) TiO_2 nanowire and TiO_2 nanowire doped Ag-PVP nanocomposite for antimicrobial and self-cleaning cotton textile. *Carbohydr Polym* 91:549–559. doi: 10.1016/j.carbpol.2012.08.068.

Higazy A, Hashem M, El Shafei A, Shaker N, Hady MA (2010) Development of antimicrobial jute packaging using chitosan and chitosan–metal complex. *Carbohydr Polym* 79:867–874. doi: 10.1016/j.carbpol.2009.10.011.

Huang L, Li D-Q, Lin Y-J, Wei M, Evans DG, Duan X (2005) Controllable preparation of nano-MgO and investigation of its bactericidal properties. *J Inorg Biochem* 99:986–993.

Ilakoon N (2014) Use of iron oxide magnetic nanosorbents for Cr(VI) removal from aqueous solution: a review. *J Eng Res Appl* 4:55–63.

Kang S, Pinault M, Pfefferle LD, Elimelech M (2007) Single-walled carbon nanotubes exhibit strong antimicrobial activity. *Langmuir* 23:8670–8673. doi: 10.1021/la701067r.

Kang YL, Zhang J, Wu G, Zhang MX, Chen SC, Wang YZ (2018) Full-biobased nanofiber membranes toward decontamination of wastewater containing multiple pollutants. *ACS Sustain Chem Eng* 6:11783–11792.

Karan S, Samitsu S, Peng X, Kurashima K, Ichinose I (2012) Ultrafast viscous permeation of organic solvent through diamond-like carbon nanosheets. *Science* 335(6067):444–447.

Karwowska E (2017) Antibacterial potential of nanocomposite-based materials–a short review. *Nanotechnol Rev* 6(2):243–254.

Kim JY, Lee C, Love DC, Sedlak DL, Yoon J, Nelson KL (2011) Inactivation of MS2 coliphage by ferrous ion and zero-valent iron nanoparticles. *Env Sci Technol* 45:6978–6984. doi: 10.1021/es201345y.

Kim JY, Park HJ, Lee C, Nelson KL, Sedlak DL, Yoon J (2010) Inactivation of Escherichia Coli by nanoparticulate zerovalent iron and ferrous ion. *Appl Environ Microbiol* 76:7668–7670.

Klačanová K, Fodran P, Šimon P, Rapta P, Boča R, Jorik V, Miglierini M, Kolek E, Čaplovič L (2013) Formation of Fe(0)-nanoparticles via reduction of Fe(II) compounds by amino acids and their subsequent oxidation to iron oxides. *J Chem* 2013:961629. doi: 10.1155/2013/961629.

Kumar A, Sharma G, Naushad M, Singh P, Kalia S (2014) Polyacrylamide/$Ni_{0.02}Zn_{0.98}O$ nanocomposite with high solar light photocatalytic activity and efficient adsorption capacity for toxic dye removal. *Ind Eng Chem Res* 53:15549–15560.

Lee MW, An S, Latthe SS, Lee C, Hong S, Yoon SS (2013) Electrospun polystyrene nanofiber membrane with superhydrophobicity and superoleophilicity for selective separation of water and low viscous oil. *ACS Appl Mat Interfaces* 5:10597–10604.

Leonard P, Hearty S, Brennan J, Dunne L, Quinn J, Chakarborty T, O'Kennedy R (2003) Advances in biosensors for detection of pathogens in food and water. *Enzyme Microb Technol* 32:3–13.

Liang J, Liu J, Yuan X, Dong H, Zeng G, Wu H, Wang H, Liu J, Hua S, Zhang S, Yu Z (2015) Facile synthesis of alumina-decorated multi-walled carbon nanotubes for simultaneous adsorption of cadmium ion and trichloroethylene. *Chem Eng J* 273:101–110.

Liu S, Zeng TH, Hofmann M, Burcombe E, Wei J, Jiang R, Kong J, Chen Y (2011) Antibacterial activity of graphite, graphite oxide, graphene oxide, and reduced graphene oxide: membrane and oxidative stress. *ACS Nano* 5:6971–6980. doi: 10.1021/nn202451x.

Lu H, Wang J, Stoller M, Wang T, Bao Y, Hao H (2016) An overview of nanomaterials for water and wastewater treatment. *Adv Mater Sci Eng.*

Machado FM, Bergman CP, Fernandez TH, Lima EC, Royer V, Calvete T, Fagan SB (2011) Adsorption of reactive red M-2BE dye from water solution by multiwalled carbon nanotubes and activated carbon. *J Hazard Mater* 192:1122–1131.

Makhluf S Dror R Nitzan Y Abramovich Y, Jelinek R, Gedanken A (2005) Microwave-assisted synthesis of nanocrystalline MgO and its use as a bacteriocide. *Adv Funct Mater* 15:1708–1171.

Maliyekkal SM, Sreeprasad TS, Krishnan D, Kouser S, Mishra AK, Waghmare UV, Pradeep T (2013) Graphene: a reusable substrate for unprecedented adsorption of pesticides. *Small* 9:273–283.

Maqbool Q, Nazar M, Naz S, Hussain T, Jabeen N, Kausar R, Anwaar S, Abbas F, Jan T (2016) Antimicrobial potential of green synthesized CeO_2 nanoparticles from Olea europaea leaf extract. *Int J Nanomed* 11:5015–5025.

Min LL, Zhong LB, Zheng YM, Liu Q, Yuan ZH, Yang LM (2016) Functionalized chitosan electrospun nanofiber for effective removal of trace arsenate from water. *Sci Rep* 6:32480.

Moussavi G, Mahmoudi M (2009) Removal of azo and anthraquinone reactive dyes from industrial wastewaters using MgO nanoparticles. *J Hazard Mater* 168:806–812.

Ni Y, Jin L, Zhang L, Hong J (2010) Honeycomb-like Ni@C composite nanostructures: synthesis, properties and applications in the detection of glucose and the removal of heavy-metal ions. *J Mater Chem* 20:6430–6436.

Pendergast MM, Hoek EMV (2011) A review of water treatment membrane nanotechnologies. *Energy Environ Sci* 4:1946–1971. doi: 10.1039/c0ee00541j.

Prado AGS, Bolzon LB, Pedroso CP, Moura AO, Costa LL (2008) Nb_2O_5 as efficient and recyclable photocatalyst for indigo carmine degradation. *Appl Catal B* 82(3–4):219–224.

Prasad R, Bhattacharyya A, Nguyen QD (2017) Nanotechnology in sustainable agriculture: recent developments, challenges, and perspectives. *Front Microbiol* 8:1014. doi: 10.3389/fmicb.2017.01014.

Prasad R, Jha A, Prasad K (2018) *Exploring the Realms of Nature for Nanosynthesis*. Springer International Publishing. https://www.springer.com/978-3-319-99570-0.

Pratap Reddy M, Venugopal A, Subrahmanyam M (2007) Hydroxyapatite-supported Ag–TiO_2 as Escherichia coli disinfection photocatalyst. *Water Res* 41:379–386. doi: 10.1016/j.watres.2006.09.018.

Prinsen P, Luque R (2019) Introduction to nanocatalysts. In: *Nanoparticle Design and Characterization for Catalytic Applications in Sustainable Chemistry*. 1–36. doi: 10.1039/9781788016292-00001.

Rabea EI, Badawy ME-T Stevens CV, Smagghe G, Steurbaut W (2003) Chitosan as antimicrobial agent: applications and mode of action. *Biomacromolecules* 4:1457–1465.

Rafique M, Tahir MB, Sadaf T (2019) Nanotechnology: An innovative way for wastewater treatment and purification: A study in large companies. In: *Risks, Relationships and Success Factors in IT Outsourcing*. doi: 10.1007/978-3-030-02381-2_5.

Rai M, Yadav A, Gade A (2009) Silver nanoparticles as a new generation of antimicrobials. *Biotechnol Adv* 27:76–83.

Ray PC, Yu H, Fu PP (2009) Toxicity and environmental risks of nanomaterials: challenges and future needs. *J Environ Health C* 27(1):1–35.

Reijnders L (2006) Cleaner nanotechnology and hazard reduction of manufactured nanoparticles. *Clean Prod* 14:124–133.

Romanos GE, Athanasekou CP, Katsaros FK, Kanellopoulos NK, Dionysiou DD, Likodimos V, Falaras P (2012) Double-side active TiO_2-modified nanofiltration membranes in continuous flow photocatalytic reactors for effective water purification. *J Hazard Mater* 211:304–316.

Sa RM, Premlatha M (2016) Application of nanotechnology in wastewater treatment: a review. *Int J Interdiscip Res* 2(11):1500–1511.

Saravanan R, Sacari E, Gracia F, Kha, MM, Mosquera E, Gupta VK (2016) Conducting PANI stimulated ZnO system for visible light photocatalytic degradation of coloured dyes. *J Mol Liq* 221:1029–1033.

Schäfer AI, Andritsos N, Karabelas AJ, Hoek EMV, Schneider R, Nyström M (2004) Fouling in nanofiltration. In: Schäfer AI, Waite TD, Fane AG (Eds) *Nanofiltration – Principles and Applications*. Elsevier. 169–239.

Seil JT, Webster TJ (2012) Antibacterial effect of zinc oxide nanoparticles combined with ultrasound. *Nanotechnology* 23(49):495101.

Shah MP (2020) *Advanced Oxidation Processes for Effluent Treatment Plants*. Elsevier.

Shah MP (2020) *Microbial Bioremediation and Biodegradation*. Springer.

Shah MP (2021) *Removal of Emerging Contaminants through Microbial Processes*. Springer.

Sharma V, Sharma A (2013) Nanotechnology: an emerging future trend in wastewater treatment with its innovative products and processes. *Int J Enhanc Res Sci Technol Eng* 2(1).

Shrivastava S, Bera T, Roy A, Dash D (2007) Characterization of enhanced antibacterial effects of novel silver nanoparticles. *Nanotechnology* 18(22):225103.

Spesia MB, Milanesio ME, Durantini EN (2008) Synthesis, properties and photodynamic inactivation of *Escherichia coli* by novel cationic fullerene C60 derivatives. *Eur J Med Chem* 43:853–861. doi: 10.1016/j.ejmech.2007.06.014.

Srivastava A, Srivastava ON, Talapatra S, Vajtai R, Ajayan PM (2004) Carbon nanotube filters. *Nat Mater* 3(9):610–614.

Suresh AK, Pelletier DA, Doktycz MJ (2013) Relating nanomaterial properties and microbial toxicity. *Nanoscale* 5:463–474.

Taghavi SM, Momenpour M, Azarian M, Ahmadian M, Souri F, Taghavi SA, Sadeghain M, Karchani M (2013) Effects of nanoparticles on the environment and outdoor workplaces. *Electron Physician* 5(4):706.

Tchobanoglous G, Stensel HD, Tsuchihashi R, Burton F (2014) *Wastewater Engineering: Treatment and Resource Recovery*, Fifth Edition. McGraw-Hill: New York.

Tsarev S, Collins RN, Fahy A, Waite TD (2016) Reduced uranium phases produced from anaerobic reaction with nanoscale zerovalent iron. *Environ Sci Technol* 50:2595–2601.

Verkman AS, Tradtrantip L, Smith AJ, Yao X (2017) Aquaporin water channels and hydrocephalus. *Pediatr Neurosurg* 52:409–416. doi: 10.1159/000452168.

Verliefde ARD, Cornelissen ER, Heijman, Verberk JQJC, Amy GL, Van der Bruggen B, Dijk van JC (2008) The role of electrostatic interactions on the rejection of organic solutes in aqueous solutions with nanofiltration. *J Membr Sci* 322(1):52–66.

Visa M, Carcel RA, Andronic L, Duta A (2009) Advanced treatment of wastewater with methyl orange and heavy metals on TiO_2, fly ash and their mixtures. *Catal Today* 144:137–142.

Vuković GD, Marinković AD, Škapin SD, Ristić MĐ, Aleksić R, Perić-Grujić AA, Uskoković PS (2011) Removal of lead from water by amino modified multi-walled carbon nanotubes. *Chem Eng J* 173:855–865.

Wang H, Gao H, Chen M, Xu X, Wang X, Pan C, Gao J (2016) Microwave assisted synthesis of reduced graphene oxide/titania nanocomposites as an adsorbent for methylene blue adsorption. *Appl Surf Sci* 360:840–848.

Wang SG, Gong WX, Liu XW, Yao YW, Gao BY, Yue QY (2007) Removal of lead (II) from aqueous solution by adsorption onto manganese oxide-coated carbon nanotubes. *Sep Purif Technol* 58:17–23.

Wu Y-Y, Xiong Z-H (2016) Multi-walled carbon nanotubes and powder-activated carbon adsorbents for the removal of nitrofurazone from aqueous solution. *J Dispers Sci Technol* 37:613–624.

Xie G, Xi P, Liu H, Chen F, Huang L, Shi Y, Hou F, Zeng Z, Shao C, Wang J (2012) A facile chemical method to produce superparamagnetic graphene oxide-Fe_3O_4 hybrid composite and its application in the removal of dyes from aqueous solution. *J Mater Chem* 22:1033–1039.

Yang W, Shen C, Ji Q, An H, Wang J, Liu Q, Zhang Z (2009) Food storage material silver nanoparticles interfere with DNA replication fidelity and bind with DNA. *Nanotechnology* 20(8):085102.

Zhang B, Misak H, Dhanasekaran PS, Kalla D, Asmatulu R (2011a) Environmental impacts of nanotechnology and its products. In: *Proceedings of the 2011 Midwest Section Conference of the American Society for Engineering Education.* 1–9.

Zhang HB, Pan XL, Liu JJ, Qian W, Wei F, Huang Y, Bao X (2011b) Enhanced catalytic activity of subnanometre titania clusters confined inside double-wall carbon nanotubes. *Chem Sus Chem* 4(7):975–980.

Zhang X, Du AJ, Lee P, Sun DD, Leckie JO (2008) TiO_2 nanowire membrane for concurrent filtration and photocatalytic oxidation of humic acid in water. *J Membr Sci* 313(1–2):44–51.

Zhao G, Ren X, Gao X, Tan X, Li J, Chen C, Huang Y, Wang X (2011) Removal of Pb(II) ions from aqueous solutions on few-layered graphene oxide nanosheets. *Dalton Trans* 40:10945–10952.

Zhao Y, Qiu C, Li X, Vararattanavech A, Shen W, Torres J, Hélix-Nielsen C, Wang R, Hu X, Fane AG et al. (2012) Synthesis of robust and high-performance aquaporin-based biomimetic membranes by interfacial polymerization-membrane preparation and RO performance characterization. *J Membr Sci* 422–428. doi: 10.1016/j.memsci.2012.08.039.

Zhao YG, Shen HY, Pan SD, Hu MQ, Xia QH (2010) Preparation and characterization of aminofunctionalized nano-Fe_3O_4 magnetic polymer adsorbents for removal of chromium (VI) ions. *J Mater Sci* 45:5291–5301.

Zheng X, Wu R, Chen Y (2011) Effects of ZnO nanoparticles on wastewater biological nitrogen and phosphorus removal. *Environ Sci Technol* 45:2826–2832.

Zhu J, Wang J, Hou J, Zhang Y, Liu J, Van der Bruggen B (2017) Graphene-based antimicrobial polymeric membranes: a review. *J Mat Chem A* 5:6776–6793. doi: 10.1039/C7TA00009J.

6 Emerging Role of Internet of Things (IoT) for Wastewater Management

Sensing, Treatment and Process Optimization

Satyam, Tarun Gangar, Risha Hazarika, and Sanjukta Patra
Indian Institute of Technology

CONTENTS

DOI: 10.1201/9781003164982-6

6.1 INTRODUCTION

Water is the most valuable natural resource with multi-usage applicability. It has a versatile domestic and commercial applications. Conversely, only about 2%–3% of the fresh water on the planet's surface is consumable. Therefore, it is compulsory to remediate wastewater as a solution to adequate global water demand. Industrial and municipal wastewater treatment is one such initiative adopted worldwide to strengthen the sustainable usage of water. Industries frequently use freshwater to improve the consistency of their commodities as a cooling medium or solvent. Wastewater discharge from industries intoxicates the aquatic ecosystem and can severely impact the local population consuming groundwater. To meet the safety norms and comply with discharge regulation, many enterprises are now taking the initiative to build up water treatment inside their premises to ensure the safe disposal of wastewater to water bodies. Wastewater from domestic sewage is also treated by local authorities, which can later be used for industrial or irrigation purposes.

The convergence and advancement of innovative technologies like microelectromechanical, microelectrochemical sensors, wireless data transmission technologies, and the internet resulted in the concept of the Internet of Things (IoT). Kevin Ashton, in 1999 outlined the idea of a system of network-connected "things" with the label "Internet of Things" (Ashton, 2009). With an estimated 75 billion devices connected over the internet by 2025, IoT has immense capabilities to become an extensive computing network where anyone and every device will be connected. Making all these systems communicate with each other via cellular or wireless technology would place a significant source of knowledge at our fingertips (Lee et al., 2019). The core principle of IoT is to facilitate sharing valuable data obtained by sensors that are then processed for a verdict, based on which functions are carried out in practical scenarios. IoT has evolved drastically over few years, from its application by selected enterprises to wide-range applications in mobile communication, waste management, point of care (POC) diagnostic and sensing devices (Geetha & Gouthami, 2016). Induction of IoT in wastewater management can address a variety of challenges faced by conventional wastewater remediation plants. IoT can help monitor wastewater parameters in real time and suggest the best measures for its remediation. The automation and analytics input from IoT can help in the seamless operation of wastewater treatment plants (Edmondson et al., 2018; Saravanan et al., 2018). Emerging role of a programmable microcontroller board has aided remarkably in the widespread use of IoT-based instruments. This chapter discusses the hurdles in wastewater management and the role of IoT in sensing and smart water management.

6.2 CHALLENGES IN CONVENTIONAL WASTEWATER MANAGEMENT/TREATMENT TECHNIQUES

The primary cause of freshwater pollution is the discharge of untreated municipal sewage, industrial wastewater, and agricultural run-offs. Industrial development and urbanization impart a heavy load on our natural water bodies. Treatment of wastewater from industrial and municipal sources is a complex procedure. Its complexity accounts for the amalgamation of various technologies and processes. Therefore, wastewater management demands the need of heavy maintenance and

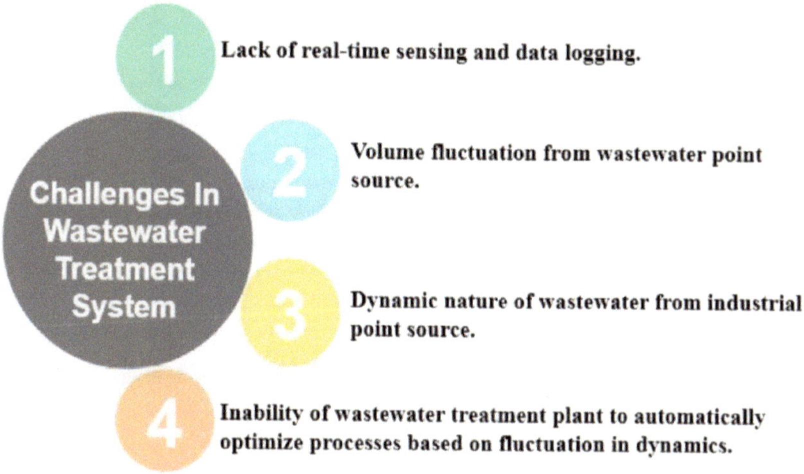

FIGURE 6.1 Challenges in conventional wastewater management and treatment techniques.

continuous upgradation and improvement for efficient applications. Heedful monitoring is required as it experiences fluctuations due to the diverse and heterogeneous chemical composition of wastewater from different sources. There have been significant varied fundamental transformations and paradigm changes related to wastewater, such as wastewater reusability, epidemiological studies, harvesting, and valuable goods production. However, there are various challenges associated with these transformations. Figure 6.1 shows various challenges associated with conventional wastewater management and treatment techniques.

Treatment of industrial and municipal wastewater requires real-time monitoring of physicochemical parameters. These parameters include pH, temperature, turbidity, total dissolved solids (TDS), conductivity, biochemical oxygen demand (BOD), and chemical oxygen demand (COD). The data gathered after assessing these parameters decides the type and extent of remediation protocols to be applied to the wastewater (Blum et al., 2020). Conventional methods of examining wastewater parameters include sampling and assessing physicochemical values. These values are then analyzed manually. The remediation protocols curated after studying the physicochemical parameters are then applied to wastewater treatment plants. Manual sensing and treatment protocols are slow and inefficient for wastewater having rapidly changing dynamics. Apart from these challenges, the lack of energy-efficient treatment plants and automation is a big hurdle in wastewater management practices. Government authorities are trying their best to provide basic sanitation and safe water to the growing population. The primary issue of water scarcity and the need to distribute sustainable water to megacities simultaneously needs to be addressed. This can be accomplished through the implementation of innovative and cost-effective real-time monitoring of wastewater management systems. Though eliminating regional and global discrepancies, executing valiant policies and bringing in effect advanced technological solutions for proper sanitation systems and wastewater reprocessing can solve the issue.

6.3 SMART WATER MANAGEMENT IN WASTEWATER TREATMENT PLANTS

Smart water management practices deliver more robust and effective water treatment. Automation in process optimization and distribution minimize operating cost and promote sustainability. Digital meters, sensors, data acquisition and controller micro-boards, and geographic information

TABLE 6.1
Major Components in Smart Water Management Practices

Module	Objective	Devices/Probes/Application
Digital/analog sensor(s)	Real-time monitoring	TDS, pH, COD, BOD, turbidity, conductivity and temperature sensors. Asset management camera. Rain gauge, pressure and leakage monitoring sensors.
Topographical information system	Infrastructural asset management and monitoring local inputs.	Mapping and management of plants asset. Monitoring inputs from local population to water bodies. Monitoring environmental and climatic parameters.
Data gathering & distribution system	Transmission of acquired data to control room over a network.	Wired/wireless data logging into a server or database management system for further analysis.
Software	For processing and storing data.	Interpretation of processed data for process optimization, real-time control and monitoring.

systems provide minute details of dynamic water parameters. These interconnected setups form smart water management systems that can deliver reliable and up-to-date data that facilitate structural and effective measures best suited for the treatment and distribution of wastewater. Major components in smart water management are represented in Table 6.1. Smart technologies can turn traditional wastewater treatment processes into highly optimized, integrated, and intelligent systems (Drenoyanis et al., 2019; Li et al., 2011; Wang et al., 2013).

6.4 ROLE OF SINGLE-BOARD COMPUTER(S) FOR DEVELOPMENT OF IoT-BASED DEVICES

The new IoT revolution is driven by the invention of single-board computers (SBC). SBC circuits have a microprocessor, memory, versatile input/output slots, and electrically erasable programmable read-only memory (EEPROM) for storage and execution of the operating system. The critical advantage of SBC is its compact size which allows universal applicability with minor mobility issues. SBC can be dynamically made to run either Windows or 'Linux' based operating systems. Some SBCs are equipped with programmable microprocessors, which can deliver the output to a specialized workbench. Latest advances have demonstrated that SBCs can also be used for database servers (Isikdag, 2015). Some of the popular SBCs easily available in the market are 'Raspberry Pi', 'Arduino' and 'Intel'-based boards.

Raspberry Pi is a small SBCs developed by 'Raspberry Pi foundation' in collaboration with 'Broadcom'. The first-generation Raspberry Pi was released in February 2012. Since then, many variants of these SBCs have been released. Raspberry Pi-based SBCs generally have a Broadcom system on a chip, a CPU with Acorn RISC Machine (ARM) compatibility, and an on-board graphics processing unit (GPU). At the time of writing this chapter, the latest compact version of Raspberry is Raspberry Pi Pico, which was released in January 2021. Pico has 264 Kilobyte (KB) of random-access memory (RAM) and 2 Megabytes (MB) of flash memory. It can be programmed by coding languages like C, CircuitPython, and MicroPython. Raspberry Pi runs on a Linux-based open-source operating system called 'Raspbian' (Johnston et al., 2018; Mitchell, 2012).

Arduino boards are also a programmable open-source SBC circuit. Arduino hardware and software are licensed under CC-BY-SA license (right to share, use and build) and GNL Lesser General Public Licence (LGPL), respectively. Arduino boards are fitted with many input/output extension pins which can be attached to other circuits/devices/sensors. These SBCs have a USB (universal serial bus) connectivity option to program the board. Arduino boards can be powered using a dedicated power pin or using a USB port. Programming languages like C and C++ are used to program

Arduino microcontroller, and the same can be done using an integrated development environment (IDE) and command-line tool. Since Arduino boars support a wide variety of sensors, potentiometers, antennas, and output devices like LCD (liquid crystal display) screens, speakers, etc., these boards are widely used for research and development of IoT-based devices monitoring platforms (Badamasi, 2014; Louis, 2016).

Intel's compact board like 'Galileo' are Arduino-certified SBCs, which consist of powerful on-board processors curated for low power consumption. These Intel boards have input/output pins for external device attachment like Arduino. Intel SBCs run on embedded Linux kernel and can be programmed using C, C++, Python, and Java coding languages. All of these gradually improving SBCs have immense potential to revolutionize the digital world. With their compact, cost-effective, and easy-to-code interface, IoT technologies are getting more efficient and reliable (Isikdag, 2015; Ramon & Ramon, 2014).

Single board computer is used with sensors to access water parameters in digital or analog form. Multi-input/output ports can access data from many probes simultaneously. The processed live feed is then transferred to a wireless network for further processing and interpretation (Zakaria & Michael, 2017).

6.5 FACTORS AFFECTING EFFECTIVE USE OF IoTs IN WASTEWATER TREATMENT PLANTS

Since IoT is a technologically emerging field, many technical, operational, ethical, and environmental factors influence its implementation feasibility. It is essential to understand the relation between these factors and how stakeholders think when implementing IoT. Figure 6.2 illustrates various factors affecting the integration of IoT technology in wastewater treatment plants. The sub-section below discusses some significant factors affecting the adoption of IoT-based technologies for wastewater management (Omoyiola, 2019).

6.5.1 Security Concerns with IoT-Integrated Wireless Communication

The security of IoT-based devices includes many levels of complexity. IoT-integrated water management technologies have components such as physical sensors, computational and networking instruments,

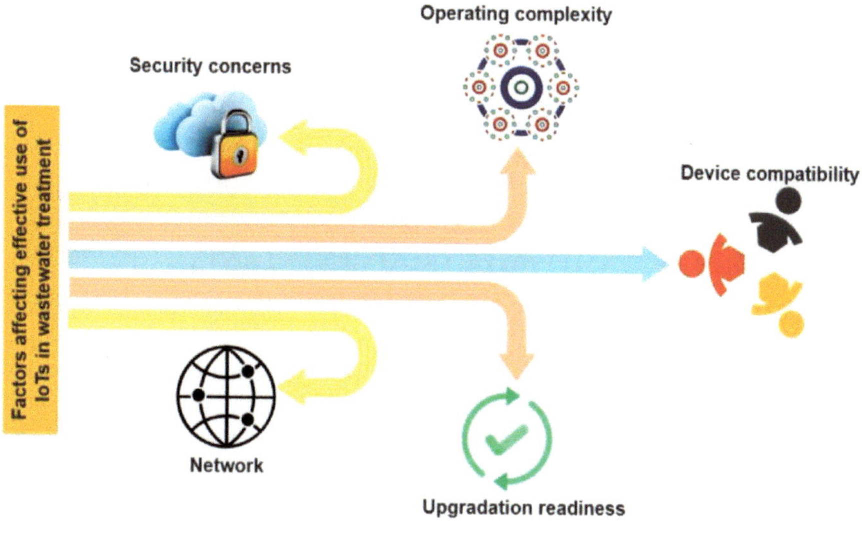

FIGURE 6.2 Factors affecting integration of IoT technology in wastewater treatment plants.

and software. The majority of security concerns are related to software and data transmission. Private firms are always concerned about their intellectual properties, data, and remediation protocol composition. A security breach in data transmission, processing, or command centers can seriously impact the reliability of IoT-based technologies. Although encryption technology can help secure data transmission, it may increase complexity and operating cost (Ammar et al., 2018; Xu et al., 2015).

6.5.2 Operating Complexity Associated with IoT-Integrated Devices

IoT-based setups have both hardware and software components integrated into each other. For a successful induction of IoT technology, the staff handling IoT devices should be technically proficient. They must have a certain level of expertise for periodic examination and calibration of installed probes at the wastewater treatment plant. Managing associated software for data processing and interpretation also requires prior training and knowledge about the domain. The success of IoT devices strongly depends on operational complexity; the lesser the complexity, the greater its adoption rate (Ng & Wakenshaw, 2017).

6.5.3 Device Compatibility Issues with IoT

Compatibility describes the degree of innovation to blend with the existing procedures or value structures. Progressive research and development have contributed to many compact and reliable sensors for monitoring wastewater parameters. Industries are pushing their limits to enhance their process optimization each day for effective remediation of wastewater. The compatibility of emerging sensors with existing IoT devices is a determining factor for the widespread adoption of IoT-based technologies (Haddud et al., 2017).

6.5.4 Network Requirement for IoT Integration

Two or more devices connected to each other in order to exchange data is called network. One of the defining features of IoT devices is their ability to communicate over a network. The data gathered by the probes/sensors are needed to be transmitted over a network/internet for further processing. The factors such as data transmission charges, network uptime, latency, and bandwidth play an essential role in the widespread adoption of IoT platforms (Omoyiola, 2019).

6.5.5 Upgradation Readiness of IoT-Connected Devices

SBCs, software, and sensors require periodic upgradation to enhance practical user experience and output. IoT-based technologies are usually dynamic and can easily be programmed as per the requirement. Upgradation provides enhanced capabilities and is economical as companies do not have to change the whole infrastructure for up-grading probes and firmware.

6.6 ROLE OF IoTs IN WASTEWATER TREATMENT PLANTS

A rapidly growing population utilizes a massive amount of water and pollutes rivers, lakes, and groundwater with industrial, agricultural, and municipal waste. Excessive organic matter, nitrogen, phosphorous, heavy metals, and dyestuff lead to the rapid depletion of oxygen in the water. One of the most sustainable approaches to combat wastewater is to treat, in which 'IoT' and 'real-time water quality monitoring' (RTWQM) can play a crucial role. The whole process of the water treatment plant is divided mainly into two parts, i.e., sensing and application of remediation protocol. IoT plays a vital role as a mediator in controlling both of these processes. IoT-based wastewater management systems include input devices or sensors, network modules, and computing and data analysis modules. Three most important components of the IoT module are illustrated in Figure 6.3.

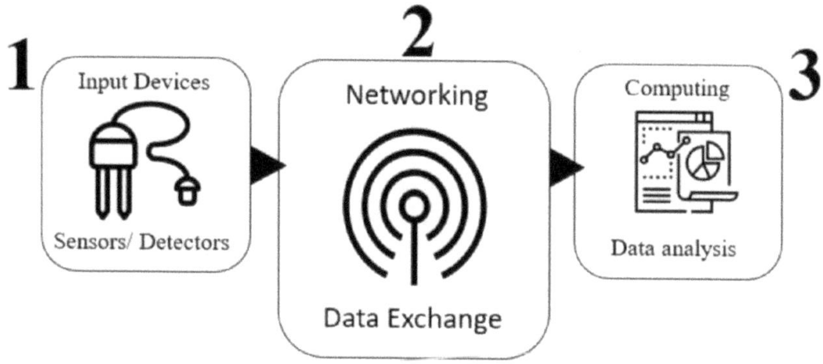

FIGURE 6.3 Three major components of IoT module.

The supervisory control and data acquisition (SCADA) control system of most wastewater treatment plants is obsolete. With the advent of the IoT software platform, a smart and savvy water quality monitoring and management system can be accomplished. In such scenarios, to remove the insoluble solids such as organic matter, sand, etc. proper, adjustments and modifications need to be made so that the solid impurities settle out prior to moving to the next phase of treatment. Thus, the flow rate of the sludge is monitored, and grit chambers are designed to slow down the water flow rate. The sludge can further burden the machinery and damage the pumps if it passes to the next phase of the wastewater treatment plant. IoT-based instruments continuously monitor the energy consumed by sewage pumps. The IoT system facilitates preventive maintenance and optimized settings that can be done in the machinery that uses maximum power by sub-metering to allow energy-efficient functioning. Other monitoring applications involve assessing critical physicochemical parameters that efficiently optimize wastewater remediation procedures by collecting and analyzing data (Martínez et al., 2020; Salam, 2020). Standard wastewater parameters and their IoT-compatible sensors are defined below.

6.6.1 Assessing Temperature in Wastewater Treatment Plants

Water temperature is a critical parameter used to calculate the quality of wastewater. All other water parameters like CO_2 concentration, conductivity, pH, compound toxicity, salinity, TDS, photosynthetic rate of aquatic plants, the metabolic rate of aquatic species, and dissolved oxygen are more-or-less temperature-dependent. The variables that affect wastewater temperature are ambient heat transfer, sunlight, and turbidity of wastewater. Various IoT-compatible sensors that can be used to measure the temperature of wastewater are listed below.

6.6.1.1 Thermistor as an IoT Sensor for Temperature Measurement

Thermistors have a temperature-dependent resistor. These devices show a change in resistance value when exposed to a temperature change. In general, the components used in these temperature sensors are temperature-dependent semiconductors. Modern thermistors are composed of a combination of oxide of nickel, copper, iron, cobalt, manganese, titanium, magnesium, and other metals and doped ceramics. Thermistors are well embedded with epoxy glass, resins, or paint, making them water-resistant, leak, and shockproof. Water-proof submersible sensors make them a suitable choice for real-time monitoring of wastewater. There are two kinds of thermistors: negative temperature coefficient (NTC) and positive temperature coefficient (PTC). In the NTC-type thermistor, with the increase in temperature, its resistance decreases, whereas in the PTC type, resistance rises with the temperature rise. Thermistors are used in various appliances like ovens, refrigerators, fire alarms and various automotive devices to measure temperature. A significant disadvantage of these instruments is that the sensors drop their accuracy in a rapid or massive temperature increase.

6.6.1.2 Thermocouple as an IoT Sensor for Temperature Measurement

Thermocouples are reported to be one of the most common temperature-sensing devices used in the industrial field. They are also used as handheld temperature-sensing tools. A thermocouple operates on the principle of the thermoelectric effect, which contributes to temperature-dependent voltage transition. It uses two distinct types of an electrical conductor material that form an electrical junction. The operating mechanism of these thermocouple sensors is based on the Seebeck effect, according to which, when the two conductors of different conductive materials undergo a temperature change in a circuit, it leads to the generation of electromotive force. This voltage change resulting from temperature change is digitally analyzed and expressed as temperature. The significant benefit of these temperature sensors is that they provide accurate and precise readings over an extensive range of temperatures. Thermocouples have been used in kilns, gas turbines, and diesel engines for assessing temperature.

6.6.1.3 Resistance Thermo-Sensors as an IoT Sensor for Temperature Measurement

Resistance thermo-sensors operate under the theory of the temperature dependence of electrical resistance. They are also often called resistance temperature detectors (RTDs). The material used in these devices to have resistance is typically gold, copper, platinum, nickel, and silver. Copper-based RTDs are comparatively cheaper than other RTDs. The sensitivity of RTDs made from various materials can vary, as the sensitivity of RTD depends on the resistance of the material employed. Resistance thermo-sensors have been used in automobiles, medical devices, computers, and cooking appliances for monitoring temperature.

6.6.1.4 Semiconductor-Based Thermo-Sensors as an IoT Sensor for Temperature Measurement

Thermo sensor-based semiconductors can accurately quantify temperatures between −55°C and +150°C and are often very inexpensive. These properties make them perfect and desirable for monitoring water temperature. The sensitivity of these sensors is around ±0.8°C. The temperature-sensitive properties of semiconductor junction devices such as transistors and diodes can be used to determine temperature. For sensing the temperature, these instruments use bandgap voltage reference proportional to the absolute temperature. Semiconductor-based thermo-sensors are generally used in amplifiers, regulators, digital signal processors, and microcontrollers.

6.6.2 Assessing Conductivity, Salinity, and TDS in Wastewater Treatment Plants

The conductivity of water can be described as its capacity to move through an electrical flow. It depends specifically on the number of dissolved ions present in water. The significant elements contributing to the conductivity and salinity of water are dissolved salts and certain inorganic materials such as chloride, alkali, carbonated compounds, and sulfides. Wastewater has a high conductivity value as it contains massive concentrations of organic compounds and other contaminants. On the other hand, pure water serves as a current insulator due to its low conductivity content. Conductance is expressed in micro- or milli-siemens per centimeter (µS/cm or mS/cm). Salinity is due to the presence of ions in water which is also responsible for the conductivity; therefore, it is not measured separately but is obtained from the conductivity measurement.

TDS can be described as measuring the total amount of ion particles smaller than 2 microns contained in water. It is measured in mg/L. TDS can be calculated by the evaporation dish or by the multiplication of empirical factors by estimated conductivity. Salinity and TDS can be easily calculated from conductance, and most of the modern digital water conductivity sensors quantify salinity and TDS with conductivity simultaneously.

6.6.2.1 Conductivity Sensor

The conductivity sensors test conductivity using the four-electrode and potentiometric system. The conductivity calculation is performed by calculating the potential between the inner pair of electrodes. This is achieved by adding an alternating current to the outer pair of electrodes. The electrodes are usually made up of platinum or other metals electroplated with platinum and are concentrically organized in the sensor. Conductivity is measured using Ohm's law by using the factors such as the potential difference, the distance between the electrodes, their surface area, and the current applied. Periodic calibration is required for precise and reliable results.

6.6.3 Assessing pH in Wastewater Treatment Plants

The pH can be described as the scale used to calculate the basicity and acidity of various aqueous solutions. The pH of a solution is the negative logarithm of hydrogen ion activity in the solution, and it is given by $-\log[H]^+$. The concentration of hydrogen ions is roughly equal to that of the activity of hydrogen ions in dilute solutions. The optimal level of drinking water is between the pH range of 6.5–8.5. Any change in the pH of aquatic water bodies can be lethal for marine organisms. Fluctuations in the pH also affect solubility and chemical properties. Heavy metal toxicity is also affected by pH. As the pH reduces, it causes leaching and increased mobility of elements and compounds, which cause the bio-accumulation of heavy metals in organisms. Although humans have excellent resistance to a wide variety of pH values in drinking water, severe pH change will negatively affect human health resulting in eye and skin irritation. pH value lower than 2.5 can contribute to irreversible damage to skin and organ lining.

6.6.3.1 pH Meter

The pH meter calculates the potential difference between the pH electrode and the reference electrode, which is directly related to the sample's pH in the test solution. Therefore, these meters are also referred to as potentiometric pH meters. The glass electrode and reference electrode are submerged into the test solution, and the electric circuit is completed; this creates a potential difference. The potential difference between the electrodes is then measured and displayed as the corresponding pH value digitally by a microprocessor attached to the probes. The pH value in the test is proportional to the electrochemical potential difference as per the Nernst equation, which relates standard electrode potential, activity, and temperature to the reduction potential of the electrochemical reaction. The advantages of pH meters are that the measurements made are reasonably precise. These instruments are therefore inexpensive, accurate, and effective. However, periodic calibration is required for a precise reading.

6.6.4 Assessing Turbidity in Wastewater Treatment Plants

Turbidity can be described as the measure of the degree of clarity or opacity of water. The particulate matter and other colored compounds are considered to be the major contributing factors to water turbidity. One of the significant effects of turbidity is that it triggers the accumulation of other pollutants such as heavy metals, ions, volatile organic compounds, and pesticides. The turbidity is generally determined by a turbidimeter/sensor and Secchi disc/tube. Turbidimeter can be handheld or submersible. Secchi disc or tube is the simplest way to assess turbidity, but its reliability depends on the person handling it. Turbidity meters are used to detect scattering light, which is quick and easy to use. These meters can be used for both static as well as real-time monitoring of the turbidity of water. However, these kinds of meters show precision only when calibration is performed routinely.

6.6.4.1 Nephelometric Turbidity Sensors

These types of optical sensors work on the principle of the scattering of light. These sensors include a light source, watertight housing, a focusing housing, and a dispersed light sensor. This sensor system meets the guidelines of EPA 180.1 and has an incandescent lamp, whereas the ISO 7027 sensor system has infrared LED. The calculation of these sensors is strongly influenced by stray light and air bubbles entrapped while reading. Such sensors are used in large-scale by wastewater treatment plants.

6.6.4.2 Backscatter Turbidity Sensors

As per the US geological survey, the sensors with a detection angle of 0°–45°, i.e., 135°–180° from transmitted light, are called backscatter sensors. This kind of turbidity sensor uses a light source and a photodetector to detect backscattered light. However, the backscatter turbidity sensor normally accepts any identification above 90° nephelometric light. Typically, sensors that use a light source of 400–600 nm use the backscatter device. Others having a wavelength of 780–900 nm use Formazin Backscatter Unit (FBU), where Formazin, synthetic material, is used to produce reproducibility and calibrate the instrument. Backscatter turbidity sensors are superior to other types because of their small size and sample volume. Also, they give reproducible results over a wide range.

6.6.4.3 Attenuation Turbidity Sensor

In such sensors, when the light beam is centered on the water sample, it detects the transmitted light passing through it. The deflected light decreases in intensity due to absorption by the water sample, which is detected by the photosensor. Colored compounds have an enormous impact on light attenuation. This is why it is not recommended to use this technique for wastewater monitoring; like in textile industries, wastewater contains high dyestuff.

6.6.5 Assessing Dissolved Oxygen in Wastewater Treatment Plants

Dissolved oxygen is the amount of oxygen available for aquatic organisms and is an essential parameter for assessing water quality. Rivers, mountain springs, and other flowing water bodies have a more generous amount of dissolved oxygen than standing water. Dissolved oxygen is influenced by factors such as pressure, salinity, and temperature. That is why there are low dissolved oxygen levels in water bodies during summer. DO can be analyzed by different types of sensors. The measuring unit for DO is milligrams per liter (mg/L) or parts per million. Therefore, modern sensors are compact and are easy to use in laboratory or water treatment plants. These sensors can process reading in digital or analog form. Some DO sensors measure the parameters like temperature, pressure, and salinity parallelly to increase their credibility because they significantly affect dissolved oxygen.

6.6.5.1 Electrochemical DO Sensors

Electrochemical DO sensors are often referred to as "amperometric or dark type" sensors. They are classified into two categories: galvanic and polarographic DO sensors. Polarographic sensors are further segmented into steady-state and fast-pulsing sensors. These types of sensors have electrode and electrolyte solutions isolated by a delicate semi-permeable membrane. The pressure of oxygen in water is proportional to the dissolved oxygen distributed through the semi-permeable membrane. The penetrated oxygen is reduced and absorbed at the cathode electrode. The current thus determined over the electrode is directly proportional to the concentration of dissolved oxygen. The electrolyte solution allows ions to pass from the cathode to the anode. The partial pressure is proportional to the current measured. The reduction of oxygen provides a current of around 2 μAmps.

6.6.5.2 Polarographic DO Sensors

Polarographic DO sensors are also a type of electrochemical sensor. They have an anode made up of silver, a cathode made up of metal such as gold or platinum, and potassium chloride (KCl) as an electrolyte. These equipment types are usually expensive than other DO sensors and need a warmup period of 5–60 minutes for the electrode polarization before calibration. The polarization from cathode to anode requires a constant voltage (usually 0.4–1.2 V). Due to the movement of electrons in a direction opposite to that of the current, the anode is positively polarized while the cathode is negatively polarized. The circuit experience increases in electric current when oxygen molecules cross the membrane and are decreased in the cathode electrode. The movement of electrons from the anode to the cathode through an internal wire circuit induces polarization. The current shift is calculated by holding the polarizing potential constant. The amount of oxygen reduced after passing the semi-permeable membrane is proportional to the current measured by the sensor. The electrons are utilized for the reduction reaction. Therefore, oxidation takes place in the anode. However, this oxidation occurs while measuring the DO; this is why darkening of anode occurs (because of AgCl coating), compromising consistency and efficiency. Though polarographic sensors are costly, they are reasonably maintenance-free. Hence, they find application in various sophisticated industrial setups.

6.6.5.3 Galvanic DO Sensors

Galvanic DO sensors have electrodes made of various metals depending on their ability to obtain or donate an electron according to their activity series. When these electrodes, made of different metals, are put in an electrolyte solution, self-polarization occurs due to these dissimilar metals' potential. The difference in potential between the electrode is maintained by 0.5 V to consume oxygen molecules without external potential. In galvanic DO sensors; usually, the anode is made up of active metals like lead or zinc, while the cathode is made up of nobel metals like silver. The electrolyte used is inert, like sodium hydroxide or sodium chloride. The working mechanism of the galvanic DO sensor is similar to that of a polarographic sensor. The only difference is that galvanic DO sensors do not require separate constant potential. A significant disadvantage of galvanic sensors is the precipitation of zinc hydroxide, which can affect the reading. This results in low or inaccurate readings, but their cost-effectiveness still makes them a preferred choice by various industries.

6.7 ROLE OF IoT IN PRIMARY WASTEWATER TREATMENT: MONITORING ENERGY USAGE, FLOW RATE, AND WATER QUALITY

Real-time monitoring of suspended solids and flow rate is crucial because rapid flow rate than settling velocity can lead to improper sedimentation. Continuous monitoring of suspended solids indicates if the rate of flow is accurate, this also stipulates the comprehensive status of water quality. Integrating industry-grade flow rate sensors in the clarifier tank along with the IoT module can give an insight into various water quality parameters. The automatic IoT-integrated system notifies the end-user if there are any irregularities in the in-line sediment tank. Energy usage at each treatment step can be monitored and fixed if any installed electronic devices are not operating at the optimal energy consumption limit. Methane production at the dewatering tank can also be monitored using greenhouse gas emission sensors (Crisan & Korodi, 2018; Zhang et al., 2018).

6.8 ROLE OF IoT IN SECONDARY WASTEWATER TREATMENT: MONITORING DISSOLVED OXYGEN AND BLOWERS IN THE AERATION CHAMBER

One of the most energy-consuming devices in the aeration phase is blowers. It is essential to monitor power consumption by aeration pumps in wastewater remediation. Aeration in the presence of activated sludge provides an appropriate environment for aerobic bacteria to degrade organic matter. IoT devices continuously monitor power distribution among aeration pumps and ensure proper distribution of air in sludge digester. Other essential parameters such as pH, ammonium, nitrate, and orthophosphate are continuously monitored during whole secondary wastewater treatment (Dogo et al., 2019; Esakki et al., 2018; Lynggaard, 2015).

6.9 ROLE OF IoT IN TERTIARY WASTEWATER TREATMENT: IoT-MEDIATED DISINFECTION PHASE

Disinfection of treated wastewater is usually done by ultra violet (UV) rays, chlorine, or ozone treatment. Controlled discharge of chlorine/ozone can be done by IoT-connected valves. Ozone is a powerful oxidizing agent that acts as a biocide that devastates bacteria, viruses, and cyst. Similarly, the intensity and exposure time of UV light can also be controlled by IoT technologies.

6.10 LIMITATION AND FUTURE RESEARCH AND DEVELOPMENT

The importance of IoT is evident from its application in diverse disciplines. However, there are numerous problems and concerns related to the implementation of IoTs in wastewater treatment plants. IoT platforms require an 'always-on' internet connection. Any network failure or data latency issue can seriously impact the working and monitoring mechanisms of IoT-assisted wastewater treatment plants. Most of the communication between the sensors, servers, and local computers is unencrypted. Unencrypted data transmission can lead to security and privacy breaches. Although encryption technologies can solve this problem, data encryption will also add a certain level of complexity at the programming level. In general, the SBC used in IoT devices has a static configuration. On-board IC (integrated circuits), memory, storage, and processors can not be replaced further. So, there can be a certain level of compatibility issues that can arise due to the upgradation of SBC used in wastewater treatment plants. The problem of mixed hardware can also occur when we use different SBCs. Recently released SBCs from popular brands have standard input/output pins for the attachment of sensors. Automation in any field can drastically reduce the need for human labor. This would have a significant influence on employability. As we transition into the IoT era, there will be a noticeable reduction in the recruiting process for professionals. Despite some drawbacks, IoT technologies are getting reliable each day. Future research and development in this field will positively contribute to a more secure IoT environment.

6.11 CONCLUSION

Water is one of the most valuable natural resources. With the urban expansion, freshwater across the world is gradually becoming sparse and on the verge of extinction. Therefore, it is crucial to make sure water treatment and management systems are empowered. For stakeholders in the water management industry, certain aspects of the water treatment systems are controlled by the SCADA systems, but in conventional wastewater management techniques, there are various challenges associated with installation and maintenance which causes a massive lag in SCADA processes. Incorporating the Internet of Things in the infrastructure lowers the operational expenses of these water treatment plants and would improve their installation in even remote areas. These technologies rely on data transfer from a physical device over a wireless channel to a local computer having control software where water parameters and data like temperature, pressure, quality, pH, residual chemical traces after treatment, and leakage can be easily monitored and analyzed remotely. Smartphones and tablets having access to application programming interface (API) can synchronize with a server or enterprise asset management (EAM) or computerized maintenance management system (CMMS) to control or retrieve data from the treatment plant. This allows technicians, mechanics, and other asset administrators to get an insight into all the processes and data in real time.

These treatment centers, therefore, have the potential to become a modern foundation for various commercial and public use applications and further aid in minimizing water pollution which will undoubtedly support us in combating water shortages in the future.

ACKNOWLEDGMENT

Satyam, Tarun Gangar, Risha Hazarika, and Sanjukta Patra acknowledge the Department of Biosciences and Bioengineering, Indian Institute of Technology, Guwahati for providing infrastructure to carry out research work. Satyam, Tarun Gangar, and Risha Hazarika acknowledge Indian Institute of Technology, Guwahati for providing research work fellowship. The authors also acknowledge Indo-EU Horizon 2020 project (BT/IN/EU-WR/60/SP/2018) funded by the Department of Biotechnology, New Delhi.

REFERENCES

Ammar, M., Russello, G., & Crispo, B. (2018). Internet of things: A survey on the security of IoT frameworks. *Journal of Information Security and Applications*, *38*, 8–27. https://doi.org/10.1016/j.jisa.2017.11.002.

Ashton, K. (2009). That 'internet of things' thing. *RFID Journal*, *22*(7), 97–114.

Badamasi, Y. A. (2014, December 23). The working principle of an Arduino. *Proceedings of the 11th International Conference on Electronics, Computer and Computation, ICECCO 2014*. https://doi.org/10.1109/ICECCO.2014.6997578.

Blum, C., Verdaguer, M., Monclús, H., & Poch, M. (2020). A new optimization model for wastewater treatment planning with a temporal component. *Process Safety and Environmental Protection*, *136*, 157–168. https://doi.org/10.1016/j.psep.2019.12.034.

Crisan, R., & Korodi, A. (2018). Noninvasive control solution for energy efficiency in wastewater treatment plants. *Proceedings of the IEEE International Conference on Industrial Technology, 2018-February*, 1604–1609. https://doi.org/10.1109/ICIT.2018.8352421.

Dogo, E. M., Salami, A. F., Nwulu, N. I., & Aigbavboa, C. O. (2019). *Blockchain and Internet of Things-Based Technologies for Intelligent Water Management System* (pp. 129–150). Springer, Cham. https://doi.org/10.1007/978-3-030-04110-6_7.

Drenoyanis, A., Raad, R., Wady, I., & Krogh, C. (2019). Implementation of an IoT based radar sensor network for wastewater management. *Sensors*, *19*(2), 254. https://doi.org/10.3390/s19020254.

Edmondson, V., Cerny, M., Lim, M., Gledson, B., Lockley, S., & Woodward, J. (2018). A smart sewer asset information model to enable an 'Internet of Things' for operational wastewater management. *Automation in Construction*, *91*, 193–205. https://doi.org/10.1016/j.autcon.2018.03.003.

Esakki, B., Ganesan, S., Mathiyazhagan, S., Ramasubramanian, K., Gnanasekaran, B., Son, B., Park, S. W., & Choi, J. S. (2018). Design of amphibious vehicle for unmanned mission in water quality monitoring using internet of things. *Sensors*, *18*(10), 3318. https://doi.org/10.3390/s18103318.

Geetha, S., & Gouthami, S. (2016). Internet of things enabled real time water quality monitoring system. *Smart Water*, *2*(1), 1–19.

Haddud, A., DeSouza, A., Khare, A., & Lee, H. (2017). Examining potential benefits and challenges associated with the internet of things integration in supply chains. *Journal of Manufacturing Technology Management*, *28*(8), 1055–1085. https://doi.org/10.1108/JMTM-05-2017–0094.

Isikdag, U. (2015). Internet of things: Single-board computers. In *Springer Briefs in Computer Science* (Vol. 0, pp. 43–53). Springer. ISBN 9783319218243. https://doi.org/10.1007/978-3-319-21825-0_4.

Johnston, S. J., Basford, P. J., Perkins, C. S., Herry, H., Tso, F. P., Pezaros, D., Mullins, R. D., Yoneki, E., Cox, S. J., & Singer, J. (2018). Commodity single board computer clusters and their applications. *Future Generation Computer Systems*, *89*, 201–212. https://doi.org/10.1016/j.future.2018.06.048.

Lee, J., Kim, D., Lee, H., Lee, Y., & Cheon, J. H. (2019). RLizard: Post-quantum key encapsulation mechanism for IoT devices. *IEEE Access*, *7*, 2080–2091. https://doi.org/10.1109/ACCESS.2018.2884084.

Li, L., Hu, X., Chen, K., & He, K. (2011). The applications of WiFi-based wireless sensor network in internet of things and smart grid. *Proceedings of the 2011 6th IEEE Conference on Industrial Electronics and Applications, ICIEA 2011*, 789–793. https://doi.org/10.1109/ICIEA.2011.5975693.

Louis, L. (2016). Working principle of Arduino and using it as a tool for study and research. *International Journal of Control, Automation, Communication and Systems (IJCACS)*, *1*(2), 21–29. https://doi.org/10.5121/ijcacs.2016.1203.

Lynggaard, P. (2015). The potential of wastewater energy recovery in smart buildings by using internet of things technology. https://vbn.aau.dk/en/publications/potentialet-for-at-genvinde-energi-fra-brugsvand-i-smarte-bygning.

Martínez, R., Vela, N., el Aatik, A., Murray, E., Roche, P., & Navarro, J. M. (2020). On the use of an IoT integrated system for water quality monitoring and management in wastewater treatment plants. *Water*, *12*(4), 1096. https://doi.org/10.3390/w12041096.

Mitchell, G. (2012). The Raspberry Pi single-board computer will revolutionise computer science teaching. *Engineering & Technology*, *7*(3), 26–26. https://doi.org/10.1049/et.2012.0300.

Ng, I. C. L., & Wakenshaw, S. Y. L. (2017). The internet-of-things: Review and research directions. *International Journal of Research in Marketing*, *34*(1), 3–21. https://doi.org/10.1016/j.ijresmar.2016.11.003.

Omoyiola, B. (2019). Factors affecting IOT adoption. *IOSR Journal of Computer Engineering*, *21*(6), 19–24.

Ramon, M. C., & Ramon, M. C. (2014). Intel Galileo and Intel Galileo Gen 2. In *Intel® Galileo and Intel® Galileo Gen 2* (pp. 1–33). Apress. https://doi.org/10.1007/978-1-4302-6838-3_1.

Salam, A. (2020). Internet of things in water management and treatment. In *Internet of Things* (pp. 273–298). Springer. https://doi.org/10.1007/978-3-030-35291-2_9.

Saravanan, K., Anusuya, E., Kumar, R., & Son, L. H. (2018). Real-time water quality monitoring using Internet of Things in SCADA. *Environmental Monitoring and Assessment*, *190*(9), 1–16. https://doi.org/10.1007/s10661-018-6914–x.

Wang, S., Zhang, Z., Ye, Z., Wang, X., Lin, X., & Chen, S. (2013). Application of environmental internet of things on water quality management of urban scenic river. *International Journal of Sustainable Development & World Ecology*, *20*(3), 216–222. https://doi.org/10.1080/13504509.2013.785040.

Xu, T., Wendt, J. B., & Potkonjak, M. (2015). Security of IoT systems: Design challenges and opportunities. *IEEE/ACM International Conference on Computer-Aided Design, Digest of Technical Papers, ICCAD*, *2015-January*(January), 417–423. https://doi.org/10.1109/ICCAD.2014.7001385.

Zakaria, Y., & Michael, K. (2017). An integrated cloud-based wireless sensor network for monitoring industrial wastewater discharged into water sources.

Zhang, D., Lindholm, G., & Ratnaweera, H. (2018). Use long short-term memory to enhance internet of things for combined sewer overflow monitoring. *Journal of Hydrology*, *556*, 409–418. https://doi.org/10.1016/j.jhydrol.2017.11.018.

7 Advanced Technological Options for Treatment of Wastewater

Tejas M. Ukarde, Preeti H. Pandey, Jyoti S. Mahale, Ayush Vasishta, Pankaj Shinde, and Hitesh S. Pawar
Institute of Chemical Technology

CONTENTS

7.1 INTRODUCTION

Earth has a plentiful supply of water, over the total surface of the earth 71% of the area is covered with water but out of this water sources 97% lies in oceans as saline water and only 2.5% is available as freshwater which is limited to human activities [1]. The demand for water is constantly rising due to the multiplying population and industrialisation in the world, and the demand for water is very high as compared to the current supply of water. Also, water has a very low self-purification capacity which cannot sustain the increased load of water contamination due to increasing human activities, rapid urbanisation and industrialisation. Thus, it is necessary to minimise water use and recycle back to its resources with minimum pollution load. Currently, India accounts for 16% of the world's population with 4% of total freshwater sources. Total available water in India is estimated

DOI: 10.1201/9781003164982-7

to be 1123 BCM and is used for different purposes like irrigation, industry, drinking, energy, etc. Out of 433 BCM of groundwater present in India, 18.1 BCM is used for domestic and industrial purposes, rest of it is diverted to the irrigation sector. As India's population is increasing at a rate of 1.9% per year, the demand for water for the domestic and industrial sectors is expected to increase up to 29.2 BCM. The per capita average annual freshwater availability in India is reduced since 1951 from 5177 to 1588 m^3, in 2010. It is forecasted to further reduction to 1341 m^3 in 2025 and 1140 m^3 in 2050. So, it is very much important to the management of available water resources through better water use and wastewater recycling by taking account of the future water needs of India [2].

The wastewater generation in India due to uncontrolled urbanisation is drastically increased in the last decade. Class 1 and class 2 cities generate about 35,558 and 2696 MLD of wastewater, respectively. Out of the total water used for domestic purposes in an urban area, 80% of water is getting back to water bodies like rivers and lakes as waste-contaminated water. The city having population of about 1 lakh leads to the generation of 17 billion litres of wastewater daily [2]. On the other hand, numerous industries in India produce a large amount of effluent out of which only a part is treated with effluent treatment technology effectively and most of the remaining part is discharged into river bodies. Only 60% (8000 MLD is treated) is treated out of a total of 13,468 MLD of effluent generated by different industries in India. As per data stated by CPCB, the quantity of industrial effluent discharged in only the Ganga river is 501 MLD from different types of industries [3]. The industrial sectors include distilleries, sugar industries, paper and textile/tannery. Industrial chemicals are major sectors contributors to discharging a large amount of wastewater into natural resources. Contribution of different industries to the generation of wastewater is shown in Figure 7.1 [4].

By considering the future estimates of water consumption, water scarcity can be a crucial dispute for humankind with associated crises of the destructive effect on human well-being, financial progress and ecological traits. Thus, wastewater recovery and management is an urgent need to preserve water resources. However, there is a huge divergence between wastewater production and wastewater management in India due to insufficient treatment capacity and low operational efficiencies. [5]. Thus, there is a nascent demand for a competent approach to the viable treatment of wastewater.

The wastewater treatment strategies can be classified as primary for the elimination of sedimentable organic and inorganic solids, secondary for the exclusion of dangled and dissolved organic contaminants and tertiary treatments for attaining the final acceptable threshold of clearance boundaries. [6]. However, current established wastewater treatment methods have several drawbacks: primary treatments are capable of eliminating only sedimentable matter, and secondary treatments have drawbacks such as soothing heavy metals removal efficiencies and bio-refractory

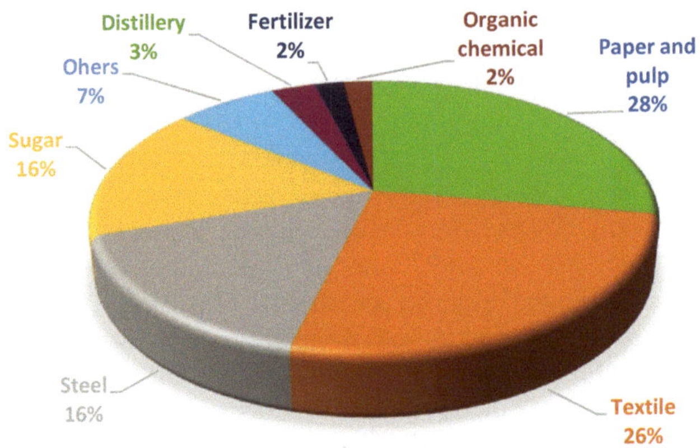

FIGURE 7.1 Contribution of different industries for generation of wastewater [4].

pollutants, the requirement of long processing period, generation of a huge amount of sludge, high cost, and land requirement [7]. Thus, the existing strategies of wastewater management are incompetent to deliver a sustainable solution for achieving high removal efficiencies for both dissolved and undissolved matter from wastewater.

In order to tackle drawbacks associated with present processes, emerging processes such as advanced oxidation processes (AOP), biological treatments, membrane processes, membrane bioreactor, coagulation, hydrodynamic cavitation, electrodialysis, photocatalysis, etc. have been explored by worldwide researchers for achieving the required sustainability and efficiency in wastewater treatment [8–10]. This chapter has highlighted the urgent need for sustainable and efficient wastewater treatment technology. It reviews the main advanced wastewater technologies, which can help humankind to tackle the crisis of water scarcity. Challenges and barriers associated with conventional and advanced wastewater treatment technologies have been discussed, which can be beneficial for future researchers. It provides a pathway to improve the efficiency of wastewater treatment processes for constructing sustainable wastewater treatment and management strategies to attain global sustainability.

7.2 ADVANCED WASTEWATER TREATMENT STRATEGIES

In industries, there are a number of processes that contribute to the increase in wastewater production. In order to prevent the increased wastewater generation, researchers are now developing advanced wastewater treatment strategies and simultaneously improving the existing processes for the acquittal of wastewater such as AOP, biological treatment, membrane processes, electrodialysis, photocatalysis, etc. The need for advanced treatment strategies is important to make wastewater treatment processes more efficient.

7.2.1 Advanced Oxidation Processes

AOPs are mostly employed for the removal of organic pollutants in wastewater. These harmful organic pollutants are perilous to both humans and marine life. Therefore, there is an urgent requirement to treat these pollutants. Advanced oxidation technologies generate strong oxidants such as hydroxyl and sulfate radicals in the reaction mixture for the oxidation of organic pollutants [11]. These radicals act as commanding oxidising agents, which can eliminate toxic pollutants, thereby solving the problem of wastewater treatment. Some of the AOPs such as hydroxyl-based, ozone-based, UV-based, Fenton and Sulfate radical-based, etc. are discussed below.

7.2.1.1 Hydroxyl Radical-Based (AOPs)

Hydroxyl radical is an extremely reactive oxidiser in water treatment, with oxidation potential between 2.8 V (pH 0) and 1.95 V (pH 14) as compared to commonly used saturated calomel electrode as a reference [12]. Hydroxyl radicals outbreak pollutants through four basic pathways: radical addition, hydrogen abstraction, electron transfer, and radical combination [13]. When hydroxyl radicals react with organic components, it produces carbon-centred radicals (R· or R·–OH) [14]. Upon oxidation, these carbon-centred radicals get converted into organic peroxyl radicals (ROO·). These radicals react again and form hydrogen peroxide (H_2O_2) and superoxide ($O2^{\bullet -}$), resulting in the chemical removal of these organic components [15]. UV photolysis of hydrogen peroxides is significant as they are an efficient process relating to environmental and economic attention. The stimulation of H_2O_2 with UV irradiation generates the hydroxyl radical species (•OH) through the homolytic cleavage of the peroxide bond [16]. These hydroxyl radicals so formed, are used for the oxidation of water pollutants. The addition of H_2O_2 is vital in the photolysis process for better removal efficiency since it can promote the degradation of micro-pollutants in wastewater [14].

7.2.1.2 Ozone-Based AOPs

Ozone (O_3) is a strong oxidant with an oxidation potential of 2.07 V as compared to SCE. Ozone oxidation is a reaction, in which O_3 favourably reacts with the ionised and dissociated form of organic compounds, rather than the neutral form [11,15]. Under certain conditions, OH· is generated from ozone to begin the oxidation. Detailed mechanisms are proposed to explain the complex OH· so produced, and the overall reaction involving OH· generation is expressed as Eq. 7.1 below [15,17].

$$3O_3 + H_2O \rightarrow 2OH\cdot + 4O_2 \tag{7.1}$$

In the existence of other oxidants, the OH·yield can be improved to a great extent. In peroxone (O_3/H_2O_2) system, the ozone decomposition and OH· generation is enhanced by hydroperoxide (HO_2^-) produced from H_2O_2 decomposition [18].

$$H_2O_2 \rightarrow HO_2^- + H^+ \tag{7.2}$$

$$HO_2^- + O_3 \rightarrow OH\cdot + O_2^- + O_2 \tag{7.3}$$

In the O_3/ultraviolet (UV) irradiation, hydrogen peroxide is produced as an additional oxidant through ozone photolysis as shown in Eq. 7.4.

$$O_3 + H_2O + hv \rightarrow H_2O_2 + O_2 \tag{7.4}$$

Therefore, OH· can be produced by three pathways: (a) ozonation, (b) O_3/H_2O_2, and (c) photolysis of H_2O_2 as shown in Eq. 7.5.

$$H_2O_2 + hv \rightarrow 2OH\cdot \tag{7.5}$$

7.2.1.3 UV-Based AOPs

Hydroxyl radicals can be initiated by photons in the presence of catalysts or oxidants [11,14]. The catalyst used is titanium dioxide (TiO_2). TiO_2 particles are excited to generate positive holes in the valence band (hv_{vb}^+) having an oxidative capacity, and negative electrons at the conduction band (e_{cb}^-) with a reductive capacity, as shown in Eq. 7.6.

$$TiO_2 + hv \rightarrow e_{cb}^- + hv_{vb}^+ \tag{7.6}$$

With the reactions of OH^-, H_2O, and O_2^- at the surface of TiO_2, these holes and electrons can form hydroxyl radicals, as shown in Eqs. 7.7–7.9 [19].

$$hv + vb + OH^-(\text{surface}) \rightarrow OH\cdot \tag{7.7}$$

$$hv + vb + H_2O(\text{absorbed}) \rightarrow OH\cdot + H^+ \tag{7.8}$$

$$e_{cb}^- + O_2(\text{absorbed}) \rightarrow O_2^- \tag{7.9}$$

In the existence of oxidants such as H_2O_2 or O_3, additional OH· may be produced under UV irradiation. H_2O_2 molecule is cleaved by UV irradiation to produce two OH· as shown in Eq. 7.10.

$$H_2O_2 + hv \rightarrow 2OH\cdot \tag{7.10}$$

In addition, at a wavelength less than 242 nm, OH· can also be generated through the photolysis of H_2O as shown in Eq. 7.11.

$$H_2O + hv \rightarrow OH\cdot + H\cdot \tag{7.11}$$

7.2.1.4 Fenton-Related AOPs

In the Fenton process, hydrogen peroxide reacts with Fe^{2+} to produce strong reactive species [20]. The Fenton radical mechanisms involve the following reactions, as shown in Eqs. 7.12–7.18.

$$Fe^{2+} + H_2O_2 \rightarrow Fe^{3+} + OH \cdot + OH^- \tag{7.12}$$

$$Fe^{3+} + H_2O_2 \rightarrow Fe^{2+} + HO_2 \cdot + H^+ \tag{7.13}$$

$$OH \cdot + H_2O_2 \rightarrow HO_2 \cdot + H_2O \tag{7.14}$$

$$OH \cdot + Fe^{2+} \rightarrow Fe^{3+} + OH^- \tag{7.15}$$

$$Fe^{3+} + HO_2 \cdot \rightarrow Fe^{2+} + O_2H^+ \tag{7.16}$$

$$Fe^{2+} + HO_2 \cdot + H^+ \rightarrow Fe^{3+} + H_2O_2 \tag{7.17}$$

$$2HO_2 \cdot \rightarrow H_2O_2 + O_2 \tag{7.18}$$

$OH \cdot$ is produced from Eq. 7.12 through electron transfer. Equation 7.13 indicates that the generated Fe^{3+} from Eq. 7.12 can be reduced to Fe^{2+}. Subsequently, Fe^{3+} forms iron sludge at wastewater treatment conditions. The sludge needs to be separately wiped of, which increases the treatment complexity and operational costs. Production of hydroxyl radicals during the Fenton reaction is effective only at an acidic pH condition. Therefore, the application of the Fenton reaction for wastewater treatment is restricted in practice [21].

Based on the Fenton treatment scheme, three modified Fenton processes are projected, including the Fenton-like system, photo-Fenton system, and electro-Fenton system. In the Fenton-like reaction, Fe^{2+} is replaced by ferric ion (Fe^{3+}), namely, the series of reactions in the Fenton system are initiated from Eq. 7.13 in the Fenton-like system, rather than from Eq. 7.12 in the traditional Fenton treatment. In the photo-Fenton reaction, UV irradiation is applied with the traditional Fenton system with the purpose of enhancing the UV-induced reduction of dissolved Fe^{3+} to Fe^{2+}. In the electro-Fenton reaction, either or both of the Fenton reagents may be generated through electrochemical methods.

7.2.1.5 Sulfate Radical-Based AOPs

$S_2O_8{}^{2-}$ is a strong oxidant with a standard oxidation potential (E^o) of 2.01 V. Once triggered by heat, ultraviolet (UV) irradiation (Eq. 7.19), transitional metals (Eq. 7.20), or elevated pH, $S_2O_8{}^{2-}$ can form more powerful sulfate radicals ($SO_4{}^{\cdot-}$, E^o=2.6 V) to initiate sulfate radical-based AOP [22].

$$S_2O_8^{2-} \rightarrow \Delta/UV\, 2SO_4^{-} \tag{7.19}$$

$$S_2O_8^{2-} + Mn+ \rightarrow SO_4^{-} + SO_4^{2-} + Mn^{+1} \tag{7.20}$$

In a thermally activated persulfate method, the temperature broadly ranges between 35°C and 130°C. From Eqs. 7.19 and 7.20, it can be noticed that with the same molar persulfate concentration, the metal activation method only produces 50% of a sulfate radical yield produced from the heat or UV-activated persulfate method. Therefore, the metal activation method is not theoretically efficient. The most frequently used metals include ferrous (Fe(II)) and ferric (Fe(III)) ions, though other metals have been demonstrated to have an activation capability, such as Cu(I) and Ag(I) [23].

Sulfate radicals are highly reactive species with a short lifespan. Hydroxyl radicals add to C=C bonds or abstract H from C–H bonds during their reactions with organic compounds. In contrast, Sulfate radicals incline to remove electrons from organic molecules that are transformed into organic radical cations. Hydroxyl radicals can also be generated from sulfate radicals through Eqs. 7.21 and 7.22 [24].

$$SO\cdot^{-4} + H_2O \rightarrow OH\cdot + SO_2^{-4} + H^+ \quad (7.21)$$

$$SO\cdot^{-4} + OH^- \rightarrow OH\cdot + SO_2^{-4} \quad (7.22)$$

Equation 7.24 shows that more hydroxyl radicals can be produced from sulfate radicals at an alkaline condition.

7.2.1.6 Other AOPs

Some other AOPs have been considered for different wastewater treatment, such as ultrasound (US) irradiation and electronic-beam irradiation. Under the US irradiation (16 kHz–100 MHz), alternate compression and rarefaction cycles of the sound waves can lead to three successive stages of cavities (i.e., nucleation, growth, and implosive collapse) that are made of vapour and gas-filled microbubble. The microbubble collapse can generate a high temperature (4200–5000 K) and a high pressure (200–500 atm). Under such conditions, water molecules in the form of gas within microbubbles are fragmented to produce hydroxyl radicals as shown in Eq. 7.23 [11].

$$H_2O \rightarrow OH\cdot + H\cdot \quad (7.23)$$

Electronic-beam irradiation generates various free radicals in water through splitting water as shown in Eq. 7.24.

$$H_2O + e^- \rightarrow 2.7OH\cdot + 2.7H_3O^+ + 2.6e^- + 0.7H_2O_2 + 0.6H\cdot + 0.45H_2 \quad (7.24)$$

7.2.2 Biological Treatment

Biological wastewater treatment is a secondary treatment process that depends on bacteria, and other small organisms to break down organic wastes remaining after primary treatment using normal cellular processes. Wastewater contains a huge amount of organic matter, such as garbage, food waste, pathogenic organisms, heavy metals, and toxins. Biological treatment of wastewater is used worldwide since its more effective and economical than chemical processes [25].

Biological treatment is mainly divided into aerobic and anaerobic processes. "Aerobic" refers to a process where oxygen is present, while "Anaerobic" refers to a biological process in which oxygen is absent. Aerobic wastewater treatment processes involve systems that require oxygen to the biomass/organic wastes. It includes aerobic tanks or ponds to create enough surface area, surface and spray mechanical aeration devices for incorporating oxygen, trickling filters, and aerobic digestor. Diffused aeration systems can be employed in order to maximise oxygen transfer and minimise odours when the wastewater is treated [25]. Aeration provides oxygen to the bacteria and other organisms so that they decompose organic compounds in the wastewater. Since we require to aerate it, aerobic systems incline to be less energy-efficient than anaerobic process.

The activated sludge process is widely used for the secondary treatment of both domestic and industrial wastewater [26]. It is employed for treating waste streams rich in organic or biodegradable compounds.

Membrane aerated biofilm reactor (MABR) is the most energy-intensive stage of biological treatment. In MABR treatment, air at atmospheric pressure is gently blown into a spirally wound membrane in a tank, with air on one side of the membrane and mixed liquor on the other in a single tank. Nitrification (biofilm oxidises ammonium compounds to nitrate using O_2 diffusing from the air through the membrane) and denitrification (suspended biomass oxidises BOD using nitrate) are achieved by a biofilm that forms on the membrane. The result is an effluent suitable for irrigation or release into the environment. MABR is much more efficient than aerobic treatment since it is less time-consuming and expensive.

In contrast, Anaerobic systems are designed to avoid the exposure of biomass sludge to air. This can be accomplished using airtight, enclosed digesters or by up-flow anaerobic sludge blanket (UASB) systems that keep the biomass layer submerged below the treated [25,26]. Anaerobic treatment uses bacteria to help organic material decompose in an oxygen-free environment. Anaerobic treatment tends to provide energy recovery as it is used to produce biogas, which is composed primarily of methane [25,26]. Anaerobic systems provide benefits over aerobic systems, including lower operational costs therefore more economical and more energy-efficient.

7.2.3 Hydrodynamic Cavitation

Cavitation has emerged as one of the new, innovative and potential technologies for wastewater treatment. Cavitation technology is eco-friendly and economical technology for the decomposition of complex compounds which are resistant to conventional disposal methods [27]. Cavitation reactor is a new form of a multiphase reactor which generates conditions for oxidation. Cavitation reactor is used to convert complex compounds, chlorinated hydrocarbon, phenolic and aromatic compounds, textiles dyes, pesticides, ester and pharmaceuticals to low molecular weight, carbon dioxide and water as the final product [28]. Cavitation is a rapid phase change phenomenon, phase changes occur because of the suddenly absorb of high energy which increases the temperature and becomes responsible for the oxidation of impurities present in the wastewater [29]. Cavitation process is classified based on the generation of cavitation i.e optic, particle, acoustic, and hydrodynamic. Hydrodynamic and acoustic cavitation is most common and efficient for wastewater and chemical treatment [30]. Optical and particle cavitation involved focusing of short laser light beam on stream which heats up the liquid and helps in creating cavitation [31]. Particle cavitation is generally used for material synthesis which is performed by creating void during material synthesis which alters the material properties [6]. Acoustic cavities (AC) are generally generated by using ultrasound waves within the frequency range of 16 kHz–2 MHz [28,30]. Hydrodynamic cavitation (HC) is generated by a variation of pressure or velocity of liquid. Drawbacks like low efficiency, high operating cost, and less feasible for scale-up HC are preferred over other cavitation. Steam Cavitation occurs by injecting direct steam into sub-cooled water creating a similar effect as HC [31]. By varying the pressure or velocity of fluid we can generate a cavitation bubble at nuclei, which collapses after a certain expansion of the bubble which creates a high temperature, pressure and turbulence in the system.

Cavitation, a physical phenomenon, has three stages: generation, evolution and downfall of vapour or gas-vapour bubbles in an exceptionally small interval of time(microseconds) within fluid due to discrepancies in local static pressure. By varying the velocity or cross-sectional area, decrease in pressure can be attained. By decreasing the pressure at a point in liquid stream, vapour pressure of liquid can be accomplished at an operating temperature which helps in the generation of gas-vapour bubbles in the liquid stream. These vapour bubble explodes with a generation of high energy as the pressure reaches back to normal which results in the increase of pressure and temperature at the point of collapse. Such extreme temperature and pressure (5000 atm and 12,000°K, intense turbulence) will be responsible for physical, chemical & biological transformations, even when the bulk conditions are ambient [34]. These extreme conditions are responsible for the breakdown of high molecular weight materials and other inert materials.

The thermodynamic concept of cavitation is very well understood by phase diagram. In water phase diagram, the curve line between the triple point and critical point represents the liquid phase and vapour phase. Crossing the curved line represents the reversible and equilibrium transformation of water. At vapour pressure of each temperature, the transformation from the liquid phase to vapour phase and vice versa takes place. Cavitation is similar to boiling, with the only difference being that boiling takes place with an increase in temperature while cavitation takes place by variation of pressure.

Cavitation is strongly influenced by water quality (quantity and quality of impurities present), occurrence of cavitation bubbles starts at several points like gas bubble or smaller impurities in water, called as bubbles nuclei, whose shape changes continuously and suddenly explodes, which break the bond between the water molecules [31].

Cavitation consists of two phases compressions and rarefactions. In the compression cycle, increase in pressure bring the molecules closer, while in the rarefaction cycle decrease in pressure pull the molecule away. Due to low pressure, cavitation bubbles occur in the rarefaction region. These bubbles grow slowly and reach unstable diameter and then explode violently [31].

In HC pressure variation is caused by changing the cross-sectional area of flow, low pressure is responsible for the initiation of cavitation bubbles. The bubbles that occur in a cavitating flow are not perfectly spherical.

The advantages of HC are as follows [34]:

1. Minimum footprint: Can be easily fitted in congested place make retrofitting possible
2. Reduce chemical usage: no pollution
3. Highly efficient: use less energy for the same work as compared to other technology
4. Save water: negligible water loss
5. Speed up the processing time: increase plant capacity with the same resources
6. Can handle large flow: Industrial applications are possible
7. Technology allows for retrofitting: make existing plants more viable
8. Can combine with other technology
9. Low cost

7.2.4 Electrodialysis (ED)

ED process is basically a membrane separation and is commonly used for wastewater treatment. In ED ion exchange membrane is applied between anode and cathode, and electric potential is used as driving force. Ion exchange membrane means it allows flow to selective ion either cation or anion. On application of electric current cation and anions move towards respective electrodes depending on the polarity. Schematic diagram of electrodialysis is shown in Figure 7.2 [33]. The main objective of ED is the removal of salts and metal contaminants by passing through the ion exchange membrane which is a separate cation and anion electrode [34]. Low concentration of salt or metals is preferred and considered the energy-efficient process. ED technology is used for the removal of heavy metal ions like nickel, zinc, chromium, cadmium, lead and organic acids, etc. [35].

7.2.4.1 Electrochemical Oxidation

Electrochemical oxidation (EO) is an AOP, which is also known as anodic oxidation is used for industrial effluents [16]. The EO can be attained in different ways (a) Direct EO, where electron transfer occurs at the electrode and molecule. (b) Indirect EO, where oxidation takes place with the help of oxidising agents like hydrogen peroxide etc. There is a substantial variety of electrode materials used for EO like noble metals gold, platinum, mixed metal oxide, boron-doped diamond, etc. [37].

The advantages of EO are that it helps in the removal of toxic chemicals and reduces the quantity of solid sludge. EO can be operated without further addition of chemicals making it environmentally friendly. EO has no restriction for the quality of water like high turbidity and coloured water. The disadvantages of the EO process are passivation, polarisation and corrosion of electrodes. Passivation is caused by the polymeric reaction that occurs at the electrode. Polarisation is caused

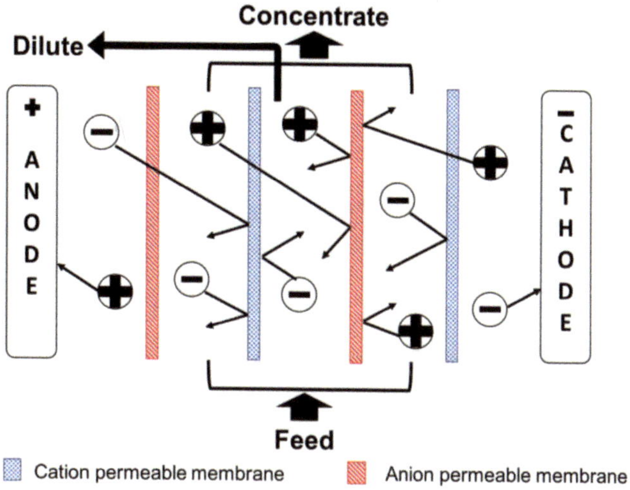

FIGURE 7.2 Schematic representation of electrodialysis.

by the synthesis of gas at the electrode which reduces the mass transfer area and coefficient of the process [36].

7.2.4.2 Microbial Electrolysis Cell

Microbial electrolysis cell (MEC) is one of the emerging technologies, which has the capability to replace conventional technology. MEC is an electrolysis process in which one of the electrodes involves reaction with electrochemical reaction with microorganism [38]. MEC is anaerobic biological process for the conversion of organic into hydrogen and methane [39]. MEC not only treat wastewater but also generate renewable energy from wastewater. In an MEC, organic pollutants are removed by oxidation at anode with the help of exo-electrogenic bacteria, simultaneously hydrogen gas is produced at the cathode on the application of a small electric voltage. The efficiency of MEC is entirely dependent on the selection of materials like anode, cathode and membrane. The following properties are required (a) high conductivity, (b) biological, physical and chemical stability (c) Economic viable (d) high surface-to-volume ratio. The only disadvantage is difficulty in scaling up and working with the help of other filtration processes [39].

7.2.5 Photocatalysis

Photocatalysis is one of the most studied and explored methods among advanced wastewater treatment strategies. Since photocatalysis is an eco-friendly process that has many advantages due to its attractive properties and can destroy the complex organic materials without using any hazardous oxidising agents at room temperature only [40]. The main advantage of photocatalysis is that it does not involve any further secondary disposal method. It is simply a light or photon-induced reaction that occurs on the surface of semiconducting materials which are known as photocatalyst. Once the photocatalyst gets activated, there occurs a transfer of electrons from the valance band to the conduction band which form the hydroxyl radical once the electron and hole get paired [41]. The basic mechanism of semiconductor-assisted photocatalysis is shown in Figure 7.3 [42]. Photocatalyst assisted with UV radiation has turned into a mature technology while solar-assisted photocatalysis is still at the demonstration stage. There are some of the pilot plants also in developed countries such as the USA, Germany, Japan, and China where the concept of nano-photocatalyst has been taken into much more consideration for the disinfection of wastewater [43]. In the past few years, photocatalysis coupled with ozone treatments has also become a promising method as it can both generate the hydroxyl radical as well as can oxidise the organic residues remaining in wastewater [44]. Both

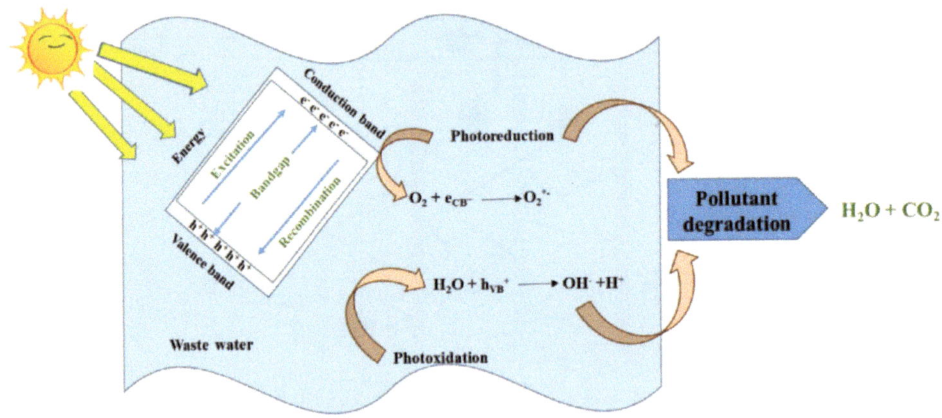

FIGURE 7.3 Mechanism of semiconductor-assisted photocatalysis [42].

homogeneous as well as heterogeneous photocatalyst are reported in the literature. Among heterogeneous catalysis, semiconductor materials are mainly utilised for the effective decomposition of organic matter to carbon dioxide and water.

TiO_2 is one of the most active photocatalysts having photon energy ranges from 300 nm<I <390 nm. It remains stable (without any loss in its catalytic activity) during the catalytic repetition cycles [45]. It has been proven as a potential photocatalyst for the treatment of wastewater either by suspending it in water or by coating it on some substrate [46]. It is reported that TiO_2 activity depends on its phase and shape in which it is prepared (nanoparticles, nanotubes, nanowires, nano ribbons, and nanosheets as well as inverse opals) as well as on its size like nano TiO_2 is much more active than the bulk TiO_2 [47,48]. In addition to this, the activity of TiO_2 also depends upon the radiation under which it is irradiated. It is reported that the band gap of TiO_2 is more suitable to absorb UV radiation as compared to that of visible light [49]. TiO_2 has been utilised for the remediation of wastewater under both UV and visible radiation as well as in doped form and with other semiconductor-based metal oxides. The section provides an outlook of TiO_2-based photocatalyst discovered so far for the treatment of wastewater.

7.2.5.1 TiO_2 under UV and Visible Light Irradiation

The wide band gap of TiO_2 makes it somewhat less efficient under visible light. If we look at the band gap of two most desired form of TiO_2 such as Anatase and Rutile which has slight different band gap which is 3.2 and 3.0 eV respectively. Therefore, rutile TiO_2 can absorb radiations till 405 nm wavelength while anatase can absorb radiations till 385 nm [50]. Most of the wastewater treatment process has been carried out under UV radiation. Nowadays, many commercial lamps are used that provide UV range radiation but they are expensive and lead to the formation of toxic substances. So, in order to make this technology economical researchers have tried to utilise solar energy as radiation source. However, solar radiation has only 5% UV light rest is visible and thus, in order to absorb visible light further modifications are required to TiO_2. This can be achieved by lowering its band gap from 3.0 eV, by increasing the mobility of charge carriers within the bands and by providing suitable potentials to conduction and valance bands. Many studies were performed by adding metal ions to TiO_2 to make it active under visible radiation [51]. It was reported that TiO_2 possess approx. 3.0 eV from hydrogen reference potential while approx. 1.2 eV from oxidation potential. Therefore, when TiO_2 is irradiated with UV light with a wavelength below 400, then its surface would get heat higher than 30,000°C which is very high and thus it can oxidise all organic contaminants [52].

The mechanism that takes place on the surface of TiO_2 while removing the pollutants under irradiation usually takes place in three steps:

i. Photoexcitation: $TiO_2 + h\nu\ e^- + h^+$
ii. Charge carriers trappings: $e^-_{CB}\ e^-_{TR}$
$h^+_{VB} h^+_{TR}$
iii. Electron-hole recombination: $e^-_{TR} + h^+_{VB}\ (h^+_{TR})\ e^-_{CB} +$ heat
iv. Photoexcitation: O_2 (ads) $+ e^-\ O_2^{\cdot -}$
v. Generation of hydroxyl radical: $H_2O + O_2^{\cdot -}\ HO_2^{\cdot -} + OH\cdot$
vi. Formation of H_2O_2: $2OH_2\cdot\ H_2O_2 + O$
vii. Reactive oxygen species: $H_2O_2 + O_2^{\cdot -} 2OH\cdot + OH^- + O_2$
viii. Photodegradation: organic $+ OH\cdot\ H_2O + CO_2$

There are many non-biodegradable organic pollutants which were degraded by TiO_2 photocatalysts such as acetaldehyde, benzene, acrolein, butyric acid, dichlorophenol, dichlorobenzidine, chloronaphthalene, chloroform, hydrazine, methylene chloride, hexachlorobenzene, tetrachloroethylene, polychlorinated biphenyl, nitroglycerine, xylene, etc. under different operating conditions [53].

7.2.5.2 Doped TiO_2/UV

In order to improve the photocatalytic activity of TiO_2, it is doped with many metals and non-metal ions. As discussed in the earlier section, doping techniques are usually done to narrow down its band gap and to make it more active towards solar radiation. This also helps in improving the thermal stability [54]. Dopants are also known to deliberate their physicochemical properties such as high surface area and high crystallinity to TiO_2. They inhibit the recombination of photogenerated holes and electrons and thus accelerate the rate of generation of hydroxyl radical which enhances the photodegradation process [55]. A detailed study has been conducted by optimising the following parameters that are important to increase the photoactivity of doped TiO_2 such as calcination temperature, reactant concentration, concentration of dopant, concentration of doped TiO_2, etc. These are some of the reported metal as well as non-metal dopants which were used to improve the photocatalytic activity of TiO_2 for the degradation of pollutants such as methyl orange, methylene blue, dye acid-7, oxalic acid, organic waste, penicillin, organophosphorus pesticides, rhodamine B, phenols, etc. [56].

7.2.5.3 Semiconductor and Other Nanocomposite TiO_2/UV

Pure semiconductor metals are also known for their photodegradation activity. But like TiO_2, they also suffer from wide band gap, and thus, researchers have found the doping techniques by which one or more semiconductors metal oxides are doped together to bring out excellent or improved photocatalytic activity. The most popular semiconductor metal oxides such as SnO_2, WO_3, ZnO, CuO, Co_2O_3, CdS, Cr_2O_3, etc. The working mechanism of semiconductors-doped TiO_2 is the same as discussed in the earlier section. The only and sole advantage of doping TiO_2 with semiconductor is to make them more active under solar radiation by narrowing the wide band gap. Many semiconductors-doped TiO_2 photocatalysts were synthesised by using different methods and have been tried for water remediation. Some of them are tabulated below in Table 7.1 [56,57].

7.2.6 Gamma Radiation

Ionising radiation technology has gained much attention in chemical and biological wastewater treatment. As it is a physical process, no such chemicals are used. The ionising radiations in the form of gamma rays and electron beams provide a solution in order to get rid of harmful microbial contaminants and thus help in the disinfection of water [58]. It is capable of killing pathogens causing without any loss in the phosphorus and nitrogen contents of useful nutrients present in water. This is very useful if the treated wastewater is to be used for agricultural purposes [59]. The rate of degradation of pollutants in agricultural wastewater with the help of the irradiation process has been presented in Table 7.2.

TABLE 7.1
Nanocomposites Used for Degradation of Pollutants [56,57]

Sr. No.	Nanocomposite	Degraded Organic Pollutants	Resulted Outcomes
1	TiO_2 and ZnO photocatalysis	Effluents from dyeing and finishing textile industry	No toxicity was observed after composite-assisted photocatalytic treatment
2	Photocatalysis with TiO_2, ZnO and Nb_2O_5	Textile effluent from jeans industrial laundry	Significant decrease in toxicity of textile wastewater after photocatalytic treatment
3	Cu-Zn-S/TiO_2NTAs	2,4-dichlorophenoxyacetic acid, anthracene-9-carboxylic acid	100% removal of (RhB) Rhodamine B
4	γ-Fe_2O_3@SiO_2@TiO_2–Ag	MO	High photocatalytic activity as compared to commercial Degussa P25
5	Ag/TiO_2–Cu	4-nitrophenol	Faster charge transfer process and 96% photocatalytic reduction of 4-nitrophenol
6	Cellulose fibre TiO_2 nanobelt –Ag NPs	MO	100% degradation of MO with better photocatalytic and antibacterial activity
7	Ag-TiO_2	MB	Decolorization of MB in visible light at lower energies than band gap of TiO_2
8	SnO_2/TiO_2	Rhodamine B (RhB)	Improved photocatalytic activity of TiO_2 under visible irradiation with decreased electron-hole pair recombination rate.
9	CdS/TiO_2	MO	Much better activity as compared to bare TiO_2 towards ofloxacin degradation
10	TiO_2–$LaFeO_3$	MB	Faster degradation rate due to inhibition of combination of electron-hole pairs
11	CeO_2/TiO_2, ZrO_2/TiO_2 $H_3PW_{12}O_{40}$-CeO_2/TiO_2, $H_3PW_{12}O_{40}$-ZrO_2/TiO_2	Methanol	Photocatalytic degradation of methanol under ultraviolet irradiation

TABLE 7.2
Percent Degradation of Agricultural Wastewater Pollutants Degradation Using Gamma Irradiation [61]

Sr. No.	Pollutants	Origin	Percent Degradation	Dose, [a]kGy
1	Polychlorinated biphenyls (PCB)	Pesticide	96%	<0.1
2	Alkali halides	Herbicides, fungicides, insecticide	98%	1
3	Trihalomethanes disinfection by-products	Disinfection by-products	95%	6
4	Nitrophenols	Degradation products	-	5
5	Carbofuran, dimethoate, imidacloprid, procloraz	Pesticide	99%	5
6	Methiocarb	Pesticide	67%	5
7	2,4-dichlorophenol pesticide	Pesticide	Complete degradation	10

[a] kGy, kilogray (SI unit of absorbed dose of ionising radiation).

Also, this technology is much superior to the conventional chlorination/de-chlorination techniques in terms of toxicity assessment of treated effluents. Therefore, it is also a technical and economical process in terms of removing persistent heavy metals. The mechanism of reducing the organic contaminants and pathogens usually takes place in the following steps:

a. Oxidation of organic molecules
b. Disturbance of structure of organic and inorganic molecules
c. Formation of colloidal system and killing of microorganisms

There are more than 15 pilot plants with large scaled irradiation treatment facilities that carry out the irradiation process with the help of electron accelerators and γ irradiators of different capacities. However, there is much need of spreading more awareness among the environmental engineering community. International Atomic Energy Agency (IAEA) has done this by providing many reports and integrated reviews on the advances in ionising radiation applications in industrial effluent treatment from technical, operational safety, and economical aspects [60]. Out of which some are based on exploring the role of ionising radiation in the degradation of azo dyes. However, some of the existing knowledge gaps need to be identified and to be addressed to make this technique fully sustainable in terms of both economically and operational safety.

7.3 CHALLENGES AND BARRIERS

The main challenge faced by conventional wastewater technology is energy consumption. Traditional secondary treatment methods are most expensive of all wastewater treatment techniques because of the large consumption of electricity and chemicals. The system involves blower, pump and aeration system to treat a large amount of water. Another drawback associated with some conventional systems is the large space requirement. The wastewater treatment plant should have a large outdoor facility that requires a large infrastructure budget. Very low energy and resource recovery can be achieved by utilising the conventional system is another most important challenge associated with conventional wastewater treatment methods.

To overcome the drawbacks of conventional wastewater treatment technologies advanced wastewater treatment technologies have been explored. Many scientists have proposed advanced wastewater treatment technologies for sustainable wastewater treatment. However, these advanced treatment strategies also have associated challenges and barriers. The huge amount of sludge generation, requirement of significant amount of chemicals are the important bottlenecks of Fenton oxidation processes, coagulation, flocculation, and chemical precipitation. In addition, the poisonousness of chemicals restricts the applicability of these treatment methods at a commercial scale. Ozonation is competent for the elimination of colour and COD which has associated drawbacks such as high cost of ozone generation and short half-life time of ozone. Recently, HC has been known to be efficient for the reduction of COD. However, its applicability at the commercial scale is confined due to complicated scheme and very huge power constraints. Most studies have stated the applicability of activated carbon-assisted adsorption for wastewater treatment; however, the short lifetime of activated carbon, necessity of high cost for thermal retrieval of adsorbent and little resource recovery during the process limits the activated carbon use [62]. Membrane filtration, membrane bioreactors are also promising options for the treatment of wastewater however membrane fouling, generation of secondary waste are the associated drawbacks [63]. The most efficient activated sludge process is associated is having the drawback of large sludge production; the large space is required to dispose this sludge and also restricted as bio scrubber for treatment of odorous air [64].

A lot of reports are available for use of these technologies in diverse areas but very limited data is available on the actual performance of these technologies at the industrial scale. Although conventional systems have been used to treat wastewater, they are not convenient from an economical and sustainable perspective. Thus, There is an urgent need to resolve these challenges and barriers associated with advanced and conventional wastewater treatment technologies by which the pathway to develop sustainable and economically and energetically efficient technology for wastewater treatment can be more realistic.

7.4 CONCLUSION AND FUTURE PERSPECTIVE

Water is one of the basic needs for the survival of humans. By considering the present rate of freshwater consumption and estimated future freshwater demand, water will become a highly prioritised need due to exponential increase in global population and industrialisation. This chapter has briefly explored the conventional technologies for wastewater treatments like primary, secondary and tertiary treatment strategies with their associated disadvantages. Mainly different types of advanced wastewater treatment technologies like AOPs, biological treatments, membrane processes, electrodialysis, HC, membrane bioreactor, gamma irradiation, etc. have been explained in detail. However, the efficiency of the wastewater treatment strategy depends on the physical and chemical characteristics of feed wastewater challenges and barriers associated with these advanced wastewater treatment technologies have been reviewed.

To conquer the challenges associated with the advance and conventional wastewater treatment strategies for achieving the highest efficiency, requirement of deep understanding of feed wastewater, in-detail knowledge of each step involves in processing etc. are the essential factors. The integration of two or more processes can be a major breakthrough in wastewater treatment for attaining higher process efficiency. There are many reports available for the integration of two or more processes but no commercial facility with expected efficiency has been established till date. Thus, there is an emerging need for sustainable, economically and environmentally efficient technology for the treatment of wastewater to maintain global balance.

REFERENCES

1. Mishra, R. K., Dubey, S. C. Fresh water availability and it's global challenge, *Int J Eng Sci Inven Res Dev* 2 (2015) 351–407.
2. Kaur, R., Wani, S. P., Singh, A. K., Lal, K. Wastewater production, treatment and use in India, *Natl. Rep. Present. 2 Nd Reg. Work. Safe Use Wastewater Agric.*, 2012.
3. Rai, B. Pollution and conservation of Ganga river in modern India, *Int J Sci Res Publ* 3 (2013) 1–4.
4. C. P. C. Board of India, Status of Water Supply, *Wastewater Generation and Treatment in Class-I Cities & Class-II Towns of India*. Ministry of Enviroment and Forest, Govt. of India, 2009.
5. How India's cities came to drown in sewage and waste. *The Guardian*, 1 August 2012. Retrieved 17 December 2019.
6. Sperling, M. Wastewater characteristics, treatment and disposal, *Water Intell* (2015) Online. https://doi.org/10.2166/9781780402086.
7. Crini, G., Lichtfouse, E. Advantages and disadvantages of techniques used for wastewater treatment, *Environ Chem Lett* 17 (2019) 145–155. https://doi.org/10.1007/s10311-018-0785-9.
8. Gogate, P. R., Pandit, A. B. A review of imperative technologies for wastewater treatment I: Oxidation technologies at ambient conditions, *Adv Environ Res* 8 (2004) 501–551. doi: 10.1016/S1093-0191(03)00032-7.
9. Ali, I., Gupta, V. K. Advances in water treatment by adsorption technology, *Nat Protoc* 1 (2007) 2661–2667. doi: 10.1038/nprot.2006.370.
10. Bruggen, B., Vandecasteele, C., Gestel, T., Doyen, W., Leysen, R. A review of pressure-driven membrane processes in wastewater treatment and drinking water production, *Environ Prog* 22 (2003) 46–56. doi: 10.1002/ep.670220116.
11. Deng, Y., Zhao, R. Advanced oxidation processes (AOPs) in wastewater treatment. *Curr Pollution Rep* 1 (2015) 167–176. https://doi.org/10.1007/s40726-015-0015-z.
12. Tchobanoglous, G., Burton, F., Stensel, H. *Wastewater Engineering*. Metcalf & Eddy Inc: New York, 2003.
13. System SE. *The UV/Oxidation Handbook*. 1994.
14. Wang, J. L., Xu, L. J. Advanced oxidation processes for wastewater treatment: Formation of hydroxyl radical and application, *Crit Rev Environ Sci Technol* 42 (2012) 251–325.

15. Ghime, D., Ghosh, P. Advanced oxidation processes: A powerful treatment option for the removal of recalcitrant organic compounds. 10 June 2020. doi: 10.5772/intechopen.90192.
16. Ameta, S. C. Chapter 1: Introduction. In: *Advanced Oxidation Processes for Waste Water Treatment*, Academic Press, 2018, 1–12. ISBN 9780128104996. https://doi.org/10.1016/B978-0-12-810499-6.00001-2.
17. Gottschalk, C., Libra, J. A., Saupe, A. *Ozonation of water and waste water: a practical guide to understanding ozone and its applications*. John Wiley & Sons, 2009.
18. Ikehata, K., Li, Y. *Chapter 5: Ozone-Based Processes Advanced Oxidation Processes for Waste Water Treatment*. Academic Press, 2018, 115–134. ISBN 9780128104996. https://doi.org/10.1016/B978-0-12-810499-6.00005-X.
19. Tang, W. Z. *Physicochemical Treatment of Hazardous Wastes*. CRC Press, 2003.
20. Pignatello, J. J., Oliveros, E., MacKay, A. Advanced oxidation processes for organic contaminant destruction based on the Fenton reaction and related chemistry, *Crit Rev Environ Sci Technol* 36 (2006) 1–84.
21. Ameta, R., Chohadia, A. K., Jain, A., Punjabi, P. B. Chapter 3: Fenton and photo-fenton processes. In: Ameta, S. C., Ameta, R. (eds.) *Advanced Oxidation Processes for Waste Water Treatment*. Academic Press, 2018, 49–87. ISBN 9780128104996. https://doi.org/10.1016/B978-0-12-810499-6.00003-6.
22. House, D. A. Kinetics and mechanism of oxidations by peroxydisulfate, *Chem Rev* 62 (1962) 185–203.
23. Anipsitakis, G. P., Dionysiou, D. D. Radical generation by the interaction of transition metals with common oxidants, *Environ Sci Technol* 38 (2004) 3705–3712.
24. Forsey, S. P. *In Situ Chemical Oxidation of Creosote/Coal Tar Residuals: Experimental and Numerical Investigation*. University of Waterloo, 2004.
25. Jung, H., Pauly, D. 4.19: Water in the pulp and paper industry. In: Wilderer, P. (ed.) *Treatise on Water Science*. Elsevier, 2011, 667–683. ISBN 9780444531995. https://doi.org/10.1016/B978-0-444-53199-5.00100-7.
26. Tirosh, U., Shechter, R. Membrane aerated biofilm reactor (Mabr)—Distributed treatment of wastewater at low energy consumption. In: *Advances in Science, Technology and Innovation*, 2020, 513–515. https://doi.org/10.1007/978-3-030-13068-8_128.
27. Dindar, E. An overview of the application of hydrodynamic cavitation for the intensification of wastewater treatment applications: A review, *Innov Energy Res* 5 (2016) 1–7.
28. Thanekar, P., Gogate, P. Application of hydrodynamic cavitation reactor for treatment of wastewater containing organic pollutants: Intensification using hybrid approaches, *Fluids* 3 (2018) 98.
29. Sun, X., Xuan, X., Song, Y., Jia, X., Ji, L., Zhao, S., Yoon, J. Y., Chen, S., Liua, J., Wang, G., Experimental and numerical studies on the cavitation in an advanced rotational hydrodynamic cavitation reactor for water treatment, *Ultrason Sonochem* 70 (2021) 105311.
30. Mishra, K. P. Gogate, P. R. Intensification of degradation of Rhodamine B using hydrodynamic cavitation in the presence of additives, *Sep Purif Technol* 75 (2010) 385–391.
31. Mancuso, G., Langone, M., Andreottola, G. A critical review of the current technologies in wastewater treatment plants by using hydrodynamic cavitation process: Principles and applications, *J Environ Health Sci Eng*.
32. Pandit, A. B. Hydrodynamic cavitation technology: Industrial applications. In: *The Mind of an Engineer*, 2015. https://doi.org/10.1007/978-981-10-0119-2_43.
33. Strathmann, H. Electrodialysis, a mature technology with a multitude of new applications, *Desalination* 264 (2010) 268–288. https://doi.org/10.1016/j.desal.2010.04.069.
34. Akhter, M., Habib, G., Qamar, S. U. Application of electrodialysis in waste water treatment and impact of fouling on process performance, *Journal of Membrane* 8 (2018). doi: 10.4172/2155-9589.1000182.
35. Gurreri, L., Cipollina, A., Tamburini, A., Micale, G. Electrodialysis for wastewater treatment—Part II: Industrial effluents, *Curr Trends Future Develop (Bio-) Membr* (2020). doi: 10.1016/B978-0-12-816823-3.00008-3.
36. Shestakova, M., Sillanpa, M. Electrode materials used for electrochemical oxidation of organic compounds in wastewater, *Rev Environ Sci Biotechnol* 16 (2017) 223–238.
37. Maronez-Huitle, A. C., Panizza, M. Electrochemical oxidation of organic pollutants for wastewater treatment, *Curr Opin Electrochem* 11 (2018) 62–71.
38. Escapa, A., Mateos, R., Martínez, E. J., Blanes, J. Microbial electrolysis cells: An emerging technology for wastewater treatment and energy recovery. From laboratory to pilot plant and beyond, *Renew Sustain Energy Rev* 55 (2016) 942–956.
39. Katuri, K. P., Ali M., Saikaly P. E. The role of microbial electrolysis cell in urban waste water treatment: Integration options, challenges, and prospects, *Curr Opin Biotechnol* 57 (2019) 101–110.
40. McCullagh, C., Robertson, P. K. J., Adams, M., Pollard, P. M., Mohhammed, A. Development of a slurry continuous flow reactor for photocatalytic treatment of industrial waste water, *J Photochem Photobiol A* 211 (2010) 42–46.

41. Fujishima, A., Rao, T. N., Tryk, D. A. Titanium dioxide photocatalysis, *J Photochem Photobiol C: Photochem Rev* 1 (2000) 1–21.
42. Bora, L. V., Mevada, R. K. Photoreactors for heterogeneous photocatalysis for wastewater treatment, *Int Res J Eng Technol (IRJET)* 3 (2016) 1011–1014.
43. Acon, A. G. Nano water worldwide nanotechnologies in water, *Drinking Water and Waste Water Worldwide*, 2008–2010–2015.
44. Sanchez, L., Peral, J., Domenech, X. Aniline degradation by combined photocatalysis and ozonation, *Appl Catal B: Environ* 19 (1998) 59–65.
45. Matilainen, A., Sillanpää, M. Removal of natural organic matter from drinking water by advanced oxidation processes, *Chemosphere* 80 (2010) 351–365.
46. Pera-Titus, M., García-Molina, V., Baños, M. A., Giménez, J., Esplugas, S. Degradation of chlorophenols by means of advanced oxidation processes: A general review, *Appl Catal B Environ* 47 (2004) 219–256.
47. Chemseddine, Moritz, T. Nanostructuring titania: Control over nanocrystal structure, size, shape, and organization, *Eur J Inorg Chem* 2 (1999) 235–245.
48. Wang, Y. Q., Hu, G., Duan, X., Sun, H., Xue, Q. Microstructure and formation mechanism of titanium dioxide nanotubes, *Chem Phys Lett* 365 (2002) 427–431.
49. Leary, R., Westwood, A. Carbonaceous nanomaterials for the enhancement of TiO_2 photocatalysis, *Carbon* 49 (2011) 741–772.
50. Egerton, T. A., Christensen, P. A. *Advanced Oxidation Processes for Water and Wastewater Treatment*, Parsons, S. (ed.). IWA Publishing: London, 2004.
51. Carp, O., Huisman, C. L., Reller, A. Photoinduced reactivity of titanium dioxide, *Prog Solid State Chem* 32 (2004) 33–177.
52. Hernández-Alonso, M. D., Fresno, F., Sua´rez, S., Coronado, J. M. Development of alternative photocatalysts to TiO_2: Challenges and opportunities, *Energy Environ Sci* 2 (2009) 1231.
53. Lee, S. Y., Park, S. J. TiO_2 photocatalyst for water treatment applications, *J Indus Eng Chem* 19 (2013) 1761–1769.
54. Ghosh Chaudhuri, R., Paria, S. Visible light induced photocatalytic activity of sulfur doped hollow TiO_2 nanoparticles, synthesized *via* a novel route, *Dalton Trans* 43 (2014) 5526.
55. Ma, Y. F., Zhang, J. L., Tian, B. Z., Chen, F., Wang, L. Z. Synthesis and characterization of thermally stable Sm, N co-doped TiO_2 with highly visible light activity, *J Hazard Mater* 182 (2010) 386.
56. Jain, A., Vaya, D. Photocatalytic activity of TiO_2 nanomaterial, *J Chil Chem Soc* 62 (2017) 3683–3690.
57. Rueda-Marquez, J. J., Levchuk, I., Fernández Ibañez, P., Sillanpää, M. A critical review on application of photocatalysis for toxicity reduction of real wastewaters, *J Clean Prod* 258 (2020) 120694.
58. Verde, S. C., Silva, T., Matos, P. Effects of gamma radiation on wastewater microbiota, *Radiat Environ Biophys* 55 (2016) 125–131.
59. Rawat, K. P., Sharma, A., Rao, S. M. Microbiological and physicochemical analysis of radiation disinfected municipal sewage, *Water Res* 32 (1998) 737–740.
60. IAEA. *Radiation Treatment of Polluted Water and Wastewater.* IAEA: Vienna, Austria, 2008.
61. Rehab, O., Rahman, A. Hung, Y. T. Application of ionizing radiation in wastewater treatment: An overview, *Water* 12 (2020) 19.
62. Margot, J., Kienle, C., Magnet, A., Weil, M., Rossi, L., de Alencastro, L. F., Abegglen, C., Thonney, D., Chèvre, N., Schärer, M., Barry, D. A. Treatment of micropollutants in municipal wastewater: Ozone or powdered activated carbon? *Sci Total Environ* 461–462 (2013) 480–498. doi: 10.1016/j.scitotenv.2013.05.034.
63. Bruggen, B., Milis, R., Vandecasteele, C., Bielen, P., San, E., Huysman, K. Electrodialysis and nanofiltration of surface water for subsequent use as infiltration water, *Water Res* 37 (2003) 3867–3874. https://doi.org/10.1016/S0043-1354(03)00296-3.
64. Dhote, J., Ingole, S., Chavhan, A. Review on waste water treatment technologies, *Int J Eng Res Technol* 1 (2012) 110.

8 Electroflotation Process
Principles and Applications

Orhan Taner Can
Bursa Technical University

Erhan Gengec
Kocaeli University

Mehmet Kobya and Erhan Demirbas
Gebze Technical University

Alireza Khataee
Gebze Technical University
University of Tabriz

CONTENTS

8.1 INTRODUCTION

In recent years, the trend toward clean technologies in the removal of pollutants from wastewater has been increasing as correct understanding. In this context, treatment processes that will minimize the use of chemicals come to the fore. It is important to remove at least some of the pollutants in wastewater by physical treatment processes such as flotation. Electroflotation is promising in water and wastewater treatment as the most efficient of classical flotation processes. Electroflotation is a physical-chemical process of water treatment from insoluble substances. The process is based on conducting the water electrolysis on insoluble electrodes and the flotation effect. While carrying out the water electrolysis, gas bubbles of oxygen and hydrogen are generated at the anode and cathode,

DOI: 10.1201/9781003164982-8

respectively. When gas bubbles rise to the water surface due to the density difference between gas bubbles and water, they are faced with dispersed particles of pollution, adhere to them, and float them to the water surface to form a stable flotosludge layer. Skimmers or other mechanical devices later remove these particles from the water surface. Nevertheless, the formation of both oxygen and hydrogen gas bubbles hydrogen bubbles generated at the cathode plays the primary role in the electroflotation process (Sillanpää and Shestakova, 2017; Shah, 2020, 2021). This chapter aims to give the background, advantages and disadvantages, significant operational parameters, and design and discuss applying the electroflotation process.

8.1.1 Background

Flotation is a unit process employed to separate solid or liquid particles from a liquid phase. Separation is accomplished by the introduction of fine gas bubbles into the liquid phase. The bubbles are connected to particles in the medium and floating force of bubbles induces the particles to rise above the surface (Lundgren, 1976; Shah, 2021). Separations of solid particles from liquid solutions are used in many environmental and industrial treatment processes. Flotation is primarily used when sedimentation is not practical because of the presence of very fine particles or globules (Zouboulis and Avranas, 2000). In addition, it is a preferred application for the thickening of the activated and flocculated chemical sludge. The wastes of various sources including papermaking, refineries, ship's bilge and ballast waste, deinking operations, metal plating, meat, chemical and soybean processing, laundries, iron, steel and glass plants, soap and plants manufacturing, barrel and drum cleaning, wash rack and equipment maintenance, mill waste and aluminum forming were treated by air flotation (Wang et al., 2010). Flotation mechanism is illustrated in Figure 8.1. Encapsulation of an air bubble into flocs or bubbles reduces the effective density of the oil or agglomerate and it causes an increase in the rate of bubble/solid or bubble/oil agglomerates over the rise in unaerated

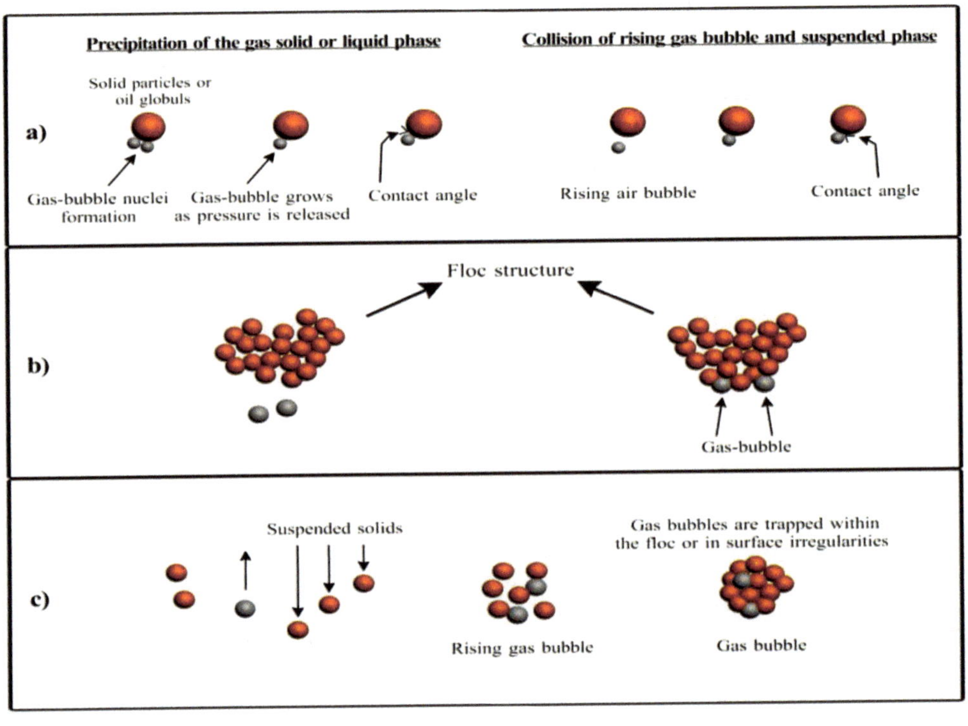

FIGURE 8.1 Mechanism of flotation. (Adapted with permission from Shammas and Bennett (2010).)

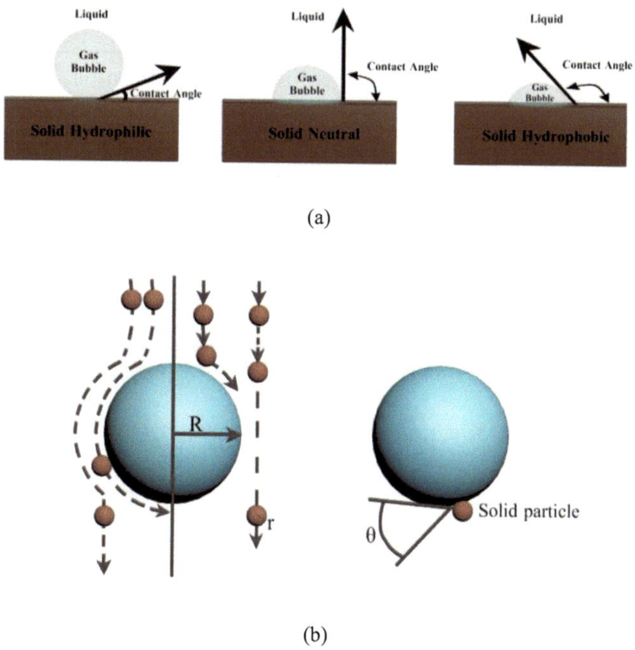

FIGURE 8.2 (a) Attachment of gas bubbles onto solids. (Adapted with permission from Lawrence K. Wang et al. (2010).) (b) Interaction of gas bubbles and solids regarding to their size. (Adapted with permission from Sillanpää and Shestakova (2017).)

systems (Shammas and Bennett, 2010). Figure 8.1 shows the attachment of solids onto gas bubbles in a flotation system. Solid surfaces that are naturally wetted by water are called hydrophilic, while a non-wettable surface is termed as hydrophobic (Shammas and Bennett, 2010).

A hydrophobic surface characteristically attracts air; hence, it is also named aerophilic (Figure 8.2a). Aerophilic surfaces are potently drawn into an air interface that easily displaces water at the surface of the solid. In a flotation system, separation of the double solid mixture can be achieved by selectively adding the hydrophobic solid particles to the gas bubbles. The other solid particles with hydrophobicity remain in the liquid. The buoyancy resulted from the difference in the density between air bubbles and water moves the hydrophobic solid particles to the surface, where they remain entrained in a froth or scum and can be drained off or mechanically skimmed away (Wang et al., 2010).

Traditionally, solid particles are removed by sedimentation; however, it does not work well for low-density particles like clay minerals, spores, and coagulated fulvic acids (Edzwald et al., 1992). In flotation, the generation of small gas bubbles is conducted at the bottom of the water column to be treated. Then, the microbubbles move upwards to the liquids' surface through buoyancy. As the bubbles rise, they collide and adsorb to particles in the solution. Consequently, the solids with low-density float to the top of the column for removal by skimming.

Air flotation and its different variations can be effectively used for the separation of light particulates and oils from wastewater. Particulates connected to an air bubble by either absorption or adsorption can be separated from the liquid phase by flotation. Polymers can be added to the wastewater for enhancing the connection of the particle to the bubble. Colloids agglomeration into floc particles with the contribution of chemicals enables the precipitation of some soluble materials, which allows the removal. When the wastewater is chemically pretreated to break the oil emulsion, most of the emulsified and free oils may be removed by the air flotation units (Shammas and Bennett, 2010). The effectiveness of surfactant-aided air flotation techniques used for the removal of solid particles from a liquid suspension is affected by different chemical and process factors. The surfactant type and concentration, hydrophobicity, zeta potential, pH, ionic strength and alkalinity are among the chemical factors; while cell concentration, growth stage, bubble size, temperature, time, flow rate, and column size are process factors (Alhattab and Brooks, 2017; Shah, 2020, 2021).

Different methods used for the production of gas bubbles cause different types of flotation processes. Flotation systems can be categorized into five distinct types based on the method of their bubble formation:

i. Dissolved air: extraction of the gas from a supersaturated solution occurs as a result of pressure reduction.
ii. Induced (dispersed) air: the liquid and gas are mechanically combined for the formation of bubbles in the liquid.
iii. Froth: the gas is injected directly into the fluid by using a sparger.
iv. Electrolytic: bubbles are formed by water electrolysis.
v. Vacuum: negative pressure causes the release of air from a saturated solution (Shammas and Bennett, 2010).

Conventional flotation cells generate large bubbles in the size of 600–2000 µm. Some spargers form a medium bubble size (100–600 µm), which is not sufficient for the flotation of ultrafine particles (<10 µm). The rise rates of larger-size bubbles are faster hence they spend less time in the reactor. On the other side, smaller bubbles rise more slowly consequently spending more time in the reactor (Yoon, 1993; Shah, 2021). Size distributions of bubbles generated by electrolysis, dispersed air, and dissolved air flotation are presented in Table 8.1. Comparisons of various flotation processes on energy consumption, retention time, and principal applications are summarized in Table 8.2 (Shammas and Bennett, 2010). EF has the potential for the generation of very fine bubbles (<100 µm) that may be effective to recover very fine particles.

The ratio between particle and gas bubble size influences the efficiency of flotation significantly because it determines the efficiency of the collision and the subsequent attraction (Sillanpää and Shestakova, 2017). Figure 8.2b shows the interaction of gas bubbles and solids regarding their size.

8.1.2 Induced (Dispersed) Air Flotation

Induced air flotation process originates from the mineral processing industry where it is generally known as froth flotation (Painmanakul et al., 2010; Zouboulis et al., 1992). Although its use for the

TABLE 8.1
Bubble Characteristics of Different Flotation Processes

Parameter	EF	IAF	DAF
d_b (µm)	5–95	170	10–300
r_b (cm/s)	0.5	1.5	0.1
N_b (m^3)	10	0.2×10^4	3.6×10^6
S_b (cm^2/cm^3)	454	293	800

d_b, average diameter of the bubble at 60 dyne/cm; r_b, rise rate of bubble; N_b, number of bubble; S_b, bubble surface area.

TABLE 8.2
Comparison of Flotation Processes in Terms of Bubble Size, Energy Consumption, Retention Time and Principal Applications

Flotation Process	d_b (μm)	C_{energy} (W/m³/h)	τ (min)	Applications
Blown air flotation	100–500	20–30	2–5	Grease
Mechanical flotation	100–1000	100–200	2–16	"Roughing" of polymer and latex or elastomer suspensions
DAF (with 20% recirculation)	10–300	45–60	20–30	Hydrocarbons solvents, fibers, fine suspension and flocculated particles
Electroflotation	5–95	150–300	-	Same applications for hot saline water as for DAF

Adapted with permission from Shammas and Bennett (2010).
d_p, size of bubble; C_{energy}, energy consumption; t, retention time; DAF, dissolved air flotation.

treatment of municipal wastewater is rare, IAF is industrially utilized for the treatment of waste or process water to remove emulsified oil and suspended solids. Nowadays, it is also used in the processing of proteins, resins, corn, clays, coal, bacteria, fats, glass, plastic, dyes, and rubber (Matis and Lazaridis, 2002).

In IAF techniques, the gas phase is directly introduced into the liquid phase by the use of a revolving impeller. A mechanical high-speed agitator or an air injection system is used for the generation of air bubbles (Jameson, 1999). In an IAF process, the bubbles have a typical diameter range of 40–1500 μm and turbulent hydrodynamic conditions (Alhattab and Brooks, 2017; Dafnopatidou and Lazaridis, 2008). The advantages of the IAF include lower capital cost, compact size, and ability to remove relatively suspended solids and free oil. On the other hand, its higher power requirement, strict hydraulic control-dependent performance and less flocculation flexibility are the disadvantages of the IAF process. In addition, the quantity of float skimming is greater than the pressurized unit (Eckenfelder, 2000).

8.1.3 Dissolved Air Flotation

There are several types of dissolved air flotation (DAF) processes used in the industry, namely vacuum flotation (Zabel and Melbourne, 1980), micro-flotation and pressure flotation (Barret, 1975). Among them, pressure flotation is the most commonly used process for the treatment of water or wastewater. DAF processes involve the use of three basic pressures that are split-flow, full-flow, and recycle flow pressure flotation. The most suitable method for water treatment applications involving the removal of a fragile floc is recycle flow pressure flotation. DAF process generates very small bubble sizes in the diameter of 30–100 μm to facilitate high treatment efficiency. However, capital cost, complicated operations, and the size of equipment limit the use of DAF systems.

DAF systems are mostly utilized to remove suspended solids and oils-greases from water or wastewater for certain treatment purposes. They are used in various processes to be as a pre-treatment step to meet sewer discharge limits or to reduce loading on downstream biological treatment systems, to reuse and recovery of certain materials, to polish effluent of biological treatment and to thicken bio-solids. The applications of DAF include clarification of refinery wastewater, separation of solids, reclamation of wastewater, thickening of sludge, separation of biological flocs, removal/separation of ions, neutralization of acid mining drainage, recovery of very small mineral particles, removal of pollutants from mining wastewaters, removal of biological and organic pollutants such as dissolved toxic organic chemicals, algae, 5–7 μm Giardia oocysts,

4–5 μm Cryptosporidium oocysts and algae from heavily algae laden waters, etc. (Gochin and Solari, 1983).

8.1.4 Froth Flotation

The froth flotation is one of the most commonly used separation systems in the mineral refining industry (Hodouin et al., 2000; McKee, 1991). It is particularly applicable for the processing of fine-grained ores not suited to the gravity concentration. On the other hand, the flotation process is not still fully understood and remains relatively inefficient despite being introduced in the early 1900s and many years of research and development efforts. It could be speculated that optimization of current froth flotation processes can provide significant economic benefits.

Froth flotation is a very adaptable technique for the physical separation of certain particles from mineral/water slurries. The separation process in this method is based on the difference of how air bubbles selectively adhere to certain mineral surfaces in the slurry. Air bubbles attach the particles to be removed and carry them to the surface where separation is conducted, while the wet particles stay in the liquid phase. A wide range of mineral separations can be performed using froth flotation by utilization of chemical treatments for selective modification of mineral surfaces to have the desired properties for separation (Haxhiaj et al., 2016).

8.1.5 Vacuum Flotation

Vacuum flotation has been used successfully to remove oil grease and suspended solids in the treatment of sewage and industrial wastes. In a vacuum flotation method, wastewater is saturated with air either directly using an aeration tank or by allowing air entrance on the suction side of a wastewater pump. In addition, a partial vacuum is applied to cause the dissolved air to emerge as minute bubbles from the solution. Air bubbles are bound to solid particles to be separated and they move up to the surface forming a scum blanket. Then using a skimming mechanism, the scum blanket is removed (Tchobanoglous et al., 2003). The bubbles are formed by the saturation of suspension with air at 1 atm and afterwards vacuum application to create relative super-saturation. The amount of air available for flotation is severely limited due to a maximum pressure difference of 1 atm, which restricts the application of the process (Atarah, 2011).

8.1.6 Electroflotation or Electrolytic Flotation

The use of electrolytic flotation or EF was first suggested by Elmore in 1904 for the separation of valuable minerals from ore (Chen, 2004). The EF is based on the generation of bubbles of oxygen and hydrogen with water electrolysis by passing a DC current between anode and cathode electrodes (Barret, 1975). EF process provides floating of pollutants to the water surface by generating small bubbles. It is an attractive alternative technology for the removal of dispersed and floatable materials or pollutants. The bubble swarm moves oil globules and suspended solids to the surface of the water where the floating sludge layer forms and is mechanically removed. Thus, free atomic oxygen is formed in the diffusion layer of the anode and then transported by convection into the water where it instantly combines and oxidizes organic and inorganic materials in the water. A large number of very small bubbles must be formed with minimal turbulence to ensure an efficient EF operation. Furthermore, the electrode grids must be arranged to provide good coverage of the tank surface so that the gas bubbles are mixed easily and uniformly (Shammas and Bennett, 2010). The cathode and anode in the EF process used for insoluble electrodes (Pt, stainless steel, Ti-coated IrO_2, RuO_2, Ta_2O_5, etc.) are generated H_2 and O_2 bubbles, respectively. Generally, the EF process consists of an EF reactor (or an electroflotator) with insoluble electrodes and tanks for chemicals and flocculant solutions if necessary (Figure 8.3). In addition, the EF process includes pumps, a power supply (or rectifier) and sludge collecting system (skimmer).

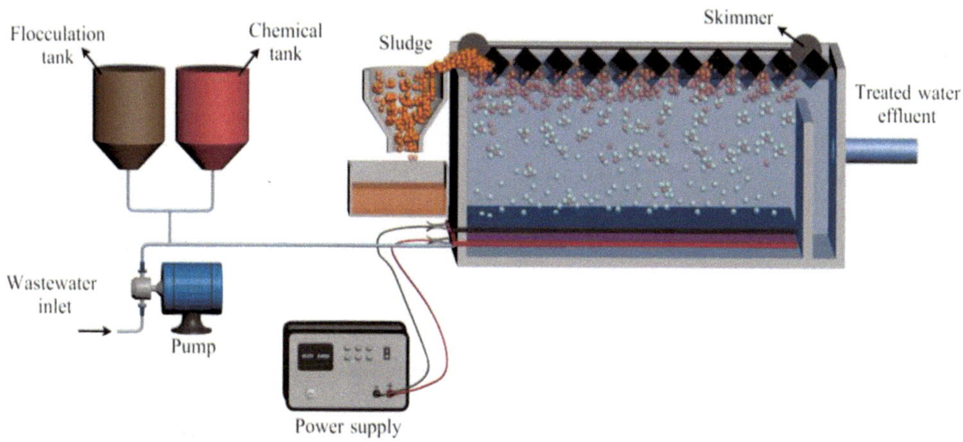

FIGURE 8.3 Schematic diagram of a typical EF process.

8.2 FUNDAMENTAL PRINCIPLES OF EF

When an aqueous solution is electrolyzed, hydrogen and oxygen are released at the cathode and anode, respectively. EF process produces hydrogen bubbles at the cathodes and oxygen bubbles at the anodes. The generation reactions of fine oxygen and hydrogen bubbles in the process are as shown in the following redox reactions (Chen et al., 2002; Kyzas and Matis, 2016; Mansour and Chalbi, 2006):

$$\text{Anode reaction}: 2H_2O \rightarrow O_2(g) + 4H^+ + 4e^- \quad (8.1)$$

$$\text{Cathode reaction}: 4H^+ + 4e^- \rightarrow 2H_2(g) \quad (8.2)$$

$$\text{The overall reaction}: 2H_2O \rightarrow 2H_2(g) + O_2(g) \quad (8.3)$$

In addition, the amount of hydrogen gas generated according to Eq. 8.3 is twice that of oxygen gas. In this case, the amount of produced gas can be measured according to the Faraday law:

$$Q_{H_2(g)} = \frac{i \times V_0}{n_{H_2} \times F} \quad (8.4)$$

$$Q_{O_2(g)} = \frac{i \times V_0}{n_{O_2} \times F} \quad (8.5)$$

where Q_{H_2} and Q_{O_2} are the generation rates of hydrogen and oxygen gases (L/s), respectively. V_o is the molar volume of gases (L/mole), F is the Faraday constant (96,487 C/mole electrons), n_{H_2} and n_{O_2} are the electron transfer number of H_2, and O_2. The total amount of gases produced (Q_g=Q_{H_2}(g)+ Q_{O_2}(g)) is calculated to be $1.74\times10^{-4}\times i$.

While these gases are dissolved at low current densities in the electrolyte solution, higher current densities enable the evolution of gas bubbles at the electrodes. The evolution of gas from the electrodes occurs in three stages; i.e. nucleation, development, and detachment. Bubbles nucleate from solutions supersaturated with product gas at the surface of the electrode. Bubbles with a radius greater than or equal to the critical radius grow, while bubbles having a radius smaller than the critical radius tend to decay. The critical bubble radius mainly depends on surface tension of liquid, molecular volume, system temperature, Boltzmann's constant, external pressure in the liquid and the dissolved gas pressure in the bubble (Lubetkin, 1995). Moreover, the pressure of dissolved gas inside the bubble is related to dissolved and saturation concentrations of the gas in the solution, the vapor pressure and volume of the water (or pure solvent), and activity coefficients of solvents (Sides, 1986). If the concentration of dissolved gas increases, the gas pressure inside the bubble will increase with all other factors being constant. This increase in the pressure of dissolved gas inside the bubble will then reduce the critical radius. In this case, the bubble nucleation rate will increase with an increase in the pressure of dissolved gas inside the bubble. As a result, the pressure of dissolved gas inside the bubble depends on the concentration of dissolved gas in the solution. The concentration of dissolved gas in the EF process depends on the applied current density. The concentration of dissolved gas increases as current density increases, thus increasing the bubble nucleation rate. The nucleated bubbles grow initially from the critical radius as a result of high internal pressure and mass transfer of dissolved gas into the growing bubble. The bubbles may also grow by coalescence with other bubbles at the electrode (Sarkar, 2012). The final stage of gas evolution in the EF process is the detachment of growing bubbles. The growing bubbles detach when the surface adhesive forces (related to surface tension and bubble contact angle) can no longer restrain them due to increasing buoyancy force. The electrostatic interaction between the bubble and the electrode double layers controls the bubble departure diameter, and the two phases are separated by a thin liquid film of electrolyte.

EF process is the formation of highly dispersed gas bubbles (hydrogen and oxygen), uniformly distributed throughout the liquid being processed during the electrolysis of a solution. In the process, the chemical composition of the liquid and the properties and state of soluble and insoluble impurities change. The interaction of gas bubbles with colloids creates the prerequisites for reliable attachment, intense coagulation of particles, vigorous sorption, adhesion, and consequently more efficient flotation. Gas bubbles possessing high buoyancy rise upwards, collide with suspended particles of contaminants, stick to them, and raise the particles to the water surface and forming a stable foam layer.

8.3 ADVANTAGES AND DISADVANTAGES OF EF PROCESS

The EF process has three main benefits. Firstly, the dispersed gas bubbles are very fine and uniform with an average bubble diameter of 5–95 μm. Second, Variable current density gives the possibility to change any concentration of gas bubbles in the flotation medium. Third, using appropriate electrode surface and solution conditions helps to obtain ideal results for a defined separation process (Hosny, 1996; Kyzas and Matis, 2016; Mansour and Chalbi, 2006).

EF processes are very flexible and competitive to settling tank techniques that require large land space. It is also competitive with other flotation processes like dispersed air flotation and DAF. EF systems are small and compact and require low maintenance and operating costs when compared to other flotation processes. The EF process is efficient at the stage of preliminary purification, for example, before sorption or membrane systems are used. Electrical energy consumption is a critical factor for the evaluation of the potential application of EF using insoluble anode electrodes. The energy consumption in the EF process significantly decreases in high-conductivity wastewater. This makes the EF process a valuable method for the removal of oil emulsions that occurred after accidents of oil spills in seawaters and floating suspended colloids from industrial wastewater. The concept of insoluble electrode application in EF is an appealing idea that can be used in oil separation as well as in other separation processes (Hosny, 1996).

There are also some disadvantages of the EF process. Compared to other flotation technologies, the energy consumption of the EF process is higher, which limits its widespread application on an industrial scale. However, the effect of this disadvantage can be lessened by using renewable energy sources. In certain applications, the EF process requires the use of specific electrodes (Pt, Ti-coated IrO_2/RuO_2/Ta_2O_5) that can be costly. Also, lifetime (or service time) and stability of the electrodes must be considered since the operating conditions of EF and properties of wastewater affects the electrode lifetime. Therefore, a lot of research has been conducted on the production of new electrodes for the lifetime and stability of the electrodes (Chen and Chen, 2005; Jia et al., 2010). The DSA (the dimensionally stable anodes) or metal mixed oxide electrodes are shown to possess high stability and lifetime, but a perfect electrode to meet all the needs for electrooxidation and electroflotation processes has not yet been made.

Sodium chloride (NaCl) is widely used to improve the conductivity of water or wastewater for electrochemical treatment. Besides its contribution to the ionic transport of electric charges, chloride ions have been found to greatly reduce the adverse effects of ions such as bicarbonate and sulfate. Also, energy consumption is reduced with the use of NaCl, thereby reducing energy cost. However, the EF process leads to greater production of chlorine gas on the anode due to high chloride ions in treated wastewaters. The current efficiency of the EF process decreases due to the evolution of Cl_2 gas. In addition, it should be noted that the generated chlorine gas is highly toxic and corrosive. The metallic materials used in the EF process can corrode, which is harmful to the public.

The hydrogen gas produced in the EF applications is a major concern due to its flammability. Therefore, the pressure exerted by hydrogen bubbles and the temperature during EF reactions should be taken into consideration for the design of the EF system.

Moreover, the formation of polymeric films (electrode fouling or passivation) due to deposition such as $CaCO_3$ and $Mg(OH)_2$ etc. on the electrode surfaces inhibits the extended use of the electrodes in the EF process.

EF process releases hydroxyl and hydrogen ions because of anode and cathode reactions. When a neutral solution is electrolyzed, in the vicinities of the hydrogen-evolving electrode (cathode) and near the oxygen-evolving electrode (anode) are set for a pH gradient because of alkaline and acid regions, respectively. The pH values at different parts of the solution depend on the amount of intermixing allowed. Thus, when hydrogen and oxygen gas are separated during electrolysis, as a result of the pH-time effect occurrence, the pulp progressively becomes more alkaline or acidic, depending on whether hydrogen or oxygen is generated. Many flotation parameters, particularly collector-mineral interaction, and bubble size may be affected by the pH changes during EF. Therefore, the changes in pH during the EF process may be difficult to control.

8.4 SIGNIFICANT OPERATIONAL PARAMETERS OF EF PROCESS

Pollutant removal efficiency and energy and/or chemical consumption determine the performance of an EF process. The efficiency of pollutant removal largely depends on the size of the produced bubbles. The power consumption relates to the cell configuration, the electrode materials, and the operating conditions such as current density and water conductivity, etc. When charging the solid particles, the opposite zeta potential is suggested for the bubbles (Chen, 2004).

8.4.1 Effect of Bubbles Size

The interactions among the process variables in the EF process influence the electrolytic gas evolution. The bubble evolution is affected by current density, electrode morphology, electrolyte and electrolyte surface tension. In a cell where the electrolyte is stirred and electrical current applied, the bubbles reduce the effective conductivity of the electrolyte thereby increasing the ohmic losses. In addition, proper mixing of the electrolyte during the process is very important.

The interactions among the process variables in the EF process influence a dynamic phenomenon called electrolytic gas evolution. The bubble evolution is affected by current density, electrode morphology, electrolyte and electrolyte surface tension. In a cell where the electrolyte is stirred and electrical current applied, the bubbles reduce the effective conductivity of the electrolyte thereby increasing the ohmic losses. In addition, proper mixing of the electrolyte during the process is very important.

In all EF applications, the size of generated bubbles is the most important parameter affecting the removal efficiency of the process. Smaller bubbles have a longer residence time in the system, a larger surface area, and are more likely to adhere to solids after a collision (de Rijk and den Blanken, 1994). Consequently, the treatment process is optimized by generating bubbles with the smallest diameter possible. The most important process variables affecting the size of the bubbles generated during EF are voltage, current density, and ionic strength. The equivalent diameter of bubbles is measured by a digital image analysis method such as high-magnification microscopy. Microbubbles produced by EF are used in separation applications like mineral processing industry and also in the treatment of water and wastewater. As the bubbles move up in a water column, particles collide with each other, attach to the bubbles, and are brought to the surface of the column. The removal efficiency of the undesirable particle is intensely affected by the bubble size. In general, smaller bubble size enables a better separation. EF process generates H_2 and O_2 gas bubbles at anode and cathode through the electrolysis of water according to Eqs. 8.1–8.3. The oxygen and hydrogen gas are dissolved into the liquid surrounding the electrodes; when the liquid becomes supersaturated with gas, bubbles begin to form on the electrode surface (Verhaart et al., 1980). The average bubble diameter increases with respect to the current density applied to the EF process. However, there is no particular trend in the bubble diameter as a function of either power or ionic strength of aqueous medium. The volume distribution of the generated bubbles indicates that EF produce bubbles with uniform diameters and narrow ranges. In addition, the zeta potential of both particles and bubbles in the separation process by flotation can be of great significance. Thus, if the particles to be removed are negatively charged, the optimum removal efficiency can be achieved by using the bubbles with a slightly positive charge (Jiménez et al., 2010).

Mansour et al. (2007) investigated the effects of the EF process on the hydrodynamic and bubble size distribution of some operating parameters. The diameter and height of a cylindrical EF reactor were 0.03 and 0.5 m, respectively. The Ti/RuO_2 anode and SS cathode screen insoluble electrodes were equipped in the EF reactor. The cathode was positioned horizontally at the top of the anode and there is 5 mm gap distance between electrodes. The mixture of distilled water glycerine at different concentrations of 0%, 5%, 20% and 20% by weight and ethanol at 5% by volume was used as the model aqueous solutions (changed viscosity from 1.0×10^{-3} to 1.39×10^{-3} kg/ms and surface tension from 72.75 to 102.1 mN/m at 0%–20%). In addition, 1 g/L Na_2SO_4 was added into all solutions, and the experiments were conducted at 20°C and 1 atm. The current density varied from 50 to 500 A/m^2. The relationships between the gas hold ups (ε_g), the bubble rise velocity (V_b), the average bubble diameter with respect to the current density (j) and the viscosity (μ) of the medium were obtained in the following equations:

$$\varepsilon_g = 1.44\times10^{-2}\times j^{0.91}\times\mu^{-0.77}\quad\left(r^2 = 0.98\right) \tag{8.6}$$

$$V_b = 0.568\times j^{0.68}\times\mu^{-1.07}\quad\left(r^2 = 0.95\right) \tag{8.7}$$

$$d_b = 6.275\times j^{0.46}\times\mu^{0.48}\quad\left(r^2 = 0.96\right) \tag{8.8}$$

Decreasing current density, surface tension and viscosity cause smaller bubbles. Gas hold up and bubble rise velocity generally increase by increasing current density and decreasing viscosity. When the current density increases from 60 to 180 A/m^2, the bubble size increases from 80 to 110 μm at the mixing solution of 10%–20%.

In a study conducted by Jiménez et al. (2010), the formation and average size of bubbles for stainless steel (SS) and titanium (Ti) electrodes in the range of 10–35 A/m^2 were compared. The EF cell was operated in batch mode. A polished stainless steel or titanium cathode is placed at the bottom of the cell and a platinum anode is above the cathodes. As shown in Figure 8.4 above, the results are affected by the electrode material and current density, but the shape of the curves is nearly identical for both electrodes. Figure 8.4a reveals that the low current densities result in the

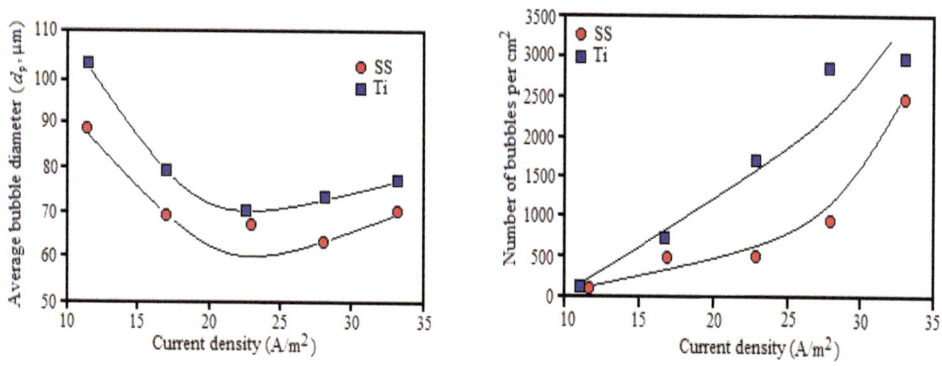

FIGURE 8.4 Effects of current density and cathode material on (a) average bubble diameter (b) production of bubbles adopted from. (Adapted with permission from Jiménez et al. (2010).)

formation of large-sized hydrogen bubbles that should be avoided for preventing non-effective EF separations. Higher current densities lead to larger numbers of small hydrogen bubbles as illustrated in Figure 8.4b. The SS cathode enables the production of smaller hydrogen bubbles compared to Ti cathode at 10–35 A/m^2; however, the Ti cathode produces a larger number of hydrogen bubbles than SS at higher current density (>20 A/m^2). The average bubble diameter at low current density (<20 A/m^2) is about >65 µm, while it is about 65–75 µm at current density above 20 A/m^2 for both electrodes. On the other hand, the change in average bubble diameter is very small in higher current densities (about >20 A/m^2), which can be explained by two different mechanisms: (a) the formation of a large number of bubbles can cause collisions thereby leading to an improvement in the coalescence process and (b) when there are too many bubbles in the system, it is possible that some bubbles are in different positions in EF reactor. At very large current densities, an apparent increase in the size of bubbles is observed, which can be simultaneously explained by actual coalescence due to the promotion of collisions and apparent coalescence due to the overlapping of bubbles in different positions.

Since the bubble evolution is generally rapid in the EF process, special techniques such as high-speed cameras and microscopy are used for its measurement. Bubble size distribution in an EF reactor with a Ti/RuO_2mesh anode and stainless steel cathode was measured by image analysis (high-speed digital camera) and the laser diffraction methods (Gonzales et al., 2012). In the study, EF conditions were to be current density of 494 A/m^2, pH of 7 and electrolyte concentration of 0.1 M. The analysis revealed that the EF process was affected by current density and pH. Higher current densities lead to the formation of a larger number of bubbles. Additionally, higher pH values also promoted the production of bubbles. The Sauter mean bubble diameter (H_2/O_2) was measured to be in the range of 82–92 µm with image analysis and 40–58 µm with laser diffraction method. It was observed that the size of hydrogen bubbles was larger in an acidic medium compared to neutral or alkaline media. Similarly, the size of oxygen bubbles was smaller in acidic medium and increase with increasing pH (Brandon et al., 1985).

The direct visualization results indicated that the hydrogen bubble size in an EF reactor with Pt wire electrodes was in the range of 15 and 23 µm in diameter at the set conditions (Pt wire diameter of 90, 120 and 190 µm, j=150–350 A/m^2, 0.2 M Na_2SO_4) (Sarkar et al., 2010). Before reaching the steady state diameter (30–50 µm) in the bulk, the bubble size showed a rapid up to an eight-fold increase in volume in the first few millimeters rise due to the transfer of dissolved hydrogen into growing bubble and passing through the supersaturated electrolyte by dissolved hydrogen gas. Moreover, Sarkar et al. (2010) found that a significant proportion of produced hydrogen on the cathodes can be dissolved in the solution. The diameter of bubble was determined as a function of hydrogen production rate, bubble nucleation rate and dissolved gas concentration.

It was found that the detached hydrogen bubble diameter varied between 15 and 23 μm at 150–350 A/m^2 for the EF reactor with Pt wire electrodes (90, 120 and 190 μm in diameter). The superficial upward liquid velocity of the bubble was in the range of 1.5–7.1 mm/s (Sarkar et al., 2010).

Consequently, the screen electrodes used in the EF process produced smaller size bubbles than plate electrodes, and the smooth screen worked better for the production of smaller bubbles compared to rough screen, while crystalline and amorphous plates produced bubbles with almost identical sizes. Average hydrogen bubble diameters increased with an increase in current density for smooth screens and plate electrodes, but decreased for rough screen electrode. In general, while the size of hydrogen bubbles produced at cathode surface during electrolysis of water varies in the range of 5–65 μm, the oxygen bubbles are larger with a size in the range of 15–95 μm.

8.4.2 Effect of PH

The size distribution and volume of formed gas bubbles in the evolution of hydrogen and oxygen in EF can be affected by the pH of the electrolyte medium (Jiménez et al., 2010). The size of hydrogen bubbles in the EF process is generally smallest at neutral pH (Chen, 2004). Larger bubbles and lower production are the results of highly acidic and alkaline media. Increasing pH causes the larger size distribution of oxygen bubbles. Bubble size which fits well with a log-normal distribution is affected by cathode and anode materials (Fukui and Yuu, 1984).

Alam et al. (2017) determined the mean bubble size by image analysis in the EF process and it was varied in the range of 32.7–68.6 μm under various operating conditions (Ti/IrO_2 anode and stainless steel 316 cathode, 10–150 A/m^2, 0.1–0.5 M KNO_3 10–30 mg/L of frother (Tennafroth 250)). The smallest mean bubble size of hydrogen with 33.9 μm was achieved at near neutral pH, but it was 65 μm at pH 2 and 50 μm at pH 12 (Figure 8.5a). Acidic environment results from the formation of hydrogen bubbles with larger volume (da Cruz et al., 2016). Recent studies have shown a similar trend for mean bubble size of hydrogen with respect to varying pH in the EF process (e.g., da Cruz et al., 2016; Jiménez et al., 2010). On the other hand, the produced oxygen on the Ti-IrO_2 plate achieved a mean bubble size of 32.7 μm in an acid medium and increased with rise in pH, which could be the result of OH^- in the solution. The trend was compatible with previous studies (e.g., Brandon et al., 1985; da Cruz et al., 2016). The size of the bubbles generated by the EF process with respect to pH for different materials of electrodes is shown in Table 8.3.

The fraction of fine oxygen bubbles increased with decreasing pH as can be seen from Figure 8.5b which illustrates the log-normal distribution of oxygen bubble size at different pH values. The production of bubbles with the uniform size is one of the benefits of EF (Ketkar et al., 1991). Figure 8.5b indicates that produced bubbles had a 50 μm difference in diameter between pH of 7 and 12. Figure 8.5a illustrates the variations of the hydrogen bubble size. The finer hydrogen bubbles were generated at a pH of 7 while they have a wider range at pH of 2. Grau and Heiskanen (2005) determined that the flotation of coarse particles could be positively affected by a broad range of bubble sizes. However, pH changes the oxidizing and reducing conditions of the pulp thereby affecting the selectivity of the flotation of minerals (Göktepe, 2002). Hence pH should be carefully adjusted to facilitate optimal recovery of the target minerals.

Llerena et al. (1996) investigated the effect of pH on the recovery of sphalerite fines using an EF process and adjusted the pH in the range of 3–4 by using a buffer solution. They indicated that smaller hydrogen bubbles (16 ± 2 μm) were produced within the pH range and the recovery of sphalerite fines was optimal. The mean hydrogen bubble diameter was 27 μm at a pH of 6 while that was 23 μm at a pH of 2 when 500 A/m^2 current density was applied to a 304 SS wire mesh. Additionally other researchers pointed out that, the distribution of bubble size is dependent on the solution as well as the electrode material used in the process (Chen, 2004). Table 8.4 shows the bubble size distribution hydrogen and oxygen with respect to pH and electrode material. In the separation of hydrogen and oxygen, the pH increases in the cathode chamber and pH decrease in the anode chamber is very fast in the absence of buffer solutions. The recovery efficiency of oxygen is half that of hydrogen in

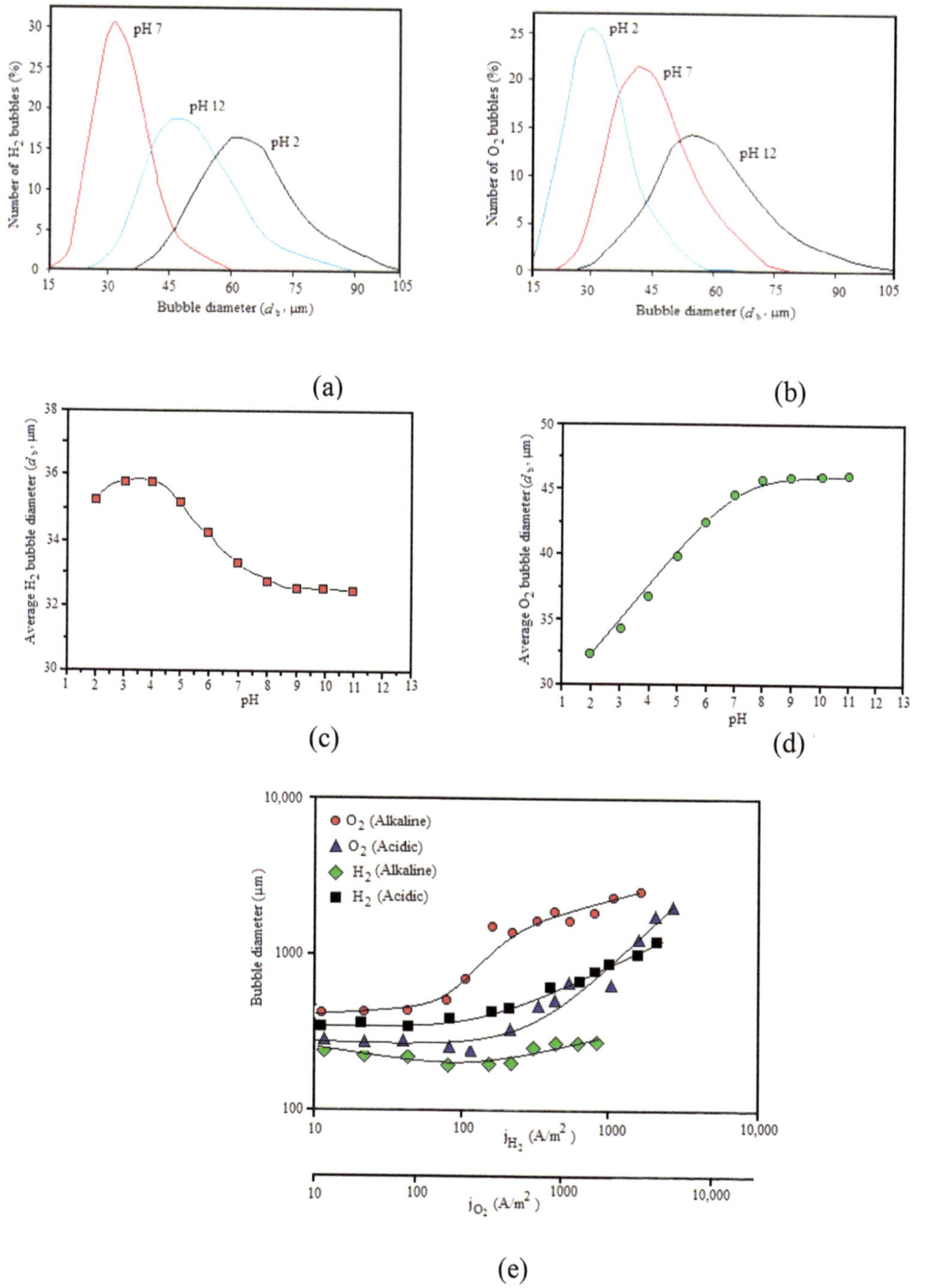

FIGURE 8.5 Log-normal distributions of: (a) H_2 (b) O_2 bubbles with different pH values. (Adapted with permission from Alam et al. (2017).) The average Sauter's diameter of O_2 and H_2 bubbles for an "U" shape EF cell with (c) meshed SS cathode and (d) Ti/RuO_2 anode. (Adapted with permission from da Cruz et al. (2016).) (e) Break off diameters of bubbles with respect to current density for Pt electrode. (Adapted with permission from Janssen and Hoogland (1973).)

TABLE 8.3
The Size of Gas Bubbles at Various pHs and Electrode Materials

Bubble	d_p (µm)	Electrode (Anode/Cathode)	j (A/m²)	pH	Electrolyte	Refs.
H_2	32	Ti-RuO_2/SS (meshed)	80	11	0.1 mol/L NaCl	da Cruz et al. (2016)
O_2	45					
H_2+O_2	30–139	Pt/SS	-	-	U=30V, i=0.9 A, pH=11, Dowfroth 200 (frother) Guar gum (depressant)	Montes-Atenas et al. (2010)
H_2+O_2	25–116	Ti-RuO_2/SS (screen)	35–500	-	0.007 M Na_2SO_4	Mansour et al. (2007)
H_2+O_2	15–45	Ti-IrO_x-Sb_2O_5-SnO_2/Ti	50-40	6.78	Conductivity=705 mS/cm	Chen et al. (2002)
O_2	17.1–37.9	Graphite/graphite	11–288 mA		0.01–0.1 M Na_2SO_4	Burns et al. (1997)
H_2	19.3–37.7		12–64 mA			
H_2	30–37	SS/SS	225–400	10	-	
H_2	28–39	SS mesh/SS mesh	125–375	9	-	Ketkar et al. (1991)
O_2	38–50	(75 mm)				
H_2	22–34	SS/SS	125–375	9	-	Ketkar et al. (1991)
O_2	42–48	Pt/Pt				
O_2	38–50	Pt plate	125–370	9	-	Ketkar et al. (1988)
O_2	42–48	Pt mesh (60 µm)			-	
H_2	17	Pt/Pt	250	7	-	
H_2	17	Fe/Fe				
H_2	25	Graphite/graphite				
O_2	30	Pt/Pt				

d_p, average bubble diameter; SS, stainless steel.

TABLE 8.4
Distribution of the Bubble Size with Respect to the Solution pH and the Material of Electrode

	Hydrogen (µm)		Oxygen (µm)	
pH	Pt	Fe	C	Pt
2	45–90	20–80	18–60	15–30
7	5–30	5–45	5–80	17–50
12	17–45	17–60	17–60	30–70

proportion to the amount of gas produced at a given current density, which was confirmed by the gas sparing of H_2 and O_2.

Ketkar et al. (1988) used a stainless steel mesh (wire diameter of 50–215 µm) and plate (mirror polished) cathodes to produce hydrogen bubbles at the current densities in the range of 125–370 A/m² and used a photographic technique to measure the size of bubbles produced. The hydrogen bubbles produced in an alkaline medium (pH of 9) were photographed at the cathode surface. The average diameters of bubbles were in the range of 28–49 and 22–34 µm produced by mesh and plate cathodes, respectively. Hydrogen bubble diameter decreased with increasing current density and decreasing wire diameter. The smallest size of the hydrogen bubble (22 µm) was observed when a stainless steel plate was used. In addition, the oxygen bubbles were larger compared to hydrogen bubbles in the alkaline media.

The average bubble size in an EF cell ("U" shape, volume of 200 mL and internal diameter of 7 mm) using Ti/RuO_2 anode and SS cathode (or mesh/plate) was measured by the laser diffraction method and image analysis (da Cruz et al., 2016). Each electrode area was 1 cm^2 and 0.10 mol/L NaCl solution was added as electrolyte. Experiments were conducted at 25°C and without agitation following the bubbling of N_2 to remove the dissolved oxygen. As can be seen in Figure 8.5c, increasing pH leads to a fast decrease in the average diameter of the H_2 bubble. The bubble diameters varied from 35.7 µm at a pH of 3 up to 32.15 µm at a pH of 11. In an alkaline environment, a smaller volume of H_2 bubbles is nucleated because of the substantial decrease in the concentration of H^+ ion, and bubble size gradually decreases with increasing pH. On the other hand, the average bubble diameter as can be seen in Figure 8.5d increases with increasing pH up to 8, after this point the bubble size remains approximately constant, with a diameter of 45 µm. In addition, the average bubble diameter of O_2 with around 32 µm is greater than the average diameter of H_2 bubbles.

At a larger-volume reactor (1 L), the bubble diameter (average of all produced bubbles on both electrodes, including H_2 on the cathode and O_2 and eventually Cl_2 on anode) decreased from about 62 µm at a pH of 2–37 µm at a pH of 12. Moreover, the H_2 volume generated in the EF cell is twice of the oxygen volume. The number of hydrogen bubbles is much greater than oxygen bubbles due to their smaller size. In addition, the mesh cathode produced smaller bubbles compared to the plate cathodes. This is due to lower surface tension on mesh-shape electrodes, which promote the generation of bubbles with smaller size.

Janssen and Hoogland (1973) reported that an EF system equipped with a platinum (Pt) disc electrode having a diameter of 11.2 mm (surface area of 100 mm^2) at the current density of 400 A/m^2 generated hydrogen bubbles with a size of 26 µm. In an acidic medium, the average hydrogen bubble diameters were 36 and 85 µm at current densities of 40 and 1000 A/m^2, respectively. However, in an alkaline medium, average hydrogen bubble diameter were similar (i.e. 26 µm at 40 A/m^2 and 28 µm at 1000 A/m^2) even if the current densities were dramatically increased from 40 to 1000 A/m^2. While the diameter of free bubbles at current density of <100 A/m^2 was founded practically independent of the current density, above 300 A/m^2 the coalescence of bubbles occurred intensely with the exception of hydrogen bubbles in an alkaline solution (Figure 8.5e). On the other hand, while the size of oxygen bubbles remained almost the same at lower current densities (typically <100 A/m^2); above current densities of 100 A/m^2 caused an increase in the average diameter of free oxygen bubbles with increasing current density both in acidic and alkaline media. While no coalescence of bubbles occurred at low current density, high current density (>300 A/m^2) often led to coalescence of bubbles. When the current density was increased from 80 to 2000 A/m^2, the bubble diameter increased from 25 to 95 µm at acidic medium and from 40 to 210 µm at alkaline medium, respectively.

8.4.3 Effect of Current Density

The most important operational parameter of an EF process is current density. The higher the current density applied, the more gas bubbles are produced (Sillanpää and Shestakova, 2017). In addition, the current density influences the formation and the properties of bubbles. A decrease in bubble size distribution was achieved by increasing current density (Shen et al., 2003). Furthermore, the bubble swarm flux, the available number of gas bubbles per second per unit cross-section area of the flotation cell, also has an important role in the recovery of fine particles and colloidal matter (Ketkar et al., 1991). As the bubbles produced an increase in size, the particle size adhering to these bubbles also increases (Sillanpää and Shestakova, 2017).

8.4.4 Effect of Surfactant and Flocculant Concentration

Hydrophobicity as well as the ratio of surfactant to metal ion in the solution is an important factor for the floatability of chemical species in liquid streams. If the surfactant has frothier properties, its amount should be at least stoichiometric (Casqueira et al., 2006). In general, a small amount of excess surfactant is incorporated into solution to ensure the maximum removal of the metallic ions.

8.5 CONCLUSION

The EF process could be better than conventional flotation processes as the design of an effective electroflotation reactor has new and superior electrode materials. Thus, the effectiveness of electroflotation in the removal of pollutant from the wastewaters of hydrometallurgical plants, chemical and petroleum industries effluents and in other fields is economically favorable. With its ability to combine with other pollutant removal processes, EF will find more applications in the future as an efficient wastewater removal process for diverse contaminants. The possibility of performing an effective flotation by controlling the parameters that affect electroflotation makes electroflotation stand out from other flotation techniques. The ease of combining electroflotation with other treatment processes will allow it to be used effectively in the coming years.

REFERENCES

Alam, R., Shang, J.Q., Khan, A.H., 2017. Bubble size distribution in a laboratory-scale electroflotation study. *Environmental Monitoring and Assessment* 189, 193. https://doi.org/10.1007/s10661-017-5888-4.

Alhattab, M., Brooks, M.S.-L., 2017. Dispersed air flotation and foam fractionation for the recovery of microalgae in the production of biodiesel. *Separation Science and Technology* 52, 2002–2016. https://doi.org/10.1080/01496395.2017.1308957.

Atarah, J.J.A., 2011. *The use of Flotation Technology in Produced Water Treatment in the Oil & Gas Industry*, University of Stavenger.

Barret, F., 1975. Electroflotation-development and application. *Water Pollution Control* 74, 59–62.

Brandon, N.P., Kelsall, G.H., Levine, S., Smith, A.L., 1985. Interfacial electrical properties of electrogenerated bubbles. *Journal of Applied Electrochemistry* 15, 485–493. https://doi.org/10.1007/BF01059289.

Burns, S.E., Yiacoumi, S., Tsouris, C., 1997. Microbubble generation for environmental and industrial separations. *Separation and Purification Technology* 11, 221–232. https://doi.org/https://doi.org/10.1016/S1383–5866(97)00024–5.

Casqueira, R.G., Torem, M.L., Kohler, H.M., 2006. The removal of zinc from liquid streams by electroflotation. *Minerals Engineering* 19, 1388–1392. https://doi.org/https://doi.org/10.1016/j.mineng.2006.02.001.

Chen, G., 2004. Electrochemical technologies in wastewater treatment. *Separation and Purification Technology* 38, 11–41. https://doi.org/http://dx.doi.org/10.1016/j.seppur.2003.10.006.

Chen, X., Chen, G., 2005. Stable Ti/RuO_2–Sb_2O_5–SnO_2 electrodes for O_2 evolution. *Electrochimica Acta* 50, 4155–4159. https://doi.org/https://doi.org/10.1016/j.electacta.2005.01.032.

Chen, X., Chen, G., Yue, P.L., 2002. Novel electrode system for electroflotation of wastewater. *Environmental Science & Technology* 36, 778–783. https://doi.org/10.1021/es011003u.

da Cruz, S.G., Dutra, A.J.B., Monte, M.B.M., 2016. The influence of some parameters on bubble average diameter in an electroflotation cell by laser diffraction method. *Journal of Environmental Chemical Engineering* 4, 3681–3687. https://doi.org/https://doi.org/10.1016/j.jece.2016.05.017.

Dafnopatidou, E.K., Lazaridis, N.K., 2008. Dyes removal from simulated and industrial textile effluents by dissolved-air and dispersed-air flotation techniques. *Industrial & Engineering Chemistry Research* 47, 5594–5601. https://doi.org/10.1021/ie071235n.

de Rijk, S.E., den Blanken, J.G., 1994. Bubble size in flotation thickening. *Water Research* 28, 465–473. https://doi.org/https://doi.org/10.1016/0043–1354(94)90284–4.

Eckenfelder, W.W., 2000. *Industrial Water Pollution Control, Environmental Engineering and Water Resources Series*, McGraw-Hill.

Edzwald, J.K., Walsh, J.P., Kaminski, G.S., Dunn, H.J., 1992. Flocculation and air requirements for dissolved air flotation. *American Water Works Association* 84, 92–100.

Fukui, Y., Yuu, S., 1984. Development of apparatus for electro-flotation. *Chemical Engineering Science* 39, 939–945. https://doi.org/https://doi.org/10.1016/0009–2509(84)87001–3.

Gochin, R.J., Solari, J., 1983. The role of hydrophobicity in dissolved air flotation. *Water Research.* https://doi.org/10.1016/0043–1354(83)90234–8.

Göktepe, F., 2002. Effect of pH on pulp potential and sulphide mineral flotation. *Turkish Journal of Engineering and Environmental Sciences* 26, 309–318.

Gonzales, L.G.V., Veneu, D.M., Monte, M.B. de M., Torem, M.L., 2012. Measurement and analysis of micro-bubbles produced in electroflotation, in: *67th ABM International Congress*, Rio de Jenario, Brazil.

Grau, R.A., Heiskanen, K., 2005. Bubble size distribution in laboratory scale flotation cells. *Minerals Engineering* 18, 1164–1172. https://doi.org/https://doi.org/10.1016/j.mineng.2005.06.011.

Haxhiaj, A., Fan, M., Haxhiaj, B., 2016. The management of lead concentrate acquisition in "Trepca" BT – Advances in molten slags, fluxes, and salts, in: *Proceedings of the 10th International Conference on Molten Slags, Fluxes and Salts 2016*, Reddy, R.G., Chaubal, P., Pistorius, P.C., Pal, U. (Eds.), Springer International Publishing, Cham, pp. 1351–1358.

Hodouin, D., Bazin, C., Gagnon, E., Flament, F., 2000. Feedforward–feedback predictive control of a simulated flotation bank. *Powder Technology* 108, 173–179. https://doi.org/https://doi.org/10.1016/S0032–5910(99)00217–X.

Hosny, A.Y., 1996. Separating oil from oil-water emulsions by electroflotation technique. *Separations Technology* 6, 9–17. https://doi.org/https://doi.org/10.1016/0956–9618(95)00136–0.

Jameson, G.J., 1999. Hydrophobicity and floc density in induced-air flotation for water treatment. *Colloids and Surfaces A: Physicochemical and Engineering Aspects* 151, 269–281. https://doi.org/https://doi.org/10.1016/S0927–7757(98)00503–2.

Janssen, L.J.J., Hoogland, J.G., 1973. The effect of electrolytically evolved gas bubbles on the thickness of the diffusion layer—II. *Electrochimica Acta* 18, 543–550. https://doi.org/https://doi.org/10.1016/0013–4686(73)85016–9.

Jia, J., Li, X., Chen, G., 2010. Stable spinel type cobalt and copper oxide electrodes for O_2 and H_2 evolutions in alkaline solution. *Electrochimica Acta* 55, 8197–8206. https://doi.org/https://doi.org/10.1016/j.electacta.2010.04.026.

Jiménez, C., Talavera, B., Sáez, C., Cañizares, P., Rodrigo, M.A., 2010. Study of the production of hydrogen bubbles at low current densities for electroflotation processes. *Journal of Chemical Technology & Biotechnology* 85, 1368–1373. https://doi.org/https://doi.org/10.1002/jctb.2442.

Ketkar, D.R., Mallikarjunan, R., Venkatachalam, S., 1988. Size determination of electrogenerated gas bubbles. *Journal of the Electrochemical Society of India* 37, 313–318.

Ketkar, D.R., Mallikarjunan, R., Venkatachalam, S., 1991. Electroflotation of quartz fines. *International Journal of Mineral Processing* 31, 127–138. https://doi.org/https://doi.org/10.1016/0301–7516(91)90009–8.

Kyzas, G.Z., Matis, K.A., 2016. Electroflotation process: A review. *Journal of Molecular Liquids* 220, 657–664. https://doi.org/https://doi.org/10.1016/j.molliq.2016.04.128.

Llerena, C., Ho, J.C.K., Piron, D.L., 1996. Effects of pH on Electroflotation of Sphalerite. *Chemical Engineering Communications* 155, 217–228. https://doi.org/10.1080/00986449608936664.

Lubetkin, S.D., 1995. The fundamentals of bubble evolution. *Chemical Society Reviews* 24, 243–250. https://doi.org/10.1039/CS9952400243.

Lundgren, H., 1976. Theory and practice of dissolved air flotation. *Journal of Filtration and Separation* 13, 24–28.

Mansour, L.B., Chalbi, S., 2006. Removal of oil from oil/water emulsions using electroflotation process. *Journal of Applied Electrochemistry* 36, 577–581. https://doi.org/10.1007/s10800-005-9109–4.

Mansour, L.B., Chalbi, S., Ksentini, I., 2007. Experimental study of hydrodynamic and bubble size distributions in electroflotation process. *Indian Journal of Chemical Technology* 14.

Matis, K.A., Lazaridis, N.K., 2002. Flotation techniques in water technology for metals recovery: Dispersed-air vs. dissolved-air flotation. *Journal of Mining and Metallurgy A: Mining* 38, 1–27.

McKee, D.J., 1991. Automatic flotation control – a review of 20 years of effort. *Minerals Engineering* 4, 653–666. https://doi.org/https://doi.org/10.1016/0892–6875(91)90055–Z.

Montes-Atenas, G., Garcia-Garcia, F.J., Mermillod-Blondin, R., Montes, S., 2010. Effect of suspension chemistry onto voltage drop: Application to electro-flotation. *Powder Technology* 204, 1–10. https://doi.org/https://doi.org/10.1016/j.powtec.2010.06.024.

Painmanakul, P., Sastaravet, P., Lersjintanakarn, S., Khaodhiar, S., 2010. Effect of bubble hydrodynamic and chemical dosage on treatment of oily wastewater by Induced Air Flotation (IAF) process. *Chemical Engineering Research and Design* 88, 693–702. https://doi.org/https://doi.org/10.1016/j.cherd.2009.10.009.

Sarkar, M.S.K.A., 2012. *Electroflotation: Its Application to Water Treatment and Mineral Processing*, The University Newcastle.

Sarkar, M.S.K.A., Evans, G.M., Donne, S.W., 2010. Bubble size measurement in electroflotation. *Minerals Engineering* 23, 1058–1065. https://doi.org/https://doi.org/10.1016/j.mineng.2010.08.015.

Shah, M.P., 2020. *Advanced Oxidation Processes for Effluent Treatment Plants*, Elsevier.

Shah, M.P., 2020. *Microbial Bioremediation and Biodegradation*, Springer.

Shah, M.P., 2021. *Removal of Emerging Contaminants through Microbial Processes*, Springer.

Shammas, N.K., Bennett, G.F., 2010. *Principles of Air Flotation Technology BT – Flotation Technology*: Volume 12, in: Wang, L.K., Shammas, N.K., Selke, W.A., Aulenbach, D.B. (Eds.), Humana Press, Totowa, NJ, pp. 1–47. https://doi.org/10.1007/978-1-60327-133-2_1.

Shen, F., Chen, X., Gao, P., Chen, G., 2003. Electrochemical removal of fluoride ions from industrial wastewater. *Chemical Engineering Science* 58, 987–993. https://doi.org/https://doi.org/10.1016/S0009–2509(02)00639–5.

Sides, P.J., 1986. *Phenomena and Effects of Electrolytic Gas Evolution BT – Modern Aspects of Electrochemistry*: Volume 18, in: White, R.E., Bockris, J.O., Conway, B.E. (Eds.), Springer US, Boston, MA, pp. 303–354. https://doi.org/10.1007/978-1-4613-1791-3_6.

Sillanpää, M., Shestakova, M., 2017. *Chapter 2: Electrochemical Water Treatment Methods BT – Electrochemical Water Treatment Methods*, Butterworth-Heinemann, pp. 47–130. https://doi.org/https://doi.org/10.1016/B978–0–12–811462–9.00002–5.

Tchobanoglous, G., Burton, F.L., Stensel, H.D., Eddy, M.&, 2003. *Wastewater Engineering: Treatment and Reuse, McGraw-Hill Series in Civil and Environmental Engineering*, McGraw-Hill.

Verhaart, H.F.A., de Jonge, R.M., van Stralen, S.J.D., 1980. Growth rate of a gas bubble during electrolysis in supersaturated liquid. *International Journal of Heat and Mass Transfer* 23, 293–299. https://doi.org/https://doi.org/10.1016/0017–9310(80)90117–9.

Wang, L.K., Shammas, N.K., Selke, W.A., Aulenbac, D.B. (Eds.), 2010. *Flotation Technology*, Humana Press. https://doi.org/10.1007/978–1–60327–133–2.

Wang, L.K, Shammas, N.K., Selke, W.A., Aulenbach, D.B., 2010. *Gas Dissolution, Release, and Bubble Formation in Flotation Systems BT – Flotation Technology*: Volume 12, in: Wang, L.K., Shammas, N.K., Selke, W.A., Aulenbach, D.B. (Eds.), Humana Press, Totowa, NJ, pp. 49–83. https://doi.org/10.1007/978-1-60327-133-2_2.

Yoon, R.-H., 1993. Microbubble flotation. *Minerals Engineering* 6, 619–630. https://doi.org/https://doi.org/10.1016/0892–6875(93)90116–5.

Zabel, T., Melbourne, J.D., 1980. Flotation, in: *Developments in Water Treatment – I*, Lewis, W.M. (Ed.), Applied Science Publishers Ltd, London, UK.

Zouboulis, A.I., Avranas, A., 2000. Treatment of oil-in-water emulsions by coagulation and dissolved-air flotation. *Colloids and Surfaces A: Physicochemical and Engineering Aspects* 172, 153–161. https://doi.org/https://doi.org/10.1016/S0927–7757(00)00561–6.

Zouboulis, A.I., Matis, K.A., Stalidis, G.A., 1992. *Flotation Techniques in Waste Water Treatment BT – Innovations in Flotation Technology*, in: Mavros, P., Matis, K.A. (Eds.), Springer, Dordrecht, Netherlands, pp. 475–497. https://doi.org/10.1007/978-94-011-2658-8_25.

9 Removal of Emerging Contaminants Present in Wastewater by Electrocoagulation Process

Ramya Sankar and V. Sivasubramanian
National Institute of Technology

CONTENTS

9.1 INTRODUCTION

Global consideration in technologies and the market requirement for energy, natural resources, and other goods are the significant contributors of environmental pollutants. Several environmental issues have prioritized emission, prevention, and mitigation intending to move towards sustainability. The pollutants are classified according to their toxicity levels and the pollutants which are highly endangered to the environment are grouped into "blacklist" pollutants. These blacklist contaminants are highly toxic, persist or bioaccumulate in the environment, and hence necessary measures should be taken to prevent contamination. These contaminants include organophosphorus, organohalogen, poisonous metals, and

DOI: 10.1201/9781003164982-9

their compounds and the treatment of effluents containing these pollutants are more complex and hard. Therefore, more efficient technologies for waste management should be adopted [1].

Waste management procedures should be prioritized by the decrease in the use of blacklist pollutants or establish a strict set of laws on their use, and the subsequent steps to be followed are the elimination, reuse, recycling, recovery and safe disposal of pollutants. Hence, to ensure environmental sustainability, it is important to treat life-threatening pollutants using appropriate treatment techniques before expelling them into the surroundings. In general, various technologies are used for the treatment processes which utilize different features of physical, chemical, biological, photolytic, and physiochemical properties for pollutant removal. However, the allowable limits of these contaminants challenge the cost-effectiveness of the process, which includes the complexity of the process, mixed pollutants, levels of micropollutants, need for recovery and recycling, etc. among various treatment technologies electrolytic treatments can deal with these challenges and often provide a straightforward solution, particularly for the industrial effluents. The electrochemical remediation techniques are effective and environmentally friendly as the main reagent involved is electron exchange [1,2]. This chapter discusses the use of electricity to treat various types of wastewater via a recognized treatment technique called electrocoagulation for pollutant recovery.

The selective separation of pollutants is important in both chemical and biochemical industries, which contribute almost 40%–70% of the total capital and operating expenditure. The separation, removal, and remediation of pollutants from complex mixed pollutant systems with cost-effectiveness and sustainability are a risk. The main problems associated with these are the dilute concentration of pollutants even at micro levels and the presence of many end products which may hinder the separation process.

Chemical coagulation and flocculation are often used in water purification systems for removing selective pollutants using specific coagulant. The principle point of this procedure is to improve molecule partition in the auxiliary treatment procedures, for example, sedimentation, filtration, or buoyancy. Compound coagulation and flocculation are utilized as a part of both civil and mechanical wastewater treatment frameworks since they help in the evacuation of various sorts of poisons from the water streams. The basis of chemical coagulation involves a two-stage process; (a) addition of a suitable coagulant that destabilizes charged or colloidal particles; (b) agglomeration of fine particles into larger structures ("flocs") which can be settled out of solution.

Coagulant dosages vary in a wide range to maximize the removal efficiency of pollutants using minimum doses at optimum pH. Generally used coagulants are iron or aluminum salts, for example, chlorides and sulfates. The destabilized contaminants and the hydrolyzed products of aluminum and iron salts agglomerate with each other and separated from the effluent either by settling or flotation. Due to destabilization the impurities are trapped and removed along with the amorphous hydroxide precipitates. Among several factors, pH and coagulant dosage have an impact on the charge neutralization and sweep coagulation mechanism for pollutant removal. The microbubbles (H_2 and O_2) released from the cathode surface assist the agglomerated pollutant particle to reach the water surface [2].

9.2 ELECTROCOAGULATION

Electrocoagulation was first proposed before the 20th century due to its simplicity in operation and high-performance efficiency. The first treatment plant was built in 1889 in London for the treatment of sewage, the sewage is first mixed with seawater and further electrolyzed. The principle of EC was first patented by A. Edietrich in 1906 for the treatment of bling water from the ship. In 1906, J.T. Harries received a patent for the treatment of wastewater using aluminum or iron electrodes as sacrificial anodes [5]. In the 1940s, a new device, electronic coagulator, was introduced for the purification of water, which dissolves aluminum ions (from anode) and hydroxyl ions from the cathode into the electrolyte. These two ions will react to form aluminum hydroxides. These hydroxides flocculate and coagulate with the suspended solids in the electrolyte and the purification of wastewater occurs. In 1956, Britain used this process for the treatment of river water using iron electrodes.

During the last decades, various water bodies have been highly polluted by industrial, mining, and urban wastewaters and by surface runoff from agricultural areas, etc. The disposal of wastewater from various industries without proper treatment and the lack of strict actions and regulations for these are also major sources of pollution. Presently, there are various treatment techniques for handling a wide range of pollutants in diverse situations. Coagulation and flocculation are the conventional methods for the treatment of wastewater. In this process, the charge destabilization of the pollutants has occurred with the help of coagulating agents(e.g. alum or ferric chloride) and other additives [6]. The dosing of these will make destabilization and charge speciation of contaminants present in the water sample and form larger aggregates that can be separated physically. This is not a single-stage operation. This process operates through various stages, and it requires extensive land area and a regular supply of chemicals. A huge amount of reaction by-products are formed and they are denser and less prone to breakage, generate a new form of contaminant concentrated waste. The disposal of the produced floc is another issue to the environment. A more clean, cost-effective, and environmentally-friendly method for the treatment of a wide variety of pollutants needs to be incorporated. The electrocoagulation technique of wastewater treatment is a better option for the above-stated problem [7–9].

Wastewater treatment technologies are classified as primary, secondary, and tertiary, which are selected depending on the type and nature of pollutants. For the treatment of organic pollutants, the biological treatment methods are sometimes impractical due to the bio-refractory characteristics of the substrates. Because of this physical-chemical methods are preferably used, but ozonation or oxidation with chlorine dioxides are not always effective for several reasons such as transportation, storage of reactants, and difficulty in safe processing [10]. An alternate to this is the use of electrochemical techniques for the treatment of wastewater. It has several advantages over traditional techniques such as simple operation, no chemical additives, not specific to the pollutants, and environmental compatibility. The electrocoagulation process is a combination of electrochemistry and physiochemical processes. In recent years, the application of electrochemistry for environmental issues has increased significantly, especially for the treatment of organic pollutants. In these studies, various factors affect the performance of electrolytic cells like the nature of electrode materials, electrolyte specifications, etc., and are explained in the next sections [11,12].

In the studies conducted after a few more years, it was further found that electrocoagulation offers many other advantages over other processes potentially suitable for remote areas or emergency conditions. EC uses solid electrodes rather than corrosive metal salts, which are much easier to handle and store. The applied voltage acts as the destabilization agent to bring about the charge speciation of the contaminants present in the wastewater. The electrochemically generated metallic ions of the electrodes form hydrolyzedpolyhydroxide and combine with the destabilized pollutants and the removal occurs. The major three stages involved in the process are coagulant formation through the dissolution of metal ions of anode reactor electrode, destabilization of pollutants, suspended particles and de-emulsification, and aggregation of unstable phases and floc-forming. Nowadays electrochemical treatments for wastewater treatments getting more attraction due to their eco-friendly nature and cost-effectiveness. Among these, electrooxidation and electrocoagulation are the most effective processes compared to other methods. Recently, electrocoagulation, owing to the increase in environmental restrictions on effluent treatments, has successfully been applied in the treatment of potable water, food and protein wastewater, leather and textile industry, agricultural wastewater, pharmaceutical wastewater, etc. The EC process is attractive due to its simplicity in operation and maintenance, no addition of chemicals, and can even remove the finest colloidal particles also [13–15].

An EC system is simply an electrolytically operated electrochemical cell, as shown in Figure 9.1. It essentially contains an anode-cathode electrode pair and an electrode source connected to it and submerged together in an electrolyte in which the effluent is to be treated. The system was operated in a constant voltage mode. The voltage was set to a constant value, and from the sacrificial metal anodes, metal hydroxides were released. From the cathode, hydrogen molecules are evolved. These processes together complete an EC system, which will release the in situ coagulants and the cathodes will release hydrogen gas, and these will help in the separation of destabilized pollutants

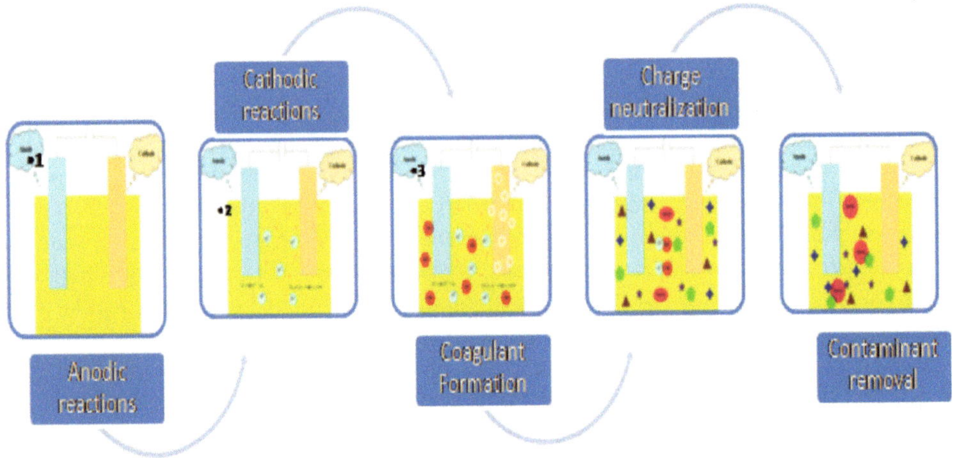

FIGURE 9.1 Chemistry of traditional electrocoagulation.

[16,17]. The process of the generation of in-situ coagulants occurs with the dissolution of the sacrificial anode. The commonly used anodes are aluminum or iron. The metal ions react with the $[OH]^-$ ions generated from the cathode and produce insoluble metal hydroxides. On the application of electricity, the contaminants present in the effluent will get dissociated, and those fragments get attached to the freshly formed metal hydroxides. The dissociation and attachment of contaminants from effluent either by complexation or by electrostatic attraction, followed by coagulation [7,18].

Emamjomeh and Sivakumar analyzed the pollutants removed by EC. The metal anodes produce cationic species of iron or aluminum in situ, which may have the same effect as the addition of coagulant salts and will destabilize charged particles, allowing for flocculation and solid–liquid separation. The entire process has three main mechanisms: electrode oxidation; gas bubble generation and flotation; and sedimentation of formed floc [19,20].

Various uses for EC were found, beginning with Kaliniichuk et al., who used EC to treat refinery wastewater. Results showed that aluminum hydroxide formed due to the dissolution of aluminum anodes and the hydroxide floc adsorbed the oil, which was then floated to the top by the hydrogen bubbles formed at the cathode. Balmer and Foulds studied the effectiveness of a sacrificial iron anode on oil separation from oil in water [21,22]. The key findings were as follows: the production rate of a reagent is controlled by the electrical current; several different anode materials are effective for treatment; energy consumption is optimized by using high electrolyte water with small electrode plate spacing; and the consumption rate of the sacrificial anode is dependent on the current applied. Ibanez et al. also treated oily wastewater and found that aluminum electrodes were more efficient than iron electrodes. This is likely due to the high adsorption capacity of hydrous aluminum oxides [23,24].

The advantages of EC over conventional coagulation (CC) consist of economic aspects (relatively low investment, maintenance, energy, and treatment costs), significantly lower volume of sludge produced, better sludge quality (lower water content, much larger and more stable flocs with better settling ability), no chemical additions, and the metals are present in their pure form, so the more active coagulants are formed, thus better performance efficiency is achieved at relatively fewer time intervals, ease of automation, simple equipment and compact size of EC systems (allowing decentralized treatment), greater functional pH range and pH neutralization effect, and the presence of electroflotation (EF) [2,25,26].

The mechanisms that can destabilize colloidal particles in water are (J.C. Crittendan et al., 2005):

i. Compression of the electrical double layer because of an increment in the concentration of particles in the bulk solution. This system is not suited for water treatment as high salt concentration will be required for the destabilization.

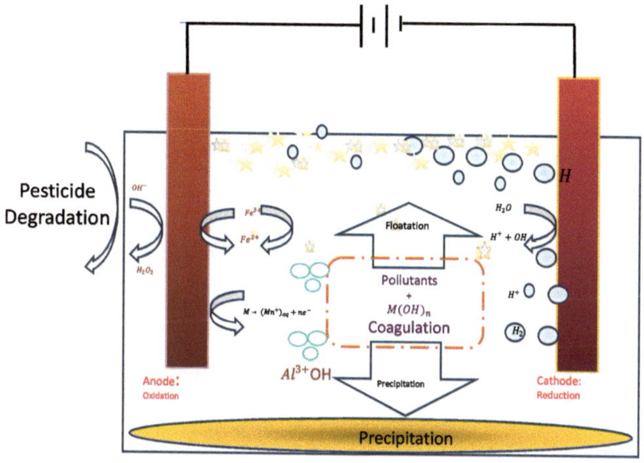

FIGURE 9.2 A schematic diagram describing the main reactions happening in an electrocoagulation process.

ii. Adsorption destabilization happens when oppositely charged ions or polymers are adsorbed onto the surface of particles. This mechanism involves diminishing the surface charge and hence the repulsive force between the particles. Be that as it may, in this situation, high coagulant dosage leads to restabilization of the particle, as the charge is reversed.
iii. Inter-particle bridging involves polymerized metal coagulants or organic chemicals forming bridges between particles. Long chained, high molecular weight organic polymers are adsorbed onto the particle surface by several mechanisms, such as dipole interactions, charge–charge interactions, hydrogen bonding, or van der Waals interaction. Bridging occurs when one polymer chain gets adsorbed on multiple particles, increasing the molecular weight.
iv. The precipitation mechanism involves high concentrations of metal salts at near-neutral pH. Under these conditions, metal salts such as ferric sulfate or alum form insoluble hydrolysis products and polymerize. Colloidal particles are then entrapped in these flocs (Figure 9.2).

9.3 COST COMPARISON OF DIFFERENT PROCESSES

Cost is one of the most important parameters to be considered while selecting a treatment technique for particular wastewater. Many studies have been conducted for the estimation of the efficiency and cost of the electrocoagulation process, compared to other conventional treatment techniques predominantly with chemical coagulation. The obtained results depend on the type of effluent, pollutant, and concentration of pollutant to be treated; the quantity of effluent; the quality of the final treated solution; the applicability of recycling; and the reuse of reclaimed water. (23 B). As a result, determining the best technology requires comparing similar types of effluents and other operating conditions [1]. Moosavirad discussed the cost estimation of greywater and concluded that electrocoagulation with aluminum electrodes is much lower than other technologies, as different authors have discussed different effluents. According to the results tabulated in Table 9.1, the insitu coagulants of aluminum formed during the electrocoagulation process have a high affinity to interact with various contaminants and result in their efficient removal. However, EC has greater effectiveness toward different types of effluents, but these applications are only limited due to the high energy demands and high current densities.

Cost comparison of EC with other following processes: electrocoagulation process using the Aluminium electrode (EC by Al electrode), electrocoagulation process using the iron electrode (EC by Fe electrode), The biological process followed by sand filtration and ozonation (biological, MF, and UV); the biological process followed by sand filtration and ozonation (biological, SF,

TABLE 9.1
Effluentstreated Using the Electrocoagulation Method

Type of Effluent	Anode/Cathode Material	Reactor Conditions	Investigated Parameters	Separation Efficiency (%)	Reference
Tannery wastewater containing organic and inorganic pollutants	Fe	Batch	COD: 4100–6700 BOD: 630–975 Cr: 11.5–14.3 TSS: 600–955 TKN: 144–170 TDS: 13,300–19,700 O &G: 638–780 Color: 3800–6330 (Pt-Co)	COD: 95 BOD: 96 Cr: 100 TSS: 96 TKN: 62 TDS: 50 O &G: 99 Color: ≈98 g (Pt-Co)	
Blue reactive (reactive blue 140),red disperse (disperse red 1), mixed dyes and real textile water	Al Fe Al-Fe	Batch	COD: 278–736 Color: n.d. TSS: 1715–6106 TS: 1800–6460	COD: ≈90 Color: ≈100 TSS: n.d TS: n.d	
Red dye solution mixture of 2- naphtoic acid & 2-naphthol	Al	Continuous	COD: 2500 Dye: 25–200 (optimum 80–100)	COD: >80 Color: 95	
Tissue paper wastewater	Al	Batch	Turbidity: 80–120 (NTU) COD: 2645 BOD: 1688 TSS: 205 TDS: 1200–1600	Turbidity: 92–97 (NTU) COD: 50 BOD: 60 TSS: n.d. TDS: n.d.	
Dairy wastewater	Fe	Batch	COD: 3900 Turbidity: 1744 (NTU) TS: 3090 TN: 113	COD: 70 Turbidity: 100 (NTU) TS: 48 TN: 93	
Pesticides contaminated groundwater	Fe/SS	Batch, Batch+UV	MB: 50–300 (Optimum 200)	MB: 89–95	
Cyanide- laden wastewater	Al Al-Fe Fe	Batch Continuous	Cyanide: 300	Cyanide (98/100)[g]	
Contaminated soil (Pb, Zn, Cd) washing solution containing EDTA	Al/SS	Batch	Pb: 1220–1620 Zn: 230–290 Cd: 8–10 EDTA: 16,500–19,700	Pb: 95 Zn: 68 Cd: 66 EDTA: ≈50	Lestan et al. (2010)

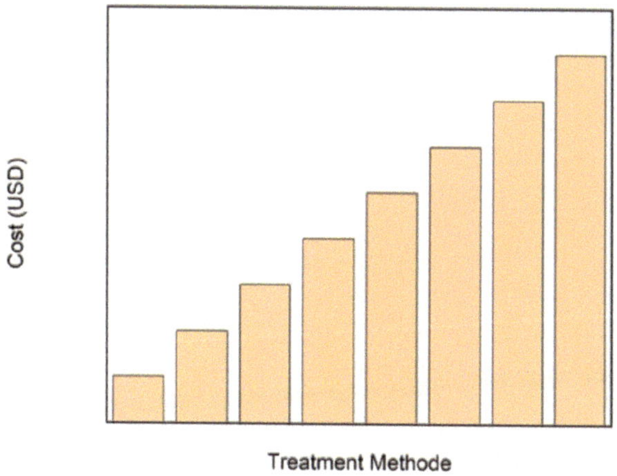

FIGURE 9.3 Cost comparison of EC with other processes.

and ozonation); the biological process followed ultrafiltration (biological and UF); coagulation by poly aluminum chloride (CC by PAC); coagulation by alum (aluminum sulfate) (CC by alum); Coagulation by ferric chloride (CC by $FeCl_3$) is shown in Figure 9.3.

9.4 APPLICATION OF ELECTROCOAGULATION FOR THE TREATMENT OF DIFFERENT EFFLUENT

Electrocoagulation can achieve greater than 90% performance efficiency within the limits of considerable energy consumption, removal of dissolved organic carbon, and 99% turbidity removal. EC can remove a wide variety of pollutants in a single process. When compared with other sorption treatment technologies such as ion exchange, EC can remove both anionic and cationic species. In addition to that, refractory pollutants such as arsenite and dichromate can also be removed without any prior treatments. Low conductivity of drinking water is required for human consumption. This is a challenge for using EC to treat drinking water since solution conductivity is an important parameter of EC. However, EC can be used for the treatment of various organic and metal contaminant removal of urban and industrial wastewater matrices [1,6].

Radioactive contaminants like strontium and cesium in water are a noteworthy risk and by utilizing diverse anode materials like aluminum, iron, magnesium, and zinc with iron as the cathode, good adsorption is found. The most extreme evacuation productivity of 97.0% and 96.8% was accomplished for strontium and cesium separately, utilizing magnesium as an anode at a present thickness of 0.08 A/dm^2 and pH of 7.0.

The electrocoagulation experiment has been performed by different researchers [10,28–31] with various process fluids. The discussions are tabulated below (Table 9.1), which includes the different parameters accounting for the EC performance and the percent efficiency.

Electrocoagulation (EC) can be used to treat wastewater containing organic matter and excess nutrients as well. The process efficiency can be varied by the use of various sacrificial electrodes such as aluminum, iron, and zinc that are oxidized to produce metal ions that can be used to coagulate and react with organic matter and nutrients in the wastewater [39,40]. A broad variety of electrode materials are suggested for EC operation. They include stable anodes(DSA, RuO_2, and ZrO_2-coated Ti), thin-film oxides (PbO_2 and SnO_2), noble metals (Platina), and carbon-based anodes. Besides these traditional electrodes like graphite electrodes(carbon felt, graphite granules, and glassy carbon), the recently developed boron-doped diamond, etc. can also perform better as metal electrodes for the removal of organic pollutants [34]. Among different anode materials in use, boron-doped diamond (BDD) has shown the best efficiency in mineralizing pollutants. In review of the electrocoagulation process, which features potable water treatment, fluoride, nitrite, air nitrate, ammonia, oily waste, heavy metal, municipal waste, industrial wastewater treatment, etc., the process was found to be applied effectively in all cases when suitably optimized [3,42].

A proposal on the improvement of the electrocoagulation process for the treatment of chemical wastewater under joined electrical associations utilizing titanium plates was progressed. The Mp Ti-Bp Al cathodes were observed to be more compelling for the treatment of material wastewater compared with Mp Ti-Bp Fe. At ideal conditions, the accompanying outcomes were acquired; 93.5% (COD), 97% (TSS), 90% (BOD5), 89% (TDS), 96% (turbidity), >99% (phenols), and 97% (phosphate) [18].

A method was identified for the preparation of nanoparticles of titanium dioxide from sludge generated during EC, choosing titanium electrodes as the sacrificial ones. This is a notable feature in the studies conducted. It ensures a solution to environmental issues in the matter of sludge disposal but also results in the production of a photocatalyst, which is a valuable product [4,38].

The improvement in the process efficiency of EC can be achieved by the modification of electrodes. One such modification is the fluidized bed electrodes. Recently, their applications are increasing due to the enhanced surface area and improved mass transport. These electrodes can provide high insitu coagulant production even at low current densities also. Since the electrodes are moving, there is very little possibility of particle agglomeration and the chances of formation of an oxide layer. The life of fluidized bed electrodes is also longer because they can provide continuous coagulants without disturbing the deposition process. The electrochemical behavior of fluidized bed electrodes is significantly influenced by bed expansion, bed height, current density, and metal concentration [39,40].

Phalakornkule et al. (2014) investigated the development of a reduced-cost electrode as an alternate to the plate or tubular, traditionally used electrode. A polyvinyl chloride (PVC) tube with an internal diameter of 28 mm was packed with machine shop turnings. The EC experiments were conducted at an optimum electrocoagulation time of 10 minutes and an applied voltage of 0.9 A and achieved a maximum color removal of 98% and a maximum COD removal of 93%. The energy consumption for the packed-bed anode (0.50 kWh/m^3) was higher than that of the traditional anode (0.38 kWh/m^3). Apart from the cost reduction, a high surface area to volume ratio is the main advantage of using PVC electrodes. It has been reported that the anode passivation is a disadvantage of this electrode. The time of onset of anode passivation is high, but once the passivation is increased, the electrode easily gets passivated. At the same applied current, the local current intensity will be lower with a higher surface area [41,42].

The studies on the removal of congo red dye, cephalothin sodium antibiotic, and malathion pesticide revealed the feasibility of the electrocoagulation process using scrap aluminum sheets as the electrodes. The main factors which affect the process were found to be reaction time, pH, current density, and rotational speed. The CR dye removal efficiency of 93% was observed at optimum conditions of 60 minutes of reaction time, pH of 8, the current density of 200 mA/cm^2, and a rotational speed of 200 rpm. The optimum conditions of antibiotic removal were found to be reaction time of 50.94 minutes, solution pH of 9, an applied voltage of 8 V, and 12.3 mL of electrolyte addition. The quantification of the pesticide degradation was carried out using HPLC, and the results are obtained as follows: removal efficiency of 96.82% and energy consumption of 0.0826 W.h. The optimum malathion pesticide removal of 96.82% and energy consumption of 0.082 kW.h was obtained with typical operating conditions–operation time, 52.12 minutes; solution pH, 9; applied voltage, 8 V; and amount of electrolyte, 11.86 mL [48,49].

9.5 COMPARISON OF REMOVAL EFFICIENCY WITH DIFFERENT ELECTRODES

Comparative batch studies between aluminum and iron as sacrificial electrodes were carried out. The removal efficiency was found to be dependent on the operating time, solution pH, operating voltage, and the amount of electrolyte. The results obtained from the batch are shown in Figure 9.4a–d, which concluded that the iron electrode was more efficient. The efficiency of the EC

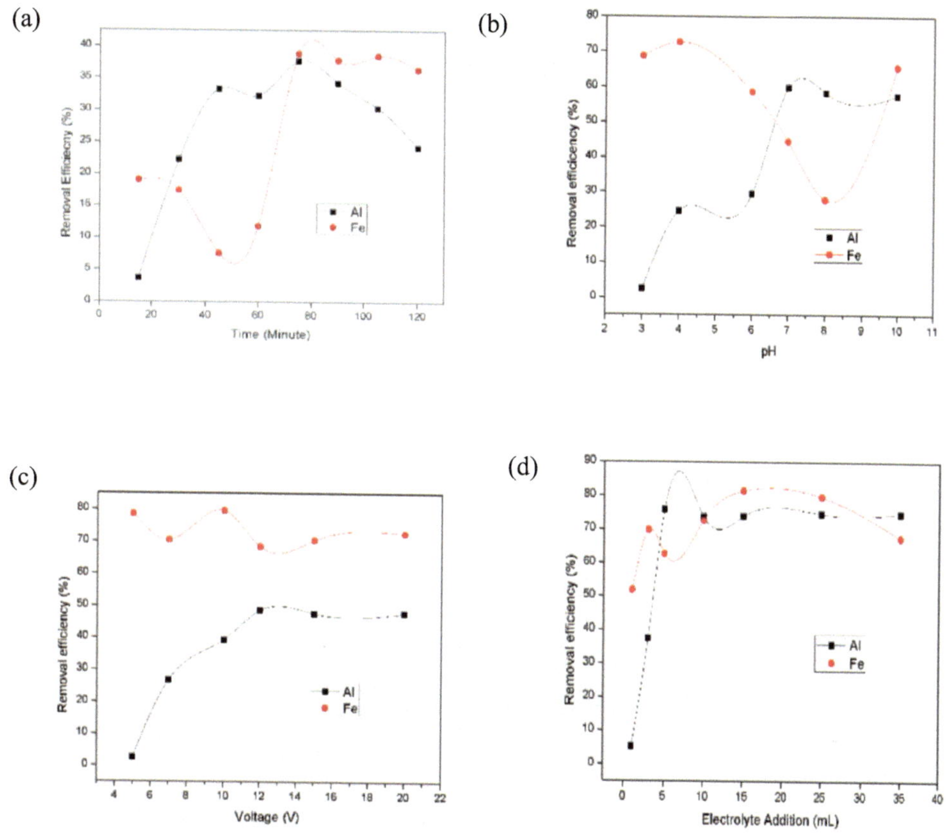

FIGURE 9.4 Comparison of iron and aluminum electrodes in the pesticide removal.

process depends on the electrodes involved in the process and the rate at which the insitu coagulants are released from the electrodes.

9.6 CHEMICAL PRINCIPLES AND ELECTROCHEMICAL REACTIONS

The electrocoagulation system is a simple electrolytic cell, the detailed figure with all the components is shown in Figure 9.5. It consists of a power source that is connected to at least one anode-cathode pair submerged together in an electrolytic solution. Water or wastewater to be treated is taken as electrolyte. It will complete the electrical circuit by allowing charges to transfer between the electrodes.

The current supply will promote the generation of metal cations by the electrolysis of sacrificial anodes. These dissolved metal ions will tend to donate hydronium ions of the respective metal cations from the neighboring hydration species. The hydroxide anions OH^- form bridging groups with the metal hydroxide, allowing further dimerization and polymerization by the liberation of hydroxide anions (Eqs. 9.4–9.14). These are more likely to carry a positive charge and form gelatinous hydroxides of sacrificial anodes. These can make bonds with the suspended matter and contaminants present in the electrolyte. This process is commonly known as sweep coagulation. The capacity of contaminants to bond with coagulants will depend on the physicochemical characteristics of the formed gelatinous hydroxides, their charge, porosity, type of bonding, etc. The careful investigation of the pollutant and the formed metal hydroxides will result in the collection of particular species. The anodic and cathodic reactions of aluminum electrodes are represented as follows [45].

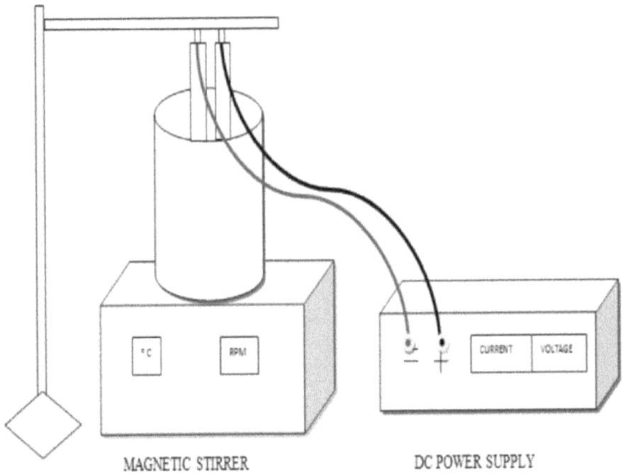

FIGURE 9.5 Schematic representation of the experimental setup.

Anodic reactions:

$$Al_{(s)} \rightarrow Al^{3+} + 3e^- \quad (9.1)$$

$$2H_2O_{(l)} \rightarrow O_{2(g)} + 4H^+ + 4e^- \quad (9.2)$$

Cathodic reactions:

$$2H_2O + 2e^- \rightarrow H_{2(g)} + 2OH^- \quad (9.3)$$

Equations 9.4–9.14 explains the possible electrochemical reactions and these charged species are responsible for the charge speciation and charge neutralization removal of various pollutants. The electrochemical reactions of both aluminum and iron are shown below.

$$(AlOH)_{3\,(s)} + 3H^+ \rightarrow Al^{3+} + 3H_2O \quad (9.4)$$

$$Al^{3+} + 1H_2O \rightarrow (AlOH)^{+2} + 1H^+ \quad (9.5)$$

$$Al^{3+} + 2H_2O \rightarrow (AlOH)_2^+ + 2H^+ \quad (9.6)$$

$$Al^{3+} + 3H_2O \rightarrow (AlOH)_3^0 + 3H^+ \quad (9.7)$$

$$Al^{3+} + 4H_2O \rightarrow (AlOH)_4^- + 4H^+ \quad (9.8)$$

$$3Fe^{3+} + H_2O \rightarrow (FeOH)^{2+} + H^+ \quad (9.9)$$

$$(FeOH)^{2+} + H_2O \rightarrow (FeOH)_2^+ + H^+ \quad (9.10)$$

$$(FeOH)_2^+ + H_2O \rightarrow (FeOH)_3^+ + H^+ \quad (9.11)$$

$$(FeOH)^{+} + H_2O_2 \rightarrow Fe^{3+} + OH^{\cdot} + OH^{-}/2OH^{-} \tag{9.12}$$

$$(SO_4)^{2-} + OH^{\cdot} \rightarrow (SO_4)^{\cdot-} + OH^{-} \tag{9.13}$$

$$(HCO_3)^{-} + OH^{\cdot} \rightarrow (CO_3)^{\cdot-} + H_2O \tag{9.14}$$

The current supply also depends on the evolution of hydrogen from the cathode. The gas evolution at the cathode can be described into three phases: nucleation, growth, and detachment. The amount of hydrogen adsorbed at the cathode surface will fill up the available reaction sites. The hydrogen desorbed from the cathode surface and combined with the hydronium ion will form hydrogen gas. Electrocoagulation process removal efficiency and energy consumption are the main performance parameters measured during the pollutant removal by the electrocoagulation process [46–48]. Those parameters were measured as follows:

$$\text{Removal efficiency} = \frac{C_i - C_f}{Ci} * 100 \tag{9.15}$$

$$\text{Energy consumption} = \text{Voltage} * \text{current} * \text{time} \tag{9.16}$$

where C_i is the concentration at zero time and C_f is the concentration of dye effluent after the preferred time interval in mg/L. Removal efficiency is represented in percentage, energy consumption in W.h, voltage in volts (V), current in amperes (A) and time in hours (h). Faraday's law will give the amount of metal cations forms, from this we can find the efficiency.

$$W = \int_0^t \left(\frac{A_i A_r}{zP} \right) dt \tag{9.17}$$

$$W = \frac{ItA_r}{zF} \tag{9.18}$$

where W is the mass of metal dissolved at the anode with the assumption of no side reactions, A_i is the total surface area of electrodes, A_r is the atomic weight of metal dissolved, z is the valance of metal ion, I is total current density, and F is the Faraday's constant.

When the applied current is increased, the potential is increased and at a particular point, the water hydrolysis will be a predominant step then the hydrolysis of electrolyte. This will lead to metal dissolution and oxygen evolution. The evolution of oxygen helps in the flotation process but the efficiency of the process will decrease considerably. The evolution of oxygen will increase the sacrificial anode dissolution rate and increased coagulant dosage, more than the results predicted by faraday's law. Industrial applications will depend on several factors such as non-Faradaic reaction, ohmic losses. Potential side reactions and the lifetime of the electrodes how frequently the electrodes should be replaced.

Coulombic efficiency is also known as Faraday efficiency. It is the ratio of the theoretical charge required for the amount of metal in a reactor to the actual charge supplied. Coulombic efficiency is found in which side reactions occur. To find the efficiency of dissolution:

$$\eta_q = \frac{zF(C_t V)}{It} * 100\% \tag{9.19}$$

where η_q is the Coulombic efficiency (%), C_t is the concentration of metal ions after time t, V is the volume of treating solution,

The lifetime of electrodes used in the electrocoagulation process can be found from the applied current density and the electrode material. The thickness and density are also important parameters. After several weeks of operation, the loss in these values will determine the lifetime of electrodes.

$$t_{\text{life}} = \frac{\rho z F l}{i A_r} \tag{9.20}$$

where ρ and l are the density and thickness of electrode plates [1].

9.7 KEY PARAMETERS

The overall efficiency of the reactor was tested using different variables: electrode materials, processing time and concentration, current density, solution pH, inter electrode distance, and the overall rotational speed.

9.7.1 Electrode Materials

Choosing the appropriate electrode can be considered as one of the most critical steps in any electrocoagulation process as it can vary the process efficiency to a great extent. Several materials such as Al, Fe, Cr, Mg, Si, Zn, etc. [10] were used. Among these, Al and Fe are the most chosen ones as they are easily available, cheap, and above all, anodically soluble. Optimal material selection depends on the chemical properties of the electrolyte and the pollutants to be removed. Apart from the expense factor, aluminum is observed to be superior compared to iron in most cases in terms of efficiency of the treatment.

Iron and a mix of aluminum and iron electrodes have likewise given high expulsion proficiency of arsenic from groundwater. As mentioned, the suitable electrode material is a mandatory factor in the electrocoagulation process. The anode material not only influences the cation introduced in the process but also can affect the current efficiency. Those materials with lower current efficiency have lower treatment efficiency and they are not preferred as electrodes like graphite, lead oxide, etc. Cleaning and replacement of the electrodes are equally important. The surface of the electrode should be cleaned regularly after each process, as the presence of sludge will inhibit the release of cations to the solution.

9.7.2 Solution pH

pH is a key parameter in the treatment process. Solution conductivity, electrode dissolution, hydroxide formation, the zeta potential of the colloidal particles formed are influenced by the same. As discussed earlier, iron and aluminum cations and hydroxides cause the destabilization of the colloids. Effective coagulant species are formed in neutral, acidic, and slightly alkaline pH. At a highly alkaline pH, $Al(OH)^{4-}$ and $Fe(OH)^{4-}$ ions, which have poor coagulation performance are formed but it is to be noted that if the species to be removed is cationic then it is favorable. It is observed that Fe (III) is effective over a wider pH range than Al (III) and also in slightly alkaline pH. Chemical coagulation is favored when pH ought to be brought down before release while EC is reasonable when higher pH (acidic) is desired. A few contaminations likewise have particular ideal pHs of treatment [49–51].

9.7.3 Current Density

Current per unit area is the current density, or in detail, it is the proportion of measure of current expended in creating an objective item to that of aggregate current consumed. The measure of metal broken up is subject to the amount of power that went through the electrolyte. At the point when current density increases, the measure of metal disintegrated additionally would build. It is to be noticed that as the current density increases, the expulsion rate increases there is an optimum at which the evacuation rate is balanced out past which promotes expulsion is impractical. Current density decreases and the amount of aluminum generated from cathode increases [52].

9.7.4 The Initial Concentration of Pollutants and Reaction Time

Variation in initial dye concentration affects the removal process which mainly depends on the kind of wastewater under consideration. If the conductivity of the wastewater solution is increased with concentration, better and efficient removal is achieved at a higher concentration whereas if the conductivity does not vary with concentration, better removal is achieved at a lower concentration only. An increase in reaction time enhances the removal or treatment by generating more flocs which aid in adsorption. For every electrocoagulation process, it is important to maintain an optimum time beyond which further removal does not take place [60].

9.7.5 Electrode Distance

Energy-saving can be achieved with optimum distance when maintained between electrodes. This can affect the size of the reactor as well as the overall cost of treatment [53]. The ohmic potential drop decreases when inter-electrode distance decreases. A narrow spacing of less than 10 mm is accompanied by low energy consumption. At higher inter-electrode spacing, an increase in cell voltage occurs, which in turn increases the power consumption [51]. A minimum distance of 3–8 mm can be maintained when handling wastewater treatment. In situations where sturdy and settable flocs are encountered, it is better to maintain a distance of at least 10 mm as this can require polarity reversal due to frequent floc deposits because of lower turbulence between the electrodes [42–64].

9.7.6 Kinetics, Isotherm, and Statistical Modeling

Another important study conducted was on kinetic and isotherm modeling. In the removal of tetracycline, the maximum removal efficiency was found to be 99% irrespective of initial concentration and for pH ranging from 3 to 10. Sips isotherm matched the experimental data. Pseudo-second-order kinetic model was followed.

Also for the removal of basic dye rhodamine B from an aqueous solution using steel electrodes kinetics, isothermal and thermodynamics studies were conducted. An increase in the removal percentage of dye with a pH up to 7 as well as contact time and current density increase resulted in a better removal. Equilibrium data went under the Langmuir isotherm equation which confirmed the monolayer adsorption capacity of 163.63 mg/g at 303 K. Pseudo-first-order kinetic model is followed. The process was found to be spontaneous and endothermic [66].

Response Surface Methodology (RSM) is a statistical tool often used for modeling and optimization where the response of interest is influenced by several operating variables. It explains the effect of the operating parameter on the response, either individually or collectively. The RSM generates an empirical model that can describe the process and analyze the influence of independent variables on a specific dependent variable (response) [68].

The interaction of various operating parameters on the removal of Acid Red 14 azo dye removal by a batch-EC (Fe/steel electrodes) process was carried out with RSM. Three independent variables were selected for the optimization studies current density, treatment time, and initial pH of the wastewater. The individual and combined effects of these on the removal efficiency were studied. The study revealed the following results for maximizing the removal efficiency of dye as 91.27% in 4 minutes, whereas 93.93% in 4.47 minutes was predicted by the model. This shows that RSM is a suitable technology for optimization studies [64].

9.8 RECENT ADVANCEMENTS IN ELECTROCOAGULATION

- Agal cultivation combined with the electrocoagulation process can treat highly contaminated waste such as animal manure and food waste. A freshwater algal sample was collected and algal assembly was cultivated on the EC-treated water sample. Anaerobic digestion can reduce the excess nutrients and turbidity in the effluent stream and this biomass can be converted into chemical and biofuel production [18,37].

- A new process to which incorporated AC-powered electrocoagulation with granular bio-char to reduce energy consumption and passivation of electrodes while achieving high treatment efficiency was a remarkable achievement.
- The process of electrocoagulation can be improved by combining with ozone, ultrasound, pulses, adsorption process. Each process is efficient depending on the application. Scale-up should be taken into consideration although bench scale and pilot plants are successful [7,65].
- Considering the total cost of the electrocoagulation process the electrical power consumption contributes more than 50% of the total expenditure. So the improvement in this enhances the application of EC in various fields. A novel current feed style is introduced the alternative pulse current. This feeding style can reduce the passivation of electrodes and can avoid extra energy consumption, which is contributed by the oxide layer formed on the electrode surface [9].

9.9 ELECTRODE CONFIGURATIONS

The operating cost and energy consumption also depend on the arrangements of the electrodes. The electrode configurations followed in the electrocoagulation process are monopolar and bipolar, as shown in Figure 9.6. In the monopolar setup, electrodes are connected to the power supply as alternative electrodes are joined to the opposite polarity terminals, giving several individual reactor units of arrangement. In a monopolar arrangement, all the units present in the system operate at the same voltage and the total current in the reactor will be the sum of the unit's currents. In the bipolar arrangement, electrodes at each extreme are connected to the power supply. Every electrode apart from one end act as an anode and the other as a cathode. Every adjacent electrode pair from the connections acts as individual units. The total voltage dropped in the system is the sum of all single voltages [14,68].

The applied voltage and the cell current in the system determine the rate of coagulant generation. The bubble production rate, its size, the distribution and mass transfer of insitu coagulants and the produced sludge will change with different electrode configurations. In the case of a bipolar arrangement, three regions of removal rate can be observed. In the initial period of the slow-moving rate of removal, for this period the solution pH got increased a little. During this period, the low

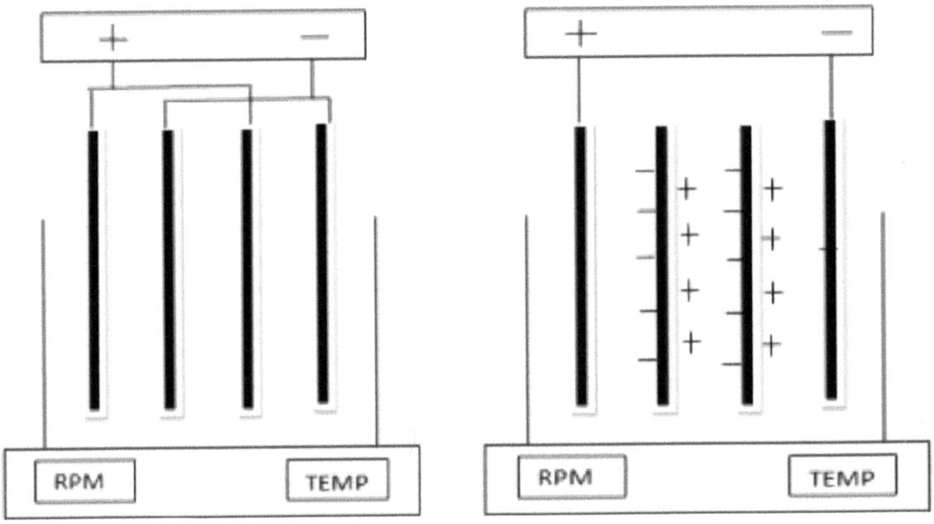

FIGURE 9.6 (a) Bench-scale EC reactor with monopolar electrodes in parallel connections and (b) bench-scale EC reactor with bipolar electrodes in series connections.

rates are due to the cathodic reduction and co-precipitation. In the second region, comparatively higher rates are found with a moderate increase in the solution pH. Further increase in the pH leading to a higher solubility of the pollutant and higher rates, this remains at a steady state. This is the third region. The same trend was observed for monopolar connections as well [69].

9.10 EFFECT OF VARIOUS ADDITIVES

The addition of different types of additives also can enhance the performance of the electrocoagulation process. Such additives are polyelectrolyte's which can create inorganic-organic composite which aid in the electrocoagulation process, in cationic or anionic other commonly used process-enhancing materials are supporting electrolytes. These directly influence the conductivity of effluent to be treated. The most commonly used supporting electrolytes are sodium chloride and sodium bicarbonate [70,71].

As the concentration of supporting electrolyte increases the conductivity of effluent also increases which will increase the removal efficiency of pollutants, all other operating conditions were kept constant. This is due to the presence of CO_3^-/Cl^- anions which have a direct effect on the solution conductivity and these will destroy the passive layer formed on the electrode surface. The formation of in-situ coagulants has been increased by the dissolution of electrodes, these anions can catalyze the dissolution of electrodes through pitting corrosion [26,72]. Therefore, EC in the presence of supporting electrolytes could improve the pollutant removal efficiency by increasing the availability of metal coagulants by the enhancement of the anodic dissolution.

Consequently, the problem of electrode passivation is partially solved when $NaHCO_3$ was used as a supporting electrolyte compared to NaCl.Cl^- ions lead to severe pitting corrosion which further will reduce the life span of the sacrificial electrode. Considering the energy consumption, as the concentration of ions increases the operating voltage also increases without any change in the cell voltage and the cell voltage decreases with increasing wastewater conductivity at constant operating voltage. Subsequently, the necessary voltage for attaining a certain operating voltage will be diminished and the consumed electrical energy will decrease.

9.11 DRAWBACKS

Since electrons are the main reagents of this process, it is referred as clean technology. The anode is continuously releasing electrons by the application of electricity, an impermeable oxide layer is formed on the surface of electrodes which may decline the performance efficiency due to the passivation effects. Gelatinous hydroxide may tend to be solubilized in some cases. Due to oxidation, a sacrificial anode is dissolved in the wastewater, so needs to be regularly replaced.The use of electricity may be expensive in some cases [56,73]. The reactors working on electrochemical principles should have good mixing for better diffusion and mass transfer. This will improve overall performance efficiency and energy consumption [48]. The conductivity of the effluent also affects the electrocoagulation process. High values of conductivity are preferably required.

9.11.1 Sludge Management

Comparing electrocoagulation with conventional coagulation involves the addition of chemicals ascoagulantsand in adjustments of pHfor the elimination of contaminants, these make the electrocoagulation process more attractive. The production of sludge is high in the case of chemical coagulation due to the added chemicals and the amount of total dissolved contents are also increased. The addition of the extra chemicals will affect water quality and it makes the disposal and recycling of water more challenging [1].

There is limited research available concerning the characteristics of flocs produced from EC. Harif et al. reported that, when comparing CC against EC, that EC has a very less amount of sludge since pure electrodes are the sources of coagulants in case of CC corrosive salts are the coagulants

[16,74]. For CC larger and denser flocs are formed, they are very hard to break. The characteristics of the floc such that the size, structure, fragileness, etc will depend on the downstream clarification methods. Relatively very little data are available on the EC flocs compared to CC. Gagnon et al. attempted to give attention to the characterization, growth and structure of flocs produced from the EC process [75]. The effect of various operating parameters on the floc characteristics are also been investigated. The results were consistent with the literature regarding chemical coagulation (CC) and may aid with the design of downstream clarification processes [18,22].

In light of an examination on the development and structure of iron precipitate flocs in saline water from a bench-scale electrocoagulation (EC) system, operation at higher currents was found to promote faster floc aggregation and having a higher concentration of particles which results in speedier shaping flocs [16]. The polymer specialist LLM 3135 improves the settling procedure at a measurement of 0.01 kg/m^3. The costfor the treatment of material wastewater utilizing the EC-electro-oxidation process was approximately1.69 US$/m^3$ [5,18].

The disposal of sludge residue produced during the electrocoagulation is a matter of discussion. The sludge obtained from this can be reused as an adsorbent for the removal of phosphates from the aqueous phase [79]. The residual sludge produced can be used as fuels in the making of blended fuel briquettes for the application of furnace fuel and also can be used for construction purposes by blending with cementation mixtures or can be utilized for soil stabilization. Soil stabilization is the process of enhancing the engineering properties of the soil thereby transforming it into a more stable component. The weak soils, which mainly comprise black cotton soils, loose sand and soft clays have poor engineering properties which make them unsuitable for construction. These types of soil experiencea larger change in volume due to the huge difference in moisture content. The black cotton soils are generally found in landslide areas and it covers about 30% of total land area in India. These types of soil have high plasticity, low strength, and more swelling [81]. Hence soil stabilization is an ideal technique to improve the soil properties and this can be undertaken by the addition of several admixtures. Thus the mixing of antibiotic wastewater sludge may be considered the same as other admixtures used to stabilize the soil. The residual sludge can be utilized for the stabilization of black cotton soil which is present in most of the land sliding areas. The main objective of this paper is to evaluate the feasibility of the electrocoagulation method with a contemporary reactor and convert it for drinking or irrigation purposes concerning public health necessity. It will be carried out efficiently by implementing the parameter optimization for maximum removal efficiency and minimum energy consumption. Along with this, the characterization and reusability of the reaction by-products are also discussed [82].

9.12 LIMITATIONS

The EC approach is a clean and environmental friendly technology. The most important limitation ofthe electrocoagulation process is that in many cases, it is not selective to the pollutants when more pollutants are present, especially in the case of real wastewater. Hence, the process removes the pollutants mixturesuccessively, failing the separation of contaminants individually. The final sludge obtained after the treatment may contain a combination of pollutants along with the insitu coagulants, usually aluminum and iron. In this context, electrocoagulation can be used as a separation process but for the recovery of valuable products other technologies must be incorporated. Even though electrocoagulation can be used for selective removal as a part of the treatment sequence. Silica can be selectively separated from the wastewater of oil and gas industries. Despite all these challenges, a few studies reported that electrocoagulation has been used for the selective removal of many valuable compounds such as gold, silver, copper (Au, Ag, and Cu), microalgae from biodiesel, TiO_2 from mining water, etc [1].

The formation of the oxide layer (passivation of electrodes) affects the mass transfer of contaminants and the insitucoagulants, this will lead to lower performance efficiency and high energy consumption. This is one of the limitations followed by EC technology [11]. The other limitation of the electrocoagulation process includes the regular replacement of sacrificial electrodes. Due to oxidation, a sacrificial anode is dissolved in the wastewater, which leads to a decrease inefficiency. Thus further control should be there in the pH and conductivity of the treating solution. High conductivity of the wastewater suspension is required for higher performance efficiency [7,11]. In the case of Fe-Al electrode combination in the acidic region, the rate of coagulant formation is very less because Fe and Al are soluble at the particular region this leads to the unsuccessful attempts of the electrocoagulation process.

A systematic approach should be considered in the optimization of the key parameters and their interactive effects on the removal efficiency and energy consumption before the scale-up of laboratory studies. However many industrial systems are currently installed and successfully operating. The design of an electrocoagulation system should ensure the effective contact between the electrodes and the effluent to be treated in case of both batch and continuous. The electrode materials and the shape, thickness can also influence the efficiency of the process commonly aluminium, iron, mildsteel, or occasionally zinc or zinc alloys are used as electrodes [1].

One of the most important steps in the reduction of the overall cost of the EC process is the decrease of energy consumption. This can be done by the reduction in the internal resistance IR drop of electrodes, which will improve the performance of the current supply. This can be mitigated by the improvement in the state of turbulence. Turbulence will improve the mass-transport which can be achieved by raising the flowrate and these may lead to the passivation of electrodes. Hydrogen and oxygen gas bubbles are evolved near anode and cathode, this accumulation of bubbles will increase the overall electrical resistance. These will affect the performance of the EC system and the overall cost also [43].

One area of research that is critically needed is the development of low-cost electrodes. Only a few studies have focused on the design of low-cost electrodes. In the study by Wei et al. (2016), low-cost steel wool was employed as the cathode. The steel wool cathode not only reduced the cost of electrodes but also provided a larger electrode surface area than iron plates. In the study by Un and Aytac (2015), wrapped iron wire netting was used as an anode in a packed bed electrochemical reactor. This unique anode was found to effectively remove color and to reduce COD in real textile wastewater.

9.13 SUMMARY

The electrocoagulation process is a very effective separation process for the removal of a broad variety of pollutants from water and wastewater. Even though the EC technology is more than 100 years old, since 20 years the technology got more attraction and considerable attention from the industry and academic researchers. The commercialization of electrocoagulation is improved and exploring new areas of applications by the suppliers. several aspects of the process are not explained in detailbecause of the complex reaction mechanism, the relationsamong electrochemistry, chemical coagulation, charged separation, flow process, mixing, and mass transport. In this chapter, the basic theory of the chemical and electrochemical principlesprevailing in the EC process and the most significant operational parameters that control the performance of the removal process are discussed.

The electrocoagulation process makes use of the principles of chemical coagulation by contributing a significant advantage over the external dosing of chemical coagulants. In electrocoagulation, the coagulants are formed insitu by the electrochemical dissolution of sacrificial anodes. The fundamental difference between chemical and electrocoagulation is the method of liberation of coagulants. Conversely, electrocoagulation avoids the addition of counter ion through the salt and the addition of an alkali to neutralize the pH shift coupled with the hydrolysis of the coagulant. In electrocoagulation, the pH of treating effluent should be in a neutral to alkali range, but very little pH change is observed during the process.

The chemical changes occurring during the electrocoagulation process depend on the electrode material and the effluent composition. The electrocoagulation performance depends on various operating parameters such as electrode geometry, electrochemical parameters such as applied current, voltage, the type of electrode arrangement, pH, conductivity, etc. The appropriate choice of electrode material, design of the electrolytic cell, systematic control on the operating variable can lead to a successful separation unit. The performance of electrocoagulation can be characterized by the removal efficiency of selected contaminants, energy consumed during the process, coulombic efficiency, dissolution of electrode materials. Eventhough electrocoagulation is an effective technology, the fouling layer formationon the electrode surface is an important challenge, as it may affect the coagulant production, resulting in a decrease in pollutant removal and an increase in energy consumption. Regardless of these challenges many industries are implementing this technology for separation as well as the removal of the target material. Electrocoagulation provides a relatively robust and compact approach mainly in the case of the end of pipe industrial effluents such as textile, leather, paper and pulp, pharmaceutical, oil mills, heavy metal-containing effluents.

Despite these advantages, further understanding of the basic reactions and processes taking place during the electrocoagulation process is required for the successive design and implementation to other applications. Although electrocoagulation has been used for the treatment of a wide variety of pollutants, not many studies are concentrated on the complex electrochemical and physiochemical reactions going on during electrocoagulation. Most studies are focused on the treatment of specific pollutants rather than the problems in mechanistic levels, this is the main issue faced by the industry during scale-up. Most of the studies are conducted in batch mode while in the industry most of them are based on continuous mode. Scale-up of the chemical process depends on the recognized models interms of operational parameters and process efficiency, however, very few models are developed for the electrocoagulation process. Modeling and scale-up studies are needed for the implementation of the electrocoagulation process in the industrial levels without expensive and complex trials. Future studies should focus on the above-stated problems of the mechanism of electrocoagulation process for different types of contaminants at an operating range, the interactive effects of the parameters on the process responses, the operations particularly in continuous modes, the development of appropriate models that can be scaled-up and finally the techno-economic analysis of electrocoagulation also should be conducted.

LIST OF ABBREVIATIONS

CC	Conventional coagulation
COD	Chemical oxygen demand
EC	Electrocoagulation
HPLC	High performance liquid chromatography
IR Drop	Internal resistance drop

LIST OF SYMBOLS

A	Current	A
A_i	Total surface area of electrode	m^2
A_r	Atomic weight of metal dissolved	g
C_f	Concentration of dye effluent after preferred time interval	mg/L
C_i	Concentration at zero time	mg/L
C_t	Concentration of metal ion after time t	mg/L
l	Thickness of electrode	m
I	Total current density	A/m^2
t	Time	hrs
t_{life}	Life time of electrode	min
V	Voltage	V
V	Volume of treating solution	L
W	Mass of metal dissolved at the anode with assumption of no side reactions	g
z	Valancy of metal ion	
η_q	Coulombic efficiency	%
ρ	Density of electrode	kg/m^3

REFERENCES

1. N. Yasri, J. Hu, M.G. Kibria, E.P.L. Roberts, Electrocoagulation separation processes, *ACS Symp. Ser.* 1348 (2020) 167–203. https://doi.org/10.1021/bk-2020-1348.ch006.
2. D.T. Moussa, M.H. El-Naas, M. Nasser, M.J. Al-Marri, A comprehensive review of electrocoagulation for water treatment: Potentials and challenges, *J. Environ. Manage.* 186 (2017) 24–41. https://doi.org/10.1016/j.jenvman.2016.10.032.
3. F. Fu, Q. Wang, Removal of heavy metal ions from wastewaters: A review, *J. Environ. Manage.* 92 (2011) 407–418. https://doi.org/10.1016/j.jenvman.2010.11.011.
4. J. Heffron, M. Marhefke, B.K. Mayer, Removal of trace metal contaminants from potable water by electrocoagulation, *Sci. Rep.* 6 (2016) 1–9. https://doi.org/10.1038/srep28478.
5. A.S. Naje, S. Chelliapan, Z. Zakaria, M.A. Ajeel, P.A. Alaba, A review of electrocoagulation technology for the treatment of textile wastewater, *Rev. Chem. Eng.* 33 (2017). https://doi.org/10.1515/revce-2016-0019.
6. M.T. Samadi, M.H. Saghi, A. Rahmani, J. Hasanvand, S. Rahimi, M.S. Syboney, Hamadan landfill leachate treatment by coagulation-flocculation process, *J. Environ. Health Sci. Eng.* 7 (2010) 253–258.
7. A.S. Naje, S.A. Abbas, Electrocoagulation technology in wastewater treatment: A review of methods and applications, *Civ. Environ. Res.* 3 (2013) 29–42. http://www.iiste.org/Journals/index.php/CER/article/view/8115.
8. P.K. Holt, G.W. Barton, M. Wark, C.A. Mitchell, A quantitative comparison between chemical dosing and electrocoagulation, *Colloids Surfaces A Physicochem. Eng.* Asp. 211 (2002) 233–248. https://doi.org/10.1016/S0927-7757(02)00285-6.
9. S. Krishnan, H. Rawindran, C.M. Sinnathambi, J.W. Lim, Comparison of various advanced oxidation processes used in remediation of industrial wastewater laden with recalcitrant pollutants, *IOP Conf. Ser. Mater. Sci. Eng.* 206 (2017). https://doi.org/10.1088/1757-899X/206/1/012089.
10. O. Sahu, B. Mazumdar, P.K. Chaudhari, Treatment of wastewater by electrocoagulation: A review, *Environ. Sci. Pollut. Res.* 21 (2014) 2397–2413. https://doi.org/10.1007/s11356-013-2208-6.
11. W.R.- Romero, L.V.F. Pino, J.L. Guerrero-guevara, J. Castro-torres, M. Luis, M.E.K. Santos, R. Yuli-posadas, Benefits of electrocoagulation in treatment of wastewater: Removal of Fe and Mn metals, oil and grease and COD : Three case studies, *Int. J. Appl. Eng. Res.* 13 (2018) 6450–6462.
12. T.S. Anantha Singh, S.T. Ramesh, New trends in electrocoagulation for the removal of dyes from wastewater: A review, *Environ. Eng. Sci.* 30 (2013) 333–349. https://doi.org/10.1089/ees.2012.0417.
13. M.A. Nasser Ghalwa, Removal of imidacloprid pesticide by electrocoagulation process using iron and aluminum electrodes, *J. Environ. Anal. Chem.* 02 (2015). https://doi.org/10.4172/2380-2391.1000154.
14. B. Palahouane, N. Drouiche, S. Aoudj, K. Bensadok, Cost-effective electrocoagulation process for the remediation of fluoride from pretreated photovoltaic wastewater, *J. Ind. Eng. Chem.* 22 (2015) 127–131. https://doi.org/10.1016/j.jiec.2014.06.033.
15. M.A. Nasser Ghalwa, Removal of imidacloprid pesticide by electrocoagulation process using iron and aluminum electrodes, *J. Environ. Anal. Chem.* 02 (2015). https://doi.org/10.4172/2380-2391.1000154.
16. S.Y. Lee, G.A. Gagnon, Growth and structure of flocs following electrocoagulation, *Sep. Purif. Technol.* 163 (2016) 162–168. https://doi.org/10.1016/j.seppur.2016.02.049.
17. M. Shirzad-siboni, A. Jonidi-jafari, M. Farzadkia, A. Esra, M. Gholami, Enhancement of photocatalytic activity of Cu-doped ZnO nanorods for the degradation of an insecticide: Kinetics and reaction pathways, *J. Environ. Manag.* 186 (2017). https://doi.org/10.1016/j.jenvman.2016.10.049.
18. A.S. Naje, S. Chelliapan, Z. Zakaria, M.A. Ajeel, K. Sopian, H.A. Hasan, Electrocoagulation by solar energy feed for textile wastewater treatment including mechanism and hydrogen production using a novel reactor design with a rotating anode, *RSC Adv.* 6 (2016) 10192–10204. https://doi.org/10.1039/c5ra26032a.
19. N.M. Abu Ghalwa, A.M. Saqer, Removal of reactive Red 24 dye by clean electrocoagulation process using iron and aluminum electrodes, *J. Chem. Eng. Process Technol.* 07 (2016) 1–7. https://doi.org/10.4172/2157-7048.1000269.
20. C. Hu, J. Sun, S. Wang, R. Liu, H. Liu, J. Qu, Enhanced efficiency in HA removal by electrocoagulation through optimizing flocs properties: Role of current density and pH, *Sep. Purif. Technol.* 175 (2017) 248–254. https://doi.org/10.1016/j.seppur.2016.11.036.
21. G. Chen, Y. Hung, Electrochemical wastewater treatment processes, in: and N.K.S.L.K. Wang, Y.-T. Hung (Eds.), *Handb. Environ. Eng.*, The Humana Press Inc., Totowa, NJ, 57, 1946.
22. P.K. Holt, G.W. Barton, C.A. Mitchell, The future for electrocoagulation as a localised water treatment technology, *Chemosphere.* 59 (2005) 355–367. https://doi.org/10.1016/j.chemosphere.2004.10.023.

23. M. Rodrigo Rodrigo, I. Sirés, E. Billas, M. Oturan, M. Pinazza, *Electrochemical Advanced Oxidation Processes*, 58 (2014) 256–275.
24. M. Bayramoglu, M. Kobya, M. Eyvaz, E. Senturk, Technical and economic analysis of electrocoagulation for the treatment of poultry slaughterhouse wastewater, *Sep. Purif. Technol.* 51 (2006) 404–408. https://doi.org/10.1016/j.seppur.2006.03.003.
25. Z. Abbas, A. Vadim, I. Ravil, P. Bedrikovetsky, Sensitivity study of low salinity water injection in Zichebashskoe Oilfield, *Chem. Eng.* 6 (2015) 10–21. https://doi.org/10.5897/JPGE.
26. M. Behloul, H. Grib, N. Drouiche, N. Abdi, H. Lounici, N. Mameri, Removal of malathion pesticide from polluted solutions by electrocoagulation: Modeling of experimental results using response surface methodology, *Sep. Sci. Technol.* 48 (2013) 664–672. https://doi.org/10.1080/01496395.2012.707734.
27. S. Acharya, S.K. Sharma, G. Chauhan, D. Shree, Statistical optimization of electrocoagulation process for removal of nitrates using response surface methodology, *Indian Chem. Eng.* 0 (2017) 1–16. https://doi.org/10.1080/00194506.2017.1365630.
28. V. Kuokkanen, T. Kuokkanen, J. Rämö, U. Lassi, Recent applications of electrocoagulation in treatment of water and wastewater—A review, *Green Sustain. Chem.* 03 (2013) 89–121. https://doi.org/10.4236/gsc.2013.32013.
29. T. Ramcharan, A. Bissessur, Treatment of laundry wastewater by biological and electrocoagulation methods, *Water Sci. Technol.* 75 (2017) 84–93. https://doi.org/10.2166/wst.2016.464.
30. C. Phalakornkule, T. Luanwuthi, P. Neragae, E.J. Moore, A continuous-flow sparged packed-bed electrocoagulator for dye decolorization, 0 (2016) 1–10. https://doi.org/10.1016/j.jtice.2016.03.046.
31. U. Tezcan Un, S. Topal, F. Ates, Electrocoagulation of tissue paper wastewater and an evaluation of sludge for pyrolysis, *Desalin. Water Treat.* 57 (2016) 28724–28733. https://doi.org/10.1080/19443994.2016.1196153.
32. E. Feilden, Additive manufacturing of ceramics and ceramic composites via robocasting, 2017. https://spiral.imperial.ac.uk/handle/10044/1/55940.
33. E. Abdelkader, L. Nadjia, V. Rose-Noëlle, Adsorption of Congo red azo dye on nanosized SnO_2 derived from sol-gel method, *Int. J. Ind. Chem.* 7 (2016) 53–70. https://doi.org/10.1007/s40090-015-0061-9.
34. C.A. Martínez-Huitle, S. Ferro, Electrochemical oxidation of organic pollutants for the wastewater treatment: Direct and indirect processes, *Chem. Soc. Rev.* 35 (2006) 1324–1340. https://doi.org/10.1039/b517632h.
35. M. Of, S. In, C. Engineering, Electrocoagulation/flotation treatment of synthetic surface water, 2011.
36. S. Yildiz, E. Oran, Sewage sludge disintegration by electrocoagulation, *Int. J. Environ. Health Res.* 29 (2019) 531–543. https://doi.org/10.1080/09603123.2018.1554181.
37. Y. Meas, J.A. Ramirez, M.A. Villalon, T.W. Chapman, Industrial wastewaters treated by electrocoagulation, *Electrochim. Acta.* 55 (2010) 8165–8171. https://doi.org/10.1016/j.electacta.2010.05.018.
38. A. Dura, M. Sc, Electrocoagulation for water treatment: The removal of pollutants using aluminium alloys, stainless steels and iron anodes table of contents, 2013.
39. D. Skban Ibrahim, P. Seethala Devi, C. Veerababhu, N. Balasubramanian, Treatment of petroleum effluent using a tubular electrochemical reactor, *Pet. Sci. Technol.* 32 (2014) 1932–1939. https://doi.org/10.1080/10916466.2012.701693.
40. N. Mohan, N. Balasubramanian, C.A. Basha, Electrochemical oxidation of textile wastewater and its reuse, *J. Hazard. Mater.* 147 (2007) 644–651. https://doi.org/10.1016/j.jhazmat.2007.01.063.
41. M. Kobya, E. Gengec, E. Demirbas, Operating parameters and costs assessments of a real dyehouse wastewater effluent treated by a continuous electrocoagulation process, *Chem. Eng. Process. – Process Intensif.* 101 (2016) 87–100. https://doi.org/10.1016/j.cep.2015.11.012.
42. E.S.Z. El-Ashtoukhy, A.A. Mobarak, Y.O. Fouad, Decolourization of reactive blue 19 dye effluents by electrocoagulation in a batch recycle new electrochemical reactor, *Int. J. Electrochem. Sci.* 11 (2016) 1883–1897.
43. M.S. Ramya Sankar, V. Sivasubramanian, E.V. Vidya Vijay, M. Jerold, J. Kanimozhi, P. Sinu, N. Shankar, Kinetic, isothermal and thermodynamic investigation on electrocoagulation of congo red dye removal from synthetic wastewater using aluminium electrodes, *Desalin. Water Treat.* 122 (2018). https://doi.org/10.5004/dwt.2018.23082.
44. E.V. Vidya Vijay, M. Jerold, M.S. Ramya Sankar, S. Lakshmanan, V. Sivasubramanian, Electrocoagulation using commercial grade aluminium electrode for the removal of crystal violet from aqueous solution, *Water Sci. Technol.* 79 (2019). https://doi.org/10.2166/wst.2019.066.
45. M.S. Ramya Sankar, V. Sivasubramanian, Application of statistical design to optimize the electrocoagulation of synthetic Congo red dye solution and predicting the mechanism, *Int. J. Environ. Sci. Technol.* 17 (2020). https://doi.org/10.1007/s13762-019-02555-5.

46. N. Daneshvar, A.R. Khataee, A.R. Amani Ghadim, M.H. Rasoulifard, Decolorization of C.I. acid Yellow 23 solution by electrocoagulation process: Investigation of operational parameters and evaluation of specific electrical energy consumption (SEEC), *J. Hazard. Mater.* 148 (2007) 566–572. https://doi.org/10.1016/j.jhazmat.2007.03.028.
47. I.A.W. Tan, A.L. Ahmad, B.H. Hameed, Adsorption isotherms, kinetics, thermodynamics and desorption studies of 2,4,6-trichlorophenol on oil palm empty fruit bunch-based activated carbon, *J. Hazard. Mater.* 164 (2009) 473–482. https://doi.org/10.1016/j.jhazmat.2008.08.025.
48. J. Nepo, B. Gourich, M. Cha, Y. Stiriba, C. Vial, P. Drogui, J. Naja, Electrocoagulation process in water treatment: A review of electrocoagulation modeling approaches, 404 (2017) 1–21. https://doi.org/10.1016/j.desal.2016.10.011.
49. S. Vasudevan, F. Epron, J. Lakshmi, S. Ravichandran, S. Mohan, G. Sozhan, Removal of NO3- from drinking water by electrocoagulation – An alternate approach, *Clean – Soil, Air, Water.* 38 (2010) 225–229. https://doi.org/10.1002/clen.200900226.
50. M.Y.A. Mollah, R. Schennach, J.R. Parga, D.L. Cocke, Electrocoagulation (EC) – Science and applications, *J. Hazard. Mater.* 84 (2001) 29–41. https://doi.org/10.1016/S0304-3894(01)00176-5.
51. S. Thakur, M.S. Chauhan, Electro-coagulation integrated with advance oxidation processes: Technical review on treatment of industrial wastewater, *Int. J. Innov. Res. Sci. Eng. Technol. An ISO Certif. Organ.* 3297 (2007) 11018–11023. https://doi.org/10.15680/IJIRSET.2015.0506159.
52. D. Ghosh, C.R. Medhi, M.K. Purkait, Treatment of fluoride containing drinking water by electrocoagulation using monopolar and bipolar electrode connections, *Chemosphere.* 73 (2008) 1393–1400. https://doi.org/10.1016/j.chemosphere.2008.08.041.
53. P. Aswathy, R. Gandhimathi, S.T. Ramesh, P.V. Nidheesh, Removal of organics from bilge water by batch electrocoagulation process, *Sep. Purif. Technol.* 159 (2016) 108–115. https://doi.org/10.1016/j.seppur.2016.01.001.
54. P. Aswathy, R. Gandhimathi, S.T. Ramesh, P.V. Nidheesh, Removal of organics from bilge water by batch electrocoagulation process, *Sep. Purif. Technol.* 159 (2016) 108–115. https://doi.org/10.1016/j.seppur.2016.01.001.
55. T. Ahmad, M. Rafatullah, A. Ghazali, O. Sulaiman, R. Hashim, A. Ahmad, Removal of pesticides from water and wastewater by different adsorbents: A review, *J. Environ. Sci. Heal. – Part C Environ. Carcinog. Ecotoxicol. Rev.* 28 (2010) 231–271. https://doi.org/10.1080/10590501.2010.525782.
56. M. Thite, P.K. Chaudhari, Electrocoagulation of waste water and treatment of rice mill waste water: A review, *J. Sci. Res. (IJSR)* 4 (2015) 233–239.
57. M.M.S.G. Eiband, K.C. Kamélia, K. Gama, J.V. De Melo, C.A. Martínez-Huitle, S. Ferro, Elimination of Pb^{2+} through electrocoagulation: Applicability of adsorptive stripping voltammetry for monitoring the lead concentration during its elimination, *J. Electroanal. Chem.* 717–718 (2014) 213–218. https://doi.org/10.1016/j.jelechem.2014.01.032.
58. A.S. Naje, M.A. Ajeel, P.A. Alaba, Treatment of textile wastewater using a novel electrocoagulation reactor design, *Wastewater Water Qual.* (2018). https://doi.org/10.5772/intechopen.76876.
59. A.I. Adeogun, R.B. Balakrishnan, Electrocoagulation removal of anthraquinone dye Alizarin Red S from aqueous solution using aluminum electrodes: Kinetics, isothermal and thermodynamics studies, *J. Electrochem. Sci. Eng.* 6 (2016) 199–213. https://doi.org/10.5599/jese.290.
60. A.I. Adeogun, R.B. Balakrishnan, Kinetics, isothermal and thermodynamics studies of electrocoagulation removal of basic dye rhodamine B from aqueous solution using steel electrodes, *Appl. Water Sci.* 7 (2017) 1711–1723. https://doi.org/10.1007/s13201-015-0337-4.
61. M.S. Bhatti, D. Kapoor, R.K. Kalia, A.S. Reddy, A.K. Thukral, RSM and ANN modeling for electrocoagulation of copper from simulated wastewater: Multi objective optimization using genetic algorithm approach, *Desalination.* 274 (2011) 74–80. https://doi.org/10.1016/j.desal.2011.01.083.
62. N. Barrak, R. Mannai, M. Zaidi, M. Kechida, A.N. Helal, Experimental design approach with response surface methodology for removal of Indigo dye by electrocoagulation, *J. Geosci. Environ. Prot.* 04 (2016) 50–61. https://doi.org/10.4236/gep.2016.411006.
63. J. Kanimozhi, I.G. Moorthy, R. Sivashankar, V. Sivasubramanian, Optimization of dextran production by Weissella cibaria NITCSK4 using response surface methodology-genetic algorithm based technology, *Carbohydr. Polym.* 174 (2017) 103–110. https://doi.org/10.1016/j.carbpol.2017.06.021.
64. M. Taheri, M.R. Alavi Moghaddam, M. Arami, Optimization of Acid Black 172 decolorization by electrocoagulation using response surface methodology, *Iran. J. Environ. Heal. Sci. Eng.* 9 (2012) 18–20. https://doi.org/10.1186/1735-2746-9-23.
65. H. Rubí-Juárez, C. Barrera-Díaz, I. Linares-Hernández, C. Fall, B. Bilyeu, A combined electrocoagulation-electrooxidation process for carwash wastewater reclamation, *Int. J. Electrochem. Sci.* 10 (2015) 6754–6767.

66. A. Dalvand, M. Gholami, A. Joneidi, N.M. Mahmoodi, Dye removal, energy consumption and operating cost of electrocoagulation of textile wastewater as a clean process, *Clean – Soil, Air, Water.* 39 (2011) 665–672. https://doi.org/10.1002/clen.201000233.
67. F.C. Moreira, R.A.R. Boaventura, E. Brillas, V.J.P. Vilar, Electrochemical advanced oxidation processes: A review on their application to synthetic and real wastewaters, *Appl. Catal. B Environ.* 202 (2017) 217–261. https://doi.org/10.1016/j.apcatb.2016.08.037.
68. T. Ntambwe Kambuyi, F. Eddaqaq, A. Driouich, B. Bejjany, B. Lekhlif, H. Mellouk, K. Digua, A. Dani, Using response surface methodology (RSM) for optimizing turbidity removal by electrocoagulation/electro-flotation in an internal loop airlift reactor, *Water Sci. Technol. Water Supply.* 19 (2019) 2476–2484. https://doi.org/10.2166/ws.2019.129.
69. J. Manoj Babu, S. Goel, Defluoridation of drinking water in batch and continuous-flow electrocoagulation systems, *Pollut. Res.* 32 (2013) 727–736.
70. S. Chellam, D.A. Clifford, A. Bagga, Venkataramanan, A., Electrocoagulation pretreatment for microfiltration: An innovative combination to enhance water quality and reduce fouling in integrated membrane systems University of Houston, 2007.
71. A. Matilainen, M. Vepsäläinen, M. Sillanpää, Natural organic matter removal by coagulation during drinking water treatment: A review, *Adv. Colloid Interface Sci.* 159 (2010) 189–197. https://doi.org/10.1016/j.cis.2010.06.007.
72. S.A. Abdel-Gawad, A.M. Baraka, K.A. Omran, M.M. Mokhtar, Removal of some pesticides from the simulated waste water by electrocoagulation method using iron electrodes, *Int. J. Electrochem. Sci.* 7 (2012) 6654–6665.
73. A.R.A. Aziz, P. Asaithambi, W.M.A.B.W. Daud, Combination of electrocoagulation with advanced oxidation processes for the treatment of distillery industrial effluent, *Process Saf. Environ. Prot.* 99 (2016) 227–235. https://doi.org/10.1016/j.psep.2015.11.010.
74. M. Ben Sasson, W. Calmano, A. Adin, Iron-oxidation processes in an electroflocculation (electrocoagulation) cell, *J. Hazard. Mater.* 171 (2009) 704–709. https://doi.org/10.1016/j.jhazmat.2009.06.057.
75. M. Franchuk, Заліт Грифа Чорного Aegypius Monachus У Прикордонну Зону України Та Білорусі В Межах Рівненської Області М. В. Франчук 1, Ю. М. Струс 2 1 –, 2 (2017) 97–98.
76. J. Lin, Pre-treatment of coal seam water with coagulation and keywords, 2017.
77. M.A. Shafii, N.M. Noh, Soft soil stabilization using sewage sludge ash, *ARPN J. Eng. Appl. Sci.* 13 (2018) 8280–8284.
78. M. Llor, A. Soler, J.F. Ortu, Nutrient removal and sludge production in the coagulation – flocculation process, 36 (2002) 2910–2919.
79. C.C. Ikeagwuani, D.C. Nwonu, Emerging trends in expansive soil stabilisation: A review, *J. Rock Mech. Geotech. Eng.* 11 (2019) 423–440. https://doi.org/10.1016/j.jrmge.2018.08.013.
80. D.A. Emarah, S.A. Seleem, Swelling soils treatment using lime and sea water for roads construction, *Alexandria Eng. J.* 57 (2018) 2357–2365. https://doi.org/10.1016/j.aej.2017.08.009.
81. M. Singh, A. Mittal, A review on the soil stabilization with waste materials, *Int. J. Eng. Res. Appl.* (2014) 2248–9622.
82. R. Sankar, A. Varsha, V. Sivasubramanian, J. Ranjitha, Electrocoagulation technique for cephalothin sodium removal with effective sludge management, *Chem. Eng. Technol.* (2021) 1–11. https://doi.org/10.1002/ceat.202000464.
83. G. Mouedhen, M. Feki, M.D.P. Wery, H.F. Ayedi, Behavior of aluminum electrodes in electrocoagulation process, *J. Hazard. Mater.* 150 (2008) 124–135. https://doi.org/10.1016/j.jhazmat.2007.04.090.

10 Emerging Innovative Technologies for Wastewater Treatment

Sarita Khaturia and Har Lal Singh
Mody University of Science & Technology

Mamta Chahar
Nalanda College of Engineering

Anjali Bishnoi
LD College of Engineering

CONTENTS

10.1 WASTEWATER INTRODUCTION

To make our life easy and comfortable we are using different commercial products and their synthesis generates various by-products. One of the important by-products of this commercial synthesis includes wastewater. Water is called contaminated when it contains impurities which make it useless for various daily life purposes like drinking, washing, cooking and swimming. Various natural conditions also affect the quality of water but it is said to be polluted when contamination is caused by human activity. The drainage of polluted wastewater into surface water or groundwater gives rise to water pollution because this water is unfit for human consumption or use and needs treatment.

DOI: 10.1201/9781003164982-10

Wastewater treatment is also known as sewage treatment in which the removal of various impurities from sewage or wastewater takes place before it reaches water bodies such as rivers, lakes, estuaries, and oceans [1–3]. Fortunately, by employing various wastewater treatment technologies we can make the wastewater potable and usable. These technologies filter and treat wastewater by removing contaminants such as sewage and chemicals.

10.2 WASTEWATER MANAGEMENT: AIM AND NEED

Wastewater contains impurities and is called contaminated or polluted water. Sewerage and industrial waste are the two main sources of polluted water. With the increasing population of India and its industrial landscape, wastewater volume is also increasing at an alarming rise. Day-by-day scarcities in freshwater sources such as rivers, wells, groundwater, etc. have now become an alarming situation and very soon water may become a luxurious asset. The main aim of this wastewater treatment is the removal of various suspended solids present in water called effluent, and then it is discharged back into the environment. Solid material in decaying uses oxygen, which is also needed by plants and water-living animals—due to which wastewater level also rises. This water is not only unfit for consumption but also after mixing with other water sources contaminates them as well.

Examples:

1. Contaminated water when trickling into rivers causes pollution.
2. Polluted water when runs downstream and mixes with other sources of water such as rivers, this contamination spreads rapidly.
3. Wastewater also seeps into the ground and causes contamination of underground water sources.

Nowadays every water source is heavily polluted, which is a very disastrous situation. We have to treat this alarming situation very seriously because; drinking water is essential for all humans and also needed for the food industry as well as farming livestock. Harmful effects of rising levels of wastewater are: less availability of potable water, release of harmful chemicals (Pb and Hg) which are chronic or toxic, groundwater adversely affected and ultimately coastal area becomes polluted. That's why it is today's demand to deal with this situation innovatively and intellectually by employing traditional as well as advanced technologies.

10.3 WASTEWATER MANAGEMENT TECHNIQUES

It is well known that there are various common ways to treat wastewater such as physical water treatment, biological water treatment, chemical treatment, and sludge treatment. In this chapter, we will discuss all these general technologies in short and our main emphasis is on innovative emerging technologies that can be used for the treatment of contaminated water.

Types of wastewater treatment technologies: (a) traditional technologies and (b) advanced technologies.

10.3.1 Traditional Technologies

10.3.1.1 Physical Water Treatment

For cleaning of wastewater physical methods are used in this type of treatment. For the removal of solid material different processes like screening, sedimentation and skimming are employed in this type of treatment (Figure 10.1). There is not any involvement of chemicals in this process. The most important technique of physical wastewater treatment is sedimentation in which insoluble and heavy particles get suspended from the wastewater. We can easily separate the pure water after

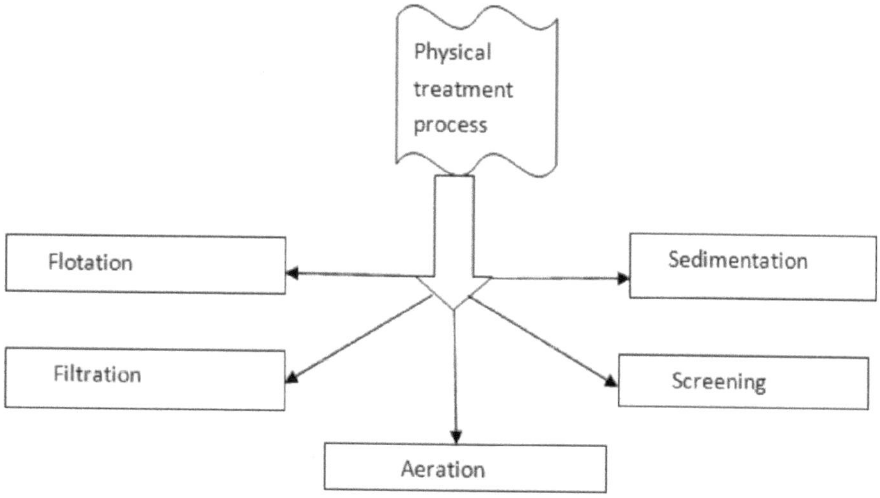

FIGURE 10.1 Physical treatment.

settling down of insoluble material at the bottom. In addition to this one more effective process involves in physical treatment technique is aeration. Aeration involves the circulation of air by the water to provide oxygen to it. Filtration is the third important method by which all the contaminants can be filtered out. Special type of filters is used for passing the wastewater and for separating the contaminants or particles present in wastewater. The most widely used filter is sand filter which is appropriate for this purpose. Mostly sticky products and grease which is found on the wastewater surface can be removed by employing this method [4,5].

10.3.1.2 Biological Water Treatment

For breaking down organic matter, such as soap, human waste, oils and food present in wastewater we can use various biological processes. In the biological treatment process, different microorganisms metabolize the organic matter present in the wastewater. Three processes are involved in this treatment can be summarized as:

i. Aerobic processes: Bacteria decompose the organic matter and convert it into carbon dioxide which is then used by plants (Figure 10.2). Aerobic treatment of wastewater involves the use of oxygen and this process is very simple and efficiently produces high-quality secondary effluent [6,7]. Sludge formed in this process is odor-free and used as potential agricultural fertilizer. It ensures complete contaminant removal when it is used in combination with anaerobic treatment. It means by applying this technique we can safely discharge the wastewater without breaking any environmental rules and regulations. For the aerobic treatment of wastewater, various different technologies are used. These include:
   a. Conventional activated sludge: In an aeration tank, organic matter is broken down by applying aerobic microorganisms which leads to the formation of biological flocs i.e. sludge. In a sedimentation tank, this sludge is now separated from the treated water.
   b. Moving bed biofilm reactor (MBBR): In MBBR growth of biofilm takes place on plastic carriers which are suspended and circulated in an aeration tank. These are kept in the tank by retention sieves.
   c. Membrane bioreactor (MBR): It is an advanced technology which combines both the processes of activated sludge and membrane filtration.

ii. Anaerobic processes [8]: In this process of fermenting, waste fermentation is used at a fixed temperature. In anaerobic process, oxygen is not used. This treatment process involves two

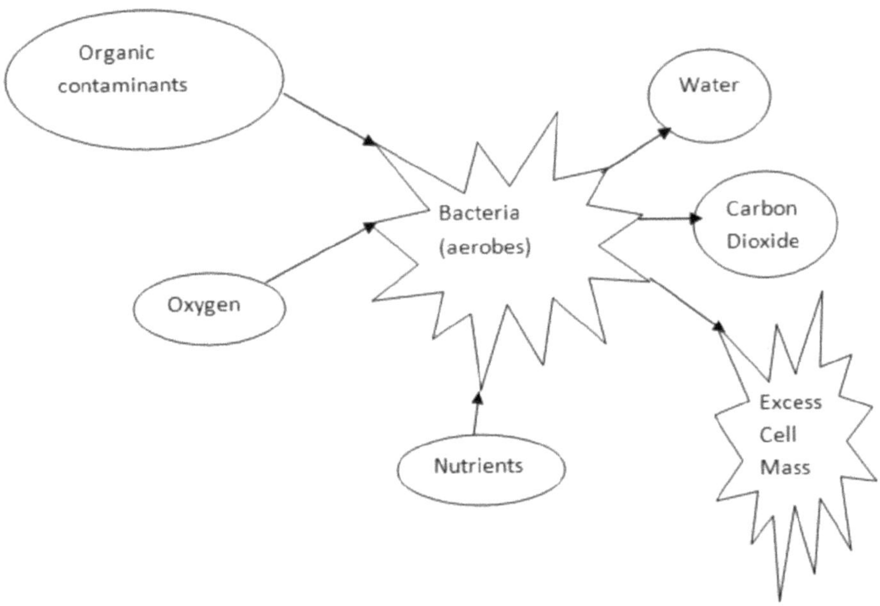

FIGURE 10.2 Aerobic versus anaerobic wastewater treatment.

stages: an acidification phase and methane production phase and both of these processes are in dynamic equilibrium.

iii. Composting: It is a type of aerobic process in which wastewater is treated by mixing it with other carbon sources.

10.3.1.3 Chemical Water Treatment

As the name suggests, this treatment involves the use of chemicals in water [9]. Chlorine is an oxidizing chemical which is used to kill bacteria which are responsible for the contamination of water. For purification purposes, another oxidizing agent can be used i.e. ozone. The natural pH of water is 7 and it can be achieved by employing a neutralization technique in which an acid or base is added. By applying chemicals we can prevent bacterial reproduction in water and stops the adulteration of water thus making the water pure. This treatment processes include chemical precipitation (coagulation, flocculation), ion exchange, neutralization, adsorption, and disinfection (chlorination, ozonization) [10,11]. With some improvements, these techniques will also be discussed in advanced methods.

i. Flocculation: This process uses flocculants for the removal of very fine particles from wastewater. These particles are very thin in size, and it is very difficult to separate them because these do not normally accumulate as larger agglomerates as a result of their electrical repulsion from having an identical charge. By adding a special type of chemicals, these particles are converted into their larger formulations which then settle out by a sedimentation process.

ii. Chemical coagulation: By applying this process destabilization of particles takes place due to which they aggregate during chemical flocculation. These fine dispersed solid particles have negative electric surface charges which prevent them from making larger groups and settling. By chemical coagulation process positively charged coagulants are introduced which destabilizes these particles and also reduce their negative particles' charge. When this charge is reduced, the particles are now free to form larger groups. After this an anionic flocculant is added to the mixture, which can react against this positively charged mixture, it either neutralizes the particle groups or creates bridges between them to bind the particles into larger groups. After the formation of larger particle groups sedimentation process is employed for the removal of particles from the mixture.

iii. Ion exchange: When the hardness of water is high, it is very difficult to use it for cleaning purposes because it leaves a gray residue on clothes. For softening of hard water an ion exchange process can be used which is quite similar to the reverse osmosis process. As we know that calcium and magnesium ions are responsible for water hardness. For softening this hard water, sodium ions which are positively charged are added in the form of dissolved sodium chloride salt or brine. The exchange of these hard calcium and magnesium ions takes place with sodium ions, and free sodium ions are released into the water.

iv. Neutralization: For adjusting the pH of the wastewater this process of neutralization is employed which involves the addition of chemicals (acids or bases) for this purpose. It uses acids (for lowering pH) or alkalis (for raising pH) and depends upon the initial pH of the influent.

v. Adsorption and chemisorption: Both of these processes adsorption and chemisorption will work on the substances which are accumulated on the surfaces of a solid body. Adsorption is the physical process and uses the van der Waal forces. In chemisorption, the same thing happens by using chemical bond.

vi. Water disinfection: This process is completely related with the removal and deactivation of disease-causing microorganisms. These microorganisms are destroyed and become deactivated, due to which their growth and reproduction are terminated. If these pathogenic microorganisms are not removed from drinking water, it will create a problem and cause people to fall ill.

vii. Chlorine and its different forms act as powerful oxidants and are used to kill or inactivate most pathogenic organisms which are harmful to human beings. Chlorination is the main disinfection process used in wastewater treatment due to its following characteristics: (a) these are readily available; (b) very effective and economical in use; (c) their application is quite simple

Various forms of chlorine used for this purpose are as follows:

a. Chlorine in elemental form
b. In hypochlorite form
c. In chlorine dioxide form

viii. Ozonization: Ozone treatment is effectively used for industrial and municipal wastewater. In various fields of wastewater treatment, we can use ozone safely due to its reaction kinetics and mechanism. All these applications are related to its pre-ozonation, post-ozonation or solely ozonation because of its powerful oxidant behavior. Also, there is not any problem in the disposal of sludge and no harmful product formation takes place which leads to increase in the DO content of wastewater. This property is very much helpful in further degradation of residual pollutants. Various types of waste such as municipal wastewater, industrial wastewater, contaminated groundwater, swimming pool water, paper industry wastewater, dye industry wastewater are effectively treated by this method. Also, this treatment is successfully employed for the removal of color, in the treatment of gaseous effluent, in the treatment of cyanide wastewater, in the treatment of heavy metals and in the elimination of phenolic compounds etc. due to its strong disinfection properties. Due to all its extensive properties, it is used as an ultimate treatment almost for all types of wastewater.

10.3.2 Advanced and Innovative Technologies for Wastewater Treatment

10.3.2.1 Requirement of Advanced Technology

In the treatment of wastewater, activated carbons are used for binding with elements which are soluble in the water and it is very difficult to remove them in earlier stages of the mechanical or biological treatment processes. Substance like colorants released from textile dying, residues of pharmaceutical industries, arsenic, and heavy metals can be effectively removed by applying these methods. But the presence of different contaminants in water such as heavy metal ions (Cr^{2+}, Cu^{2+}, Cd^{2+}, Zn^{2+}, arsenic, etc.), waterborne bacteria (*E. coli*), virus (bacteriophage MS2), organic dye effluents cause water pollution and becomes a major threat to the ecology and human health and removal of these contaminants from wastewater is very essential. Different conventional methods have been developed and employed for this purpose but all these methods of wastewater treatment are time-consuming and not cost-effective method. To resolve these issues, new technologies have come into picture. All these technologies have been employed as better alternatives with respect to traditional technologies and also protect the environment and public health.

10.3.3 Five Advanced Technologies Employed for Water Treatment

10.3.3.1 Membrane Separation

Membrane separation is one of the popular emerging techniques and is used to treat mainly pharmaceutical wastewater. Membranes which are used for this separation can be of two types (Figure 10.3):

Membranes which are used for pressure-driven separation processes (microfiltration, ultrafiltration, nanofiltration and reverse osmosis) (Figure 10.4) are made up of synthetic organic polymers such as polyethylene (PE), polytetrafluorethylene (PTFE), polypropylene, and cellulose acetate [14]. Inorganic membranes are prepared from ceramics, metals, zeolites, or silica. These are very stable (chemically/thermally) and used for industrial applications i.e. hydrogen separation, ultrafiltration, microfiltration, etc. [13,15]. In all these processes the main difference is their membrane pore sizes [16].

This technique has various advantages like (a) It controls mass transport in the membrane (b) It also controls fouling tendency of the membrane, as well as separation of solid-liquid particles passing for treatment through the membrane (c) Removal of organic matter and inorganic contaminants in control manner. This type of separation also works as pre-treatment for example distillery

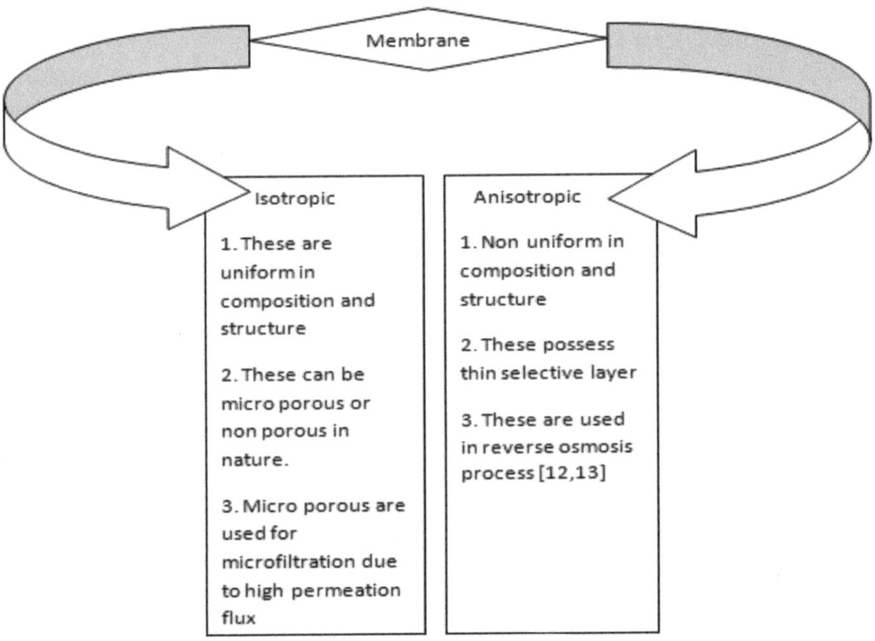

FIGURE 10.3 Types of membrane.

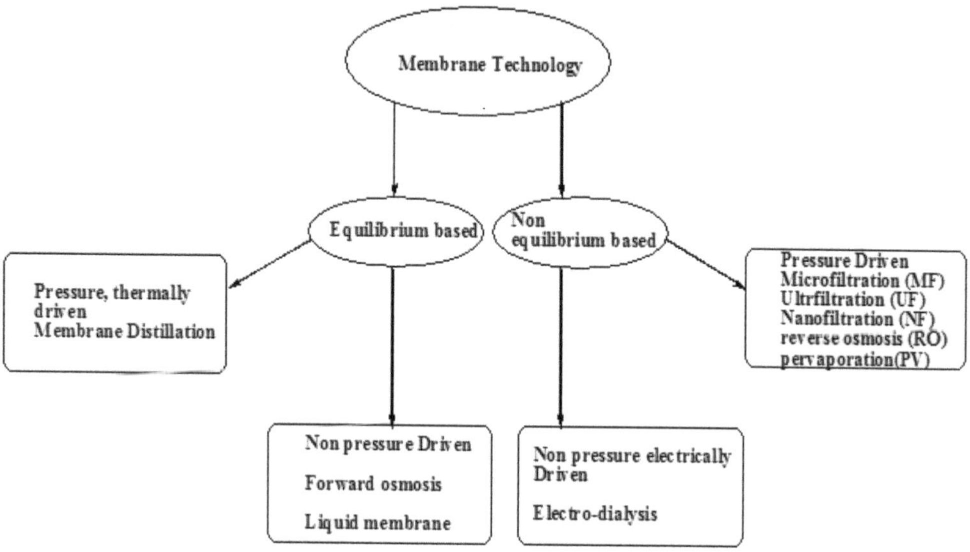

FIGURE 10.4 Schematic representation of some membrane processes.

wastewater is successfully treated by combining NF and RO processes and the removal of contaminants takes place [17]. Contaminants such as urea, reactive dye and oxidizing agents were successfully removed by using UF and RO in combination to treat reactive dye wastewater [18].

10.3.3.2 Irradiation

Another advanced popular technique is irradiation in which wastewater is exposed to radiation, by this organic contaminants are killed and get separated from inorganic one. Different types of irradiation sources (gamma rays, UV light) are used by various scientists for removing pollutants like hormones and antibiotics from water. This technique is not just applicable to organic compounds but its usefulness applies to every kind of contaminant. Ionizing irradiation technique is almost 100% effective for the elimination of pollutants from wastewater.

By using a UV disinfection system contamination of microbes in water is effectively treated by exposing them to UV-C light in the proper amount in order to effectively disinfect the water [20,21]. UV Disinfection Systems are used in many different applications ranging from the purification of drinking water in individual homes to disinfecting water supply of entire townships to industrial wastewater treatment. This technology is recognized as a safer and more cost-effective way to disinfect water for industrial applications. In this UV technology which is used for water disinfection purpose, UV light of wavelength 253.7 nm is used. This wavelength of light is appropriate and suitable for disinfecting bacteria, viruses, molds, algae, and some other microorganisms, which can multiply and grow and are the main cause of contamination. This advanced technology of UV disinfection destroys the DNA of microorganisms and leaves them dead due to which they are unable to grow further. Advanced and innovative UV disinfection technology can be used for various purposes of drinking water disinfection, process water disinfection, wastewater disinfection, and surface disinfection. This technique is not only useful in disinfection applications, but can also be used for TOC (total organic carbon) removal and ozone destruction. In hospitals, factories and offices UV sterilizer is also used for sterilization purposes.

Working: In this UV water disinfection technology, UV light is used for disinfection purposes which disinfects the contaminated water by penetrating microorganisms and destroying their DNA. DNA is very important for the functioning and reproduction of organisms hence destroying the DNA prevents the organism from being active and multiplying. This UV energy (wavelength of 240–280 nm) is also naturally found in sunlight in very small quantities. The same wavelength energy is produced in stronger intensities by using high mercury discharge lamps, commonly known as UV lamps [21]. If the correct dose of UV light is applied then no bacteria, viruses, molds or their spores can survive. Therefore we can utilize this UV technology as the best solution for water and room sterilization, there is a UV mobile sterilizer machine that can be used.

10.3.3.3 Industrial Applications of UV Disinfection System

This Ultraviolet disinfection system is not a simple lamp situated inside a pipe but the UV reactor must be designed in a way that all the microbes must have sufficient exposure to the UV light (dose). According to the hydraulic properties of water, the reactor needs to be optimized so that it would be able to guide the flow in maximizing residence time. UV water disinfection systems which are used at the industrial level are well-designed and produce consistently exceptional results in their industrial applications.

Several Examples belong to this category are:

1. Food and beverage industries: This UV disinfection system is very helpful for achieving high-quality water as per standards set and desired by the FDA (Food and Drug Administration)
2. Bio-pharmaceutical industries: It is essential that water which is used in pharmaceutical and healthcare products and as well as used for cleaning in place (CIP) must be chemical free; like chlorine, ozone, and pathogens. Mostly, pharmaceutical industries and companies must depend upon UV systems for water disinfection purposes.
3. Cosmetics industries: The use of cosmetics is a very sensitive issue, and the water used for their preparation must be free from microorganisms and toxins which ensures its quality and enhance the shelf life of cosmetics. UV Sterilization is the preferred choice for the cosmetic industry across the globe.
4. Supply of centralized drinking water: For the supply of pure water in each and every tap of your home or offices UV drinking water disinfection system is an easy and affordable solution.
5. Wastewater disinfection and reuse: To deal with the water scarcity problem and the rising cost of fresh water, this system of UV disinfection can treat wastewater in the tertiary stage. UV systems that are specially designed for wastewater can thus disinfect wastewater so that the water can be reused for secondary purposes such as flushing and gardening.
6. In swimming pools: Traditionally for cleaning water of swimming pools chlorine has been used. But as we are aware that by using this method of chemical disinfection, chemical reacts with many other organic matters and form hundreds of new chemicals which are harmful. In this case, UV is recognized as a safer and more cost-effective method for the disinfection of swimming pools.

10.3.3.4 Advantages of Disinfection System Using UV

This method is an environmentally friendly and natural way of purification because there are no by-products and toxic compound formation takes place. This technique is very effective due to the susceptibility of all the microorganisms toward UV light. Among all the disinfection systems due to their lowest operating cost, it is economical. In this technique, we are not using any chemicals hence there is no danger of overdosing and it will act as a safer and chemical-free technique. Technique is very easy to manage, fast and instant. Due to various advanced features like CFD (Computational Fluid Dynamics), high-efficiency electronic ballasts and extremely precise UV intensity monitors etc. well designed Alfaa UV systems worked as highly effective apparatus used in this technique.

10.3.3.5 Treatment Technology Using Nanoparticles

Nanotechnology is one of the most exciting developed branches in science working with objects smaller than 100nm [22]. This special branch of nanotechnology mainly shows its usefulness in medicine and engineering, but nowadays it is also used for wastewater cleaning purposes. Nanoparticles are successfully employed for the absorption of pollutants from contaminated water. People have been using special things like activated charcoal for the purification of water for a long time which is based on the adsorption process. Nanoparticle treatment uses this same concept comparatively on a smaller scale. It uses nanotubes which are made of absorbent material like carbon to pull even the smallest particles and pollutants out of water. Although nanotechnology is considered as an expensive technique but with continued development, it will become more affordable and more viable.

The use of nanoparticles is very much helpful in the removal of pesticides and heavy metals from wastewater. Nanoparticles possess various useful characteristics like their direct band gap and layered structure, their high optical absorption coefficient, low cost, and low toxicity. This chapter addresses different properties of nanoparticles and nanoadsorbents contributing to water treatment and used for the removal of numerous pollutants from groundwater. Exceptional characteristics (improved catalysis and adsorption properties) of nanomaterials are the result of their nanoscale size. These are effectively used for the removal of various pollutants in water and thus act as an important advanced technology for wastewater treatment. For this purpose, various kinds of nanomaterial such as zero-valent metal nanoparticles (Ag, Fe, and Zn), metal oxide nanoparticles (TiO_2, ZnO, and iron oxides), carbon nanotubes (CNTs), and nanocomposites can be employed. Mechanical, electrical, optical, and magnetic properties of these nanomaterials are significantly different from conventional materials due to their nanoscale size and large specific surface areas and these can show exceptional characteristics of catalysis, adsorption, and high reactivity and high mobility in solution [23]. Heavy metals [24], organic pollutants [25], inorganic anions [26], and bacteria [27] have been successfully removed by employing this technology.

10.3.3.5.1 Zero-Valent Metal Nanoparticles for Wastewater Treatment

In this category, silver nanoparticles (Ag NPs) act as highly toxic material toward microorganisms and also possess strong antibacterial effects against a wide range of microorganisms, including bacteria [27], viruses [28], and fungi [29]. As a good antimicrobial agent, silver nanoparticles have been widely used for the disinfection of water.

Mechanism of Ag NPs: When Ag NPs are in contact with bacteria, free radicals can be generated and show their ability to damage the cell membrane and cause cell deaths [30]. In addition, as DNA contains abundant sulfur and phosphorus elements, Ag NPs can act with it and thus destroy it. This is another explanation for the death of cells caused by Ag NPs [31]. These Ag NPs attached to filter materials and water disinfection due to their high antibacterial activity and cost-effectiveness [32].

Similar to this, various zero-valent metal nanoparticles, such as Fe, Zn, Al, and Ni, are successfully used in water pollution treatment. Due to its high reductive ability, nano-zero-valent Al is thermodynamically unstable in the presence of water, which favors the formation of oxides/hydroxides on the surface, impeding (completely) the transfer of electrons from the metal surface to the contaminants [33]. With a moderate standard reduction potential, nano-zero-valent Fe or Zn holds the good potential to act as reducing agents relative to many redox-labile contaminants. Despite a weaker reduction ability, Fe possesses many prominent advantages over Zn for applications in water pollution treatment, including excellent adsorption properties, precipitation and oxidation (in the presence of dissolved oxygen), and low cost. Therefore, zero-valent iron nanoparticles have been the most extensively studied zero-valent metal nanoparticles.

10.3.3.5.2 Metal Oxides Nanoparticles

In this category of nanomaterials TiO_2 act as the most exceptional photocatalyst and shows its applications in wastewater treatment also. The large band gap energy (3.2 eV) of TiO_2 requires ultraviolet (UV) excitation to induce charge separation within the particles. By UV irradiation, TiO_2 will generate reactive oxygen species (ROS) and used for the complete degradation of contaminants in a very short reaction time. Besides, TiO_2 NPs show little selectivity and thus are suitable for the degradation of all kinds of contaminants, such as chlorinated organic compounds [34], polycyclic aromatic hydrocarbons [35], dyes [36], phenols [37], pesticides [38], arsenic [39], cyanide [40], and heavy metals [41]. Hydroxyl radicals generated under UV irradiation (nm) enable TiO_2 NPs to damage the function and structure of various cells. The photocatalytic properties of TiO_2 NPs are able to kill a wide array of microorganisms, such as Gram-negative and Gram-positive bacteria, as well as fungi, algae, protozoa, and viruses [42]. In the field of photocatalysis, apart from TiO_2 NPs, ZnO NPs have emerged as another efficient candidate in water and wastewater treatment because of their unique characteristics, such as direct and wide band gap in the near-UV spectral region, strong oxidation ability, and good photocatalytic property [43–45]. ZnO NPs are environment-friendly as they are compatible with organisms [46], which makes them suitable for the treatment of wastewater. In recent years, there is a growing interest in the use of iron oxide nanoparticles for the removal of heavy metals due to their simplicity and availability. Three oxides of iron like magnetic magnetite (Fe_3O_4) and magnetic maghemite (γ-Fe_2O_4) and nonmagnetic hematite (α-Fe_2O_3) are used as nanoadsorbents. To increase adsorption efficiency and to avoid interference from other metals ions, iron oxides nanoparticles have been functionalized to tune their adsorption properties by adding various ligands (e.g., ethylenediamine tetraacetic acid (EDTA), L-glutathione (GSH), mercaptobutyric acid (MBA), α-thio-ω-(propionic acid) hepta(ethylene glycol) (PEG-SH), and meso-2,3-dimercaptosuccinic acid (DMSA)) [47] or polymers (e.g., copolymers of acrylic acid and crotonic acid) [48]. A flexible ligand shell has been reported to facilitate the incorporation of a wide array of functional groups into the shell and ensured the properties of Fe_3O_4 nanoparticles are intact [49]. Besides, a polymer shell has been found to be able to prevent the aggregation of particles and improve the dispersion stability of the nanostructures [48]. Polymer molecules could act as binders for metal ions and thus became a "carrier" of metal ions from treated water [49]. Hematite (α-Fe_2O_3) has been considered as a stable and cheap material in sensors, catalysis, and environmental applications [50]. Moreover, nanohematite has also been demonstrated to be an effective adsorbent for the removal of heavy metal ions from spiked tap water [51].

10.3.3.5.3 Carbon Nanotubes

Unique structure and electronic properties of carbon nanomaterials (CNMs) make them attractive for fundamental studies as well as diverse applications, especially in sorption processes. Their large specific surface area, fast kinetics and great capacity to absorb a wide range of contaminants make them useful for wastewater treatment [23]. CNMs are available in various forms such as carbon beads, carbon fibers, and nanoporous carbon [23]. Among all these CNTs have attracted the most attention and progressed rapidly in recent years. The diameter of carbon nanotubes is very small as 1 nm and in these graphene sheets are rolled up in cylinders [52] and act as an emerging adsorbent due to their unique properties. For removal of different contaminants, like dichlorobenzene [53], ethyl benzene [54], Zn^{2+} [55], Pb^{2+}, Cu^{2+}, and Cd^{2+} [56], and dyes [57] these CNTs are successfully employed because of its porous structures, exceptional adsorption capabilities and high adsorption efficiencies. Based on their (super) structures, CNTs can be classified into two types:

a. Multiwalled carbon nanotubes (MWCNTs): These are multiple layers of concentric cylinders have spacing of about 0.34 nm between the adjacent layers.
b. Single-walled carbon nanotubes (SWCNTs): These are made up of single layers of graphene sheets rolled into cylindrical tubes [58]. Both of these MWCNTs [55–57] and SWCNTs [59] have been very well applied for contaminants removal from water.

10.3.3.5.4 Nanocomposites

There are different problems associated with the removal efficiency of various nanomaterials like:

a. nZVI suffers from aggregation, oxidation, and separation difficulty
b. Light adsorption of TiO_2 and ZnO NPs is limited in the ultraviolet light region due to their big band gap energies.

c. Nanofiltration membranes are troubled by the problem of membrane fouling carbon nanotubes are mainly limited by their low volume of production and high cost as well as the need for supporting medium or matrix it is a common and effective strategy to fabricate nanocomposites for wastewater treatment. Various nanocomposites are used in the nanomaterial field and these novel nanoscale adsorbent are synthesized via chemical deposition of nZVI on CNTs which has the potential for effective removal of nitrate from water. These substances possess unique magnetic properties and are easily separated from solution by the use of magnet [60]. Besides, thin film nanocomposite (TFN) nanofiltration membranes have been prepared via in situ interfacial incorporation of TiO_2 NPs along with the fabrication of copolyamide network on a polyimide support. To improve the compatibility of TiO_2 NPs inside the polymer matrix, both amine and chloride compounds were utilized to functionalize TiO_2 NPs. TFN membranes exhibited higher methanol flux and dye rejection in spite of lower swelling degree. The loading of TiO_2 NPs turned out to be a crucial factor in the NF membrane performance [61].

10.4 BIOAUGMENTATION

For the purification of wastewater, the involvement of biological processes is very well utilized in pharmaceutical industries and one popular and promising biological approach in this category is bioaugmentation. In this biological process mixture of microorganisms (enzymes and bacterial strains) is introduced in polluted liquid which causes breaking followed by the removal of contaminants. By applying this organic method oil and carbon impurities are naturally degraded by these organisms. This process not only removes harmful pollutants but also releases nutrients as a by-product which is to be considered as potential and healthier benefit of this method. The potential of this option can now be enhanced in order to take advantage of important advances in the fields of microbial ecology, molecular biology, immobilization techniques and advanced bioreactor design. In municipal wastewater treatment, this process is commonly used to restart activated sludge bioreactors. Most cultures available contain microbial cultures, already containing all necessary microorganisms (*B. licheniformis*, *B. thuringiensis*, *P. polymyxa*, *B. stearothermophilus*, *Penicillium* sp., *Aspergillus* sp., *Flavobacterium*, *Arthrobacter*, *Pseudomonas*, *Streptomyces*, *Saccharomyces*, etc.). Activated sludge systems are generally based on microorganisms like bacteria, protozoa, nematodes, rotifers, and fungi, which are capable of degrading biodegradable organic matter. There are many positive outcomes from the use of bioaugmentation, such as the improvement in efficiency and speed of the process of breaking down substances and the reduction of toxic particles in an area [62]. It is an attractive strategy for the removal of recalcitrant pollutants from wastewater. This approach has proved to be successful in laboratory investigations, but some challenges still exist, especially in scaling up these processes. This process is successfully employed for the removal of chlorinated compounds by Dehalococcoides bacteria from groundwater. There are some limitations to using this process on a broad scale. One is difficulty in maintaining sufficient numbers of biodegrading microorganisms (at least as high as 106–107 cells/mL) in the environment during the bioaugmentation process.

10.5 HYBRID TECHNOLOGY

Due to the high pollutant contents in the water, clean, pure and freshwater resources are in high demand in today's world. In this connection recycling of wastewater will act as a new source of water supply. For this purpose hybrid wastewater treatment system is developed and used for the effective removal of pollutants. As we are aware that wastewater contains energy in the form of organic matter which is biodegradable. To generate and utilize this energy as well as for the removal of pollutants, a new trend of hybrid wastewater treatment system comes into the picture. By employing these hybrid systems energy gained is both sustainable and environmental friendly and becomes good source of bio-energy to compliment the power of a treatment plant. There are different combinations of hybrid wastewater systems such as: physical–biological hybrid which is made by combination of physical and biological system and similar to this physical–chemical hybrid, chemical–biological hybrid and physical–chemical–biological hybrid systems can also occur which are formed by respective combinations of physical and chemical, chemical and biological and last one is made by combinations of physical, chemical and biological hybrid systems. These hybrid systems possess various potential advantages as compared to stand-alone systems like they are more stable and sustainable in the generation of voltage and these are better in overall treatment efficiency and energy savings [63,64]. Some of the most effective modern methods of water purification are hybrid solutions. More and more pharma companies opt for membrane-bioaugmentation hybrid technology instead of picking one part of the process over the other. Membrane bioreactors are an increasingly widespread example of this technology. Membrane bioreactors first feed wastewater through a device called a bioreactor. These containers host active biological elements, like the bacteria and enzymes found in bioaugmentation. After the bioreactor, water moves through a membrane, which filters out the microorganisms as well as any leftover contaminants. The water membrane bioreactors produce is among the cleanest of any treatment process. In addition to that, these systems also allow chemists to recycle some of the removed materials from the water. That added benefit makes hybrid solutions some of the most economically viable options for pharma companies.

10.6 CONCLUSIONS AND PERSPECTIVES

As we all are aware that different textile, food and beverages agricultural, pharmaceutical industries can produce a lot of harmful pollutants in its water usage. Thankfully, however, water treatment methods improve all the time, so all these industries can use water without endangering the environment. These purification techniques are also economically enticing for the pharma industry. By thoroughly cleaning their water, industrial and pharmaceutical companies can recycle it in their processes, saving money. These modern purification methods are useful enough that they can produce the near-perfect level of purity that pharmaceutical processes require. By recycling water, these companies also reduce their amount of water waste, further protecting the environment. This industrial wastewater, if untreated, poses a threat to both humans and nature. The use of advanced treatment technologies makes pharmaceutical and industrial water environmentally safe and economically viable. Here we have discussed various traditional and advanced technologies in detail. Their current development speed and successful employment look extremely promising and make them useful for the treatment of wastewater.

Current approaches to waste management will have to change in the future since waste will have to be considered as an alternative resource rather than an inconvenient burden to dispose of Wastewater, in particular, represents an important resource of nutrients, energy (as energy contained in chemical bonds of organic matter), and water itself. In spite of the great need for improvement, the hybrid system proposed in this research is undoubtedly a promising technology that offers a vast range of applications. Bioelectrochemical wastewater treatment via the hybrid system is a novel and promising approach to wastewater treatment and simultaneous production of renewable electricity and hydrogen. The application of principles of MFCs and MECs is the main focus of this research. Design and operation parameters have to be defined in order to scale up the hybrid system to practical applications other than powering small sensor devices. Additionally, the hybrid system proposed in this work has to be competitive when compared with traditional wastewater treatment and needs to be more cost-effective than existing technologies.

REFERENCES

1. A.A. Al-Raad, M.M. Hanafiah, A.S. Naje, M.A. Ajeel, A.O Basheer, T.A. Ali, M. Ekhwan Toriman, Treatment of saline water using electrocoagulation with combined electrical connection of electrodes. *Processes* 7: 242, 2019.

2. I.H. Ali, M.K. Al Mesfer, M.I. Khan, M. Danish, M.M. Alghamdi, Exploring adsorption process of lead(II) and chromium(VI) ions from aqueous solutions on acid activated carbon prepared from Juniperus procera leaves. *Processes* 7: 217, 2019.
3. A.N. Amro, M.K. Abhary, M.M. Shaikh, S. Ali, Removal of lead and cadmium ions from aqueous solution by adsorption on a low-cost phragmites biomass. *Processes* 7: 406, 2019.
4. M.F. Hamoda, I. Al-Ghusain, N.Z. Al-Mutairi, Sand filtration of wastewater for tertiary treatment and water reuse. *Desalination* 164(3): 203–211, 2004.
5. T.H.Y. Tebbutt, Primary sedimentation of wastewater. *Water Pollution Control Federation*: 2858–2867, 1979.
6. M. Samer, GHG emission from livestock manure and its mitigation strategies, In: *Climate Change Impact on Livestock: Adaptation and Mitigation*, V. Sejian, J. Gaughan, L. Baumgard & C. Prasad (Eds.), Springer International: Germany, ISBN 978-81-322-2264-4, 2015.
7. M. Samer, E. Mostafa, A.M. Hassan, Slurry treatment with food industry wastes for reducing methane, nitrous oxide and ammonia emissions. *Journal of Agricultural Engineering* 31(4): 1523–1548, 2014.
8. M. Samer, Biogas plant constructions, In: *Biogas*, S. Kumar (Ed.), pp. 343–368, InTech: Rijeka, Croatia, ISBN 978-953-51-0204-5, 2012.
9. D.L. Russell, *Practical Wastewater Treatment*, John Wiley & Sons, Inc.: Hoboken, NJ, ISBN-13:978-0-471-78044-1, 2006.
10. M.L. Davis, *Water and Wastewater Engineering: Design Principles and Practice*, McGraw-Hill Companies, Inc.: New York, NY, ISBN 978-0-07-171385-6, 2010.
11. D.A. Clifford, Ion exchange and inorganic adsorption, In: *Water Quality and Treatment* (5th Edition), R.D. Letterman (Ed.), pp. 9.1–9.91, American Water Works Association, McGraw-Hill: New York, NY, 1999.
12. A. Sagle, B. Freeman, Fundamentals of membranes for water treatment. *Future of Desalination in Texas* 2: 137, 2004.
13. R.W. Baker, *Membrane Technology and Applications*, Wiley Online Library: Hoboken, NJ, 2012.
14. U.M. Aliyu, S. Rathilal, Y.M. Isa, Membrane desalination technologies in water treatment: a review. *Water Practice and Technology* 13: 738–752, 2018.
15. R. Mallada, M. Menéndez, *Inorganic Membranes: Synthesis, Characterization and Applications*, Elsevier: Amsterdam, The Netherlands, 2008.
16. M. Chollom, S. Rathilal, V.L. Pillay, Treatment and reuse of reactive dye effluent from textile industry using membrane technology. Ph.D. Thesis, Durban University of Technology, Durban, South Africa, 2014.
17. S.K. Nataraj, K.M. Hosamani, T.M. Aminabhavi, Distillery wastewater treatment by the membrane-based nanofiltration and reverse osmosis processes. *Water Research* 40: 2349–2356, 2006.
18. S. Šostar-Turk, M. Simonič, I. Petrinić, Wastewater treatment after reactive printing. *Dyes and Pigments* 64: 147–152, 2005.
19. J.H. Jhaveri, Z.V.P. Murthy, A comprehensive review on anti-fouling nanocomposite membranes for pressure driven membrane separation processes. *Desalination* 379: 137–154, 2016.
20. B.A. Wols, C.H.M. Hofman-caris, Review of photochemical reaction constants of organic micropollutants required for UV advanced oxidation processes in water. *Water Research* 46: 2815–2827, 2012.
21. https://www.alfaauv.com/blog/uv-disinfection-system-water-treatment.
22. C. Buzea, I.I. Pacheco, K. Robbie, Nanomaterials and nanoparticles: sources and toxicity. *Biointerphases* 2(4): MR17–MR71, 2007.
23. M.M. Khin, A.S. Nair, V.J. Babu, R. Murugan, S. Ramakrishna, A review on nanomaterials for environmental remediation. *Energy & Environmental Science* 5(8): 8075–8109, 2012.
24. W.-W. Tang, G.-M. Zeng, J.-L. Gong et al., Impact of humic/fulvic acid on the removal of heavy metals from aqueous solutions using nanomaterials: a review. *Science of the Total Environment* 468–469: 1014–1027, 2014.
25. J. Yan, L. Han, W. Gao, S. Xue, M. Chen, Biochar supported nanoscale zerovalent iron composite used as persulfate activator for removing trichloroethylene. *Bioresource Technology* 175: 269–274, 2015.
26. F. Liu, J.H. Yang, J. Zuo et al., Graphene-supported nanoscale zero-valent iron: removal of phosphorus from aqueous solution and mechanistic study. *Journal of Environmental Sciences* 26(8): 1751–1762, 2014.
27. R.S. Kalhapure, S.J. Sonawane, D.R. Sikwal et al., Solid lipid nanoparticles of clotrimazole silver complex: an efficient nano antibacterial against Staphylococcus aureus and MRSA. *Colloids and Surfaces B: Biointerfaces* 136: 651–658, 2015.
28. B. Borrego, G. Lorenzo, J.D. Mota-Morales et al., Potential application of silver nanoparticles to control the infectivity of Rift Valley fever virus in vitro and in vivo. *Nanomedicine: Nanotechnology, Biology and Medicine* 12(5): 1185–1192, 2016.

29. C. Krishnaraj, R. Ramachandran, K. Mohan, P.T. Kalaichelvan, Optimization for rapid synthesis of silver nanoparticles and its effect on phytopathogenic fungi. *Spectrochimica Acta—Part A: Molecular and Biomolecular Spectroscopy* 93: 95–99, 2012.
30. M. Danilczuk, A. Lund, J. Sadlo, H. Yamada, J. Michalik, Conduction electron spin resonance of small silver particles. *Spectrochimica Acta—Part A: Molecular and Biomolecular Spectroscopy* 63(1): 189–191, 2006.
31. K.I. Dhanalekshmi, K.S. Meena, DNA intercalation studies and antimicrobial activity of Ag@ZrO_2 core–shellnanoparticles in vitro. *Materials Science and Engineering: C* 59: 1063–1068, 2016.
32. D.V. Quang, P.B. Sarawade, S.J. Jeon et al., Effective water disinfection using silver nanoparticle containing silica beads. *Applied Surface Science* 266: 280–287, 2013.
33. M. Rivero-Huguet, W.D. Marshall, Reduction of hexavalent chromium mediated by micron- and nanoscale zerovalent metallic particles. *Journal of Environmental Monitoring* 11(5): 1072–1079, 2009.
34. T. Ohsaka, K. Shinozaki, K. Tsuruta, K. Hirano, Photoelectrochemical degradation of some chlorinated organic compounds on n-TiO_2 electrode. *Chemosphere* 73(8): 1279–1283, 2008.
35. M. Guo, W. Song, T. Wang, Y. Li, X. Wang, X. Du, Phenylfunctionalization of titanium dioxidenanosheets coating fabricated on a titanium wire for selective solid-phase microextraction of polycyclic aromatic hydrocarbons from environment water samples. *Talanta* 144: 998–1006, 2015.
36. Y. Lee, S. Kim, P. Venkateswaran, J. Jang, H. Kim, J. Kim, Anion co-doped titania for solar photocatalytic degradation of dyes. *Carbon Letters* 9(2): 131–136, 2008.
37. A.T. Nguyen, C.-T. Hsieh, R.-S. Juang, Substituent effects on photodegradation of phenols in binary mixtures by hybrid H_2O_2 and TiO_2 suspensions under UV irradiation. *Journal of the Taiwan Institute of Chemical Engineers* 62: 68–75, 2016.
38. M.G. Alalm, A. Tawfik, S. Ookawara, Comparison of solar TiO_2 photocatalysis and solar photo-Fenton for treatment of pesticides industry wastewater: operational conditions, kinetics, and costs. *Journal of Water Process Engineering* 8: 55–63, 2015.
39. G. Moon, D. Kim, H. Kim, A.D. Bokare, W. Choi, Platinum-like behavior of reduced graphene oxide as a cocatalyst on TiO_2 for the efficient photocatalytic oxidation of arsenite. *Environmental Science & Technology Letters* 1(2): 185–190, 2014.
40. S.H. Kim, S.W. Lee, G.M. Lee, B.-T. Lee, S.-T. Yun, S.-O. Kim, Monitoring of TiO_2-catalytic UV-LED photo-oxidation of cyanide contained in mine wastewater and leachate. *Chemosphere* 143: 106–114, 2016.
41. Z.P. Chen, Y. Li, M. Guo et al., One-pot synthesis of Mn-doped TiO_2 grown on graphene and the mechanism for removal of Cr(VI) and Cr(III). *Journal of Hazardous Materials* 310: 188–198, 2016.
42. H.A. Foster, I.B. Ditta, S. Varghese, A. Steele, Photocatalytic disinfection using titanium dioxide: spectrum and mechanism of antimicrobial activity. *Applied Microbiology and Biotechnology* 90(6): 1847–1868, 2011.
43. A. Janotti, C.G. Van de Walle, Fundamentals of zinc oxide as a semiconductor. *Reports on Progress in Physics* 72(12): Article ID 126501, 2009.
44. D.C. Reynolds, D.C. Look, B. Jogai, C.W. Litton, G. Cantwell, W.C. Harsch, Valence-band ordering in ZnO. *Physical Review B—Condensed Matter and Materials Physics* 60(4): 2340–2344, 1999.
45. Y. Chen, D.M. Bagnall, H.-J. Koh et al., Plasma assisted molecular beam epitaxy of ZnO on c-plane sapphire: growth and characterization. *Journal of Applied Physics* 84(7): 3912–3918, 1998.
46. M.L. Schmidt, D.J.L. MacManus, ZnO-nanostructures, defects, and devices. *Materials Today* 10: 40–48, 2007.
47. C.L. Warner, R.S. Addleman, A.D. Cinson et al., High performance, superparamagnetic, nanoparticle-based heavy metal sorbents for removal of contaminants from natural waters. *ChemSusChem* 3(6): 749–757, 2010.
48. F. Ge, M.-M. Li, H. Ye, B.-X. Zhao, Effective removal of heavy metal ions Cd^{2+}, Zn^{2+}, Pb^{2+}, Cu^{2+} from aqueous solution by polymer-modified magnetic nanoparticles. *Journal of Hazardous Materials* 211–212: 366–372, 2012.
49. R.A. Khaydarov, R.R. Khaydarov, O. Gapurova, Water purification from metal ions using carbon nanoparticleconjugated polymer nanocomposites. *Water Research* 44(6): 1927–1933, 2010.
50. H. Liang, B. Xu, Z. Wang, Self-assembled 3D flower-like α-Fe_2O_3 microstructures and their superior capability for heavy metal ion removal. *Materials Chemistry and Physics* 141(2–3): 727–734, 2013.
51. H.J. Shipley, K.E. Engates, V.A. Grover, Removal of Pb(II), Cd(II), Cu(II), and Zn(II) by hematite nanoparticles: effect of sorbent concentration, pH, temperature, and exhaustion. *Environmental Science and Pollution Research* 20(3): 1727–1736, 2013.

52. A. Chatterjee, B.L. Deopura, Carbon nanotubes and nanofibre: an overview. *Fibers and Polymers* 3(4): 134–139, 2002.
53. X. Peng, Y. Li, Z. Luan et al., Adsorption of 1,2-dichlorobenzene from water to carbon nanotubes. *Chemical Physics Letters* 376(1–2): 154–158, 2003.
54. C. Lu, F.S. Su, S. Hu, Surface modification of carbon nanotubes for enhancing BTEX adsorption from aqueous solutions. *Applied Surface Science* 254(21): 7035–7041, 2008.
55. H.-H. Cho, K. Wepasnick, B.A. Smith, F.K. Bangash, D.H. Fairbrother, W.P. Ball, Sorption of aqueous Zn[II] and Cd[II] by multiwall carbon nanotubes: the relative roles of oxygen-containing functional groups and graphenic carbon. *Langmuir* 26(2): 967–981, 2010.
56. Y.-H. Li, J. Ding, Z. Luan et al., Competitive adsorption of Pb^{2+}, Cu^{2+} and Cd^{2+} ions from aqueous solutions by multiwalled carbon nanotubes. *Carbon* 41(14): 2787–2792, 2003.
57. T. Madrakian, A. Afkhami, M. Ahmadi, H. Bagheri, Removal of some cationic dyes from aqueous solutions using magnetic-modified multi-walled carbon nanotubes. *Journal of Hazardous Materials* 196: 109–114, 2011.
58. Y.-L. Zhao, J.F. Stoddart, Noncovalent functionalization of single-walled carbon nanotubes. *Accounts of Chemical Research* 42(8): 1161–1171, 2009.
59. C.-M. Yang, J.S. Park, K.H. An et al., Selective removal of metallic single-walled carbon nanotubes with small diameters by using nitric and sulfuric acids. *Journal of Physical Chemistry B* 109(41): 19242–19248, 2005.
60. M. Herrero, D.C. Stuckey, Bioaugmentation and its application in wastewater treatment: a review. *Chemosphere* 140: 119–128, 2015.
61. A. Nzila, S.A. Razzak, J. Zhu, Bioaugmentation: an emerging strategy of industrial wastewater treatment for reuse and discharge. *International Journal of Environmental Research and Public Health* 13(9): 846, 2016.
62. C.M. Huban, R.D. Plowman, Bioaugmentation: put microbes to work. *Chemical Engineering* 104(3), 1997.
63. P.F. Tee, M.O. Abdullah, I.A.W. Tan, N.K.A. Rashid, M.A.M. Amin, C. Nolasco-Hipolito, K. Bujang, Review on hybrid energy systems for wastewater treatment and bio-energy production. *Renewable and Sustainable Energy Reviews* 54: 235–246, 2016.
64. F.I. Hai, K. Yamamoto, K. Fukushi, Hybrid treatment systems for dye wastewater. *Critical Reviews in Environmental Science and Technology* 37(4): 315–377, 2007.

11 Bio Strategies for the Removal of Contaminants of Emerging Concern from Wastewater

Cristina Quintelas, Daniela Mesquita, and Eugénio Campos Ferreira
University of Minho

CONTENTS

11.1 INTRODUCTION

According to the literature, contaminants of emerging concern (CEC) are chemical substances or compounds characterised by a perceived or veridical threat to the environment or human health with a lack of published health criteria and can also be compounds identified from an unknown source, a new exposure to humans, or a novel detection approach or technology (Lei et al., 2015; Shah, 2020, 2021). CEC include many types of contaminants such as pesticides, pharmaceuticals and personal care products, fragrances, plasticisers, hormones, flame retardants, nanoparticles, perfluoroalkyl compounds, chlorinated paraffins, siloxanes, algal toxins, various trace elements including rare earths and radionuclides (Sauvé and Desrosiers, 2014), polychlorinated naphthalenes, biocides such as pentachlorophenol (PCP), lindane, dichlorodiphenyltrichloroethane (DDT), pyrethroids, etc. for outdoor and indoor applications, and plasticisers, as polychlorinated biphenyls and polybrominated biphenyl ethers (Salthammer, 2020).

Due to CEC's increasing use in industry, transport, agriculture, and urbanisation, these chemicals are entering the environment at increasing levels causing threats to ecological systems and human health. The focus of this chapter is to assess the bio strategies for the removal of CEC from wastewater. The content of this chapter is structured as follows: In Section 11.2, an overview of the occurrence of CEC around the world will be performed. Aspects such as main sources, seasonal changes, range of concentrations found and ecological/human health effects will be addressed. In Section 11.3, several biological methods to remove contaminants of emerging concern in wastewater will be presented and the advantages/disadvantages of each one will be analysed. The main methods presented included activated sludge, granular activated sludge, microalgae-based treatments, membrane bioreactors, and anaerobic digestion, among others (Figure 11.1). In Section 11.4, the challenges and future perspectives to overcome the contamination by CEC will be presented.

DOI: 10.1201/9781003164982-11

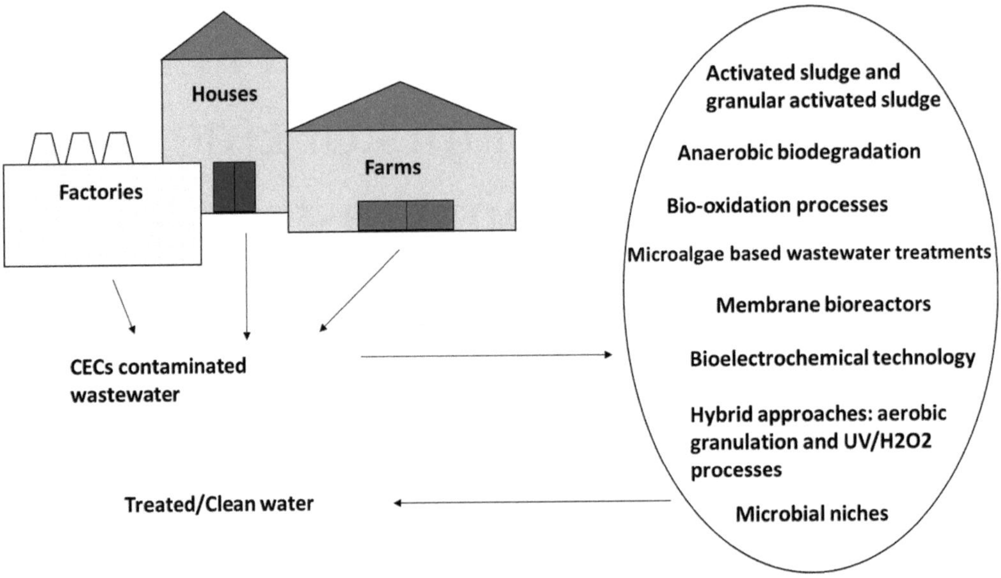

FIGURE 11.1 CECs sources and main methods to remove these compounds from wastewater.

11.2 OCCURRENCE OF CONTAMINANTS OF EMERGING CONCERN IN WASTEWATER

The presence of CEC is widespread throughout the world. In 2013, a team of 16 researchers led by Robert Loos (Loos et al., 2013; Shah, 2021) performed an EU-wide monitoring survey on emerging polar organic contaminants in wastewater treatment plant (WWTPs) effluents. These authors analysed 90 European WWTPs and found 156 different CEC. The list of CEC found included pharmaceuticals and personal care products, veterinary (antibiotic) drugs, perfluoroalkyl substances, organophosphate ester flame retardants, pesticides (and some metabolites), industrial chemicals such as benzotriazoles (corrosion inhibitors), iodinated X-ray contrast agents, and gadolinium magnetic resonance imaging agents. 80% of the target compounds are present in concentrations ranging from low nanograms to milligrams per litre. The most relevant compounds in the WWTPs with the highest median concentration levels were the artificial sweeteners acesulfame and sucralose, benzotriazoles (corrosion inhibitors), several organophosphate ester flame retardants and plasticisers, pharmaceutical compounds such as carbamazepine, tramadol, telmisartan, venlafaxine, irbesartan, fluconazole, oxazepam, fexofenadine, diclofenac, citalopram, codeine, bisoprolol, eprosartan, the antibiotics trimethoprim, ciprofloxacine, sulfamethoxazole, and clindamycine, the insect repellent N, N0-diethyltoluamide (DEET), the pesticides MCPA and mecoprop, perfluoroalkyl substances (such as PFOS and PFOA), caffeine, and gadolinium. Some examples of the highest concentrations found include 1H-benzotriazole (221 mg/L), iopromide (150 mg/L), TBEP (43 mg/L), methylbenzotriazole (24 mg/L), PFHxA (23.9 mg/L), TCPP (21 mg/L), irbesartan (17.9 mg/L), PFOA (15.9 mg/L), DEET (N, N0-diethyltoluamide) (15.8 mg/L), sucralose (12.9 mg/L), iomeprol (12.0 mg/L), amidotrizoic acid (8.4 mg/L), iohexol (7.7 mg/L), eprosartan (6.8 mg/L), iopamidol (6.1 mg/L), EHDPP (5.4 mg/L), carbamazepine (4.6 mg/L), caffeine (3.0 mg/L), ibuprofen (2.1 mg/L) and sulfamethoxazole (1.7 mg/L).

Looking at these findings, the authors concluded that the European WWTPs should be upgraded with additional tertiary or complementary treatment steps such as ozonation and/or powdered activated carbon adsorption to remove micropollutants from WWTP effluents to improve water quality. The increased water stress in many regions of the world, due to the increase in population, become the reuse of treated water and wastewater recycling one of the more important options for water supply. The increasing worldwide contamination of freshwater systems with thousands of industrial and natural chemical compounds is one of the key environmental problems facing humanity, despite the fact of these compounds are present at low concentrations, many of them raise considerable toxicological concerns, particularly when present in complex mixtures.

Two important works were published in 2014. Fernández and his colleagues (2014) analysed the seasonal occurrence and removal of six pharmaceuticals from different therapeutic classes, in municipal wastewaters. The pharmaceuticals chosen were an analgesic- paracetamol, three anti-inflamatory- diclofenac ibuprofen and naproxen, an antiepileptic- carbamazepine and a stimulant compound- caffeine. The authors collected influent and effluent water samples from two conventional WWTPs from the North of Spain, at different seasons. Influent concentrations ranges were 2.3–42 mg/L for ibuprofen and naproxen, 0.04–7.8 mg/L for caffeine and paracetamol, and 0.03–0.4 mg/L for carbamazepine and diclofenac. The highest concentrations were found for ibuprofen in the untreated municipal wastewaters. Effluent concentrations were always below 5.7 mg/L. Diclofenac and carbamazepine persisted in WWTP effluents, whereas paracetamol, ibuprofen, naproxen and caffeine showed removal efficiencies between 75% and 99%. The authors concluded that the range of concentrations found was similar to levels reported in other studies worldwide. At the entrance of the WWTPs no seasonal changes were found and the results also showed that traditional treatment for municipal wastewaters was not effective in removing diclofenac and carbamazepine, whereas paracetamol, ibuprofen, naproxen and caffeine were removed in between 75% and 99% during the process. At the exit of WWTPs some seasonal changes were found indicating that, in general, the treatment throughout the WWTPs was more effectively for warmer temperatures. Golovko and his research team (Golovko et al., 2014) conducted a similar work and analysed the seasonal changes in antibiotics, antidepressants/psychiatric drugs, antihistamines and lipid regulators in a wastewater treatment plant from České Budějovice (Czech Republic). The authors analysed the concentration of 21 pharmaceuticals over 1 year. The target compounds included ten antibiotics, seven antidepressants and psychiatric drugs, two antihistamines and two lipid regulators. During the year, a total of 272 wastewater samples were collected, 136 from influents and 136 from effluents, and analysed for the target compounds. All studied pharmaceuticals were detected in both, influent and effluent wastewater samples, except for the antihistamine- meclozine, which was only found in influent samples. The mean concentration varied from the 0.006 μg/L for memantine, a psychiatric drug, to 1.48 μg/L for clarithromycin (antibiotic), for influent samples, and from 0.003 μg/L for sertraline (antidepressant) to 0.93 μg/L for clarithromycin (antibiotic), in effluent samples. The authors detected obvious seasonal changes in the CEC concentrations with higher values during winter and this can be attributed to increased human consumption of pharmaceuticals during winter and less efficient removal of these compounds during this period.

More recently, in 2016, Beek and his colleagues (2016) performed a a comprehensive literature review of 1516 research and review articles about the presence of CEC for human and veterinary usage reported worldwide in surface water, groundwater, tap/drinking water, manure, soil, and other environmental matrices. This work reveals that pharmaceuticals or their transformation products were detected in 71 countries covering all 5 United Nations (UN) regional groups. A total of 631 pharmaceuticals were detected revealing distinct regional patterns. Sixteen pharmaceutical substances were detected in the surface, drinking, and groundwater of all the UN regions. As an example, the anti-inflammatory drug diclofenac has been detected in environmental matrices in 50 countries, and concentrations found in several locations exceeded predicted no-effect concentrations. Urban wastewater seems to be the dominant emission pathway for pharmaceuticals globally, although emissions from industrial production, hospitals, agriculture, and aquaculture are important locally. The authors conclude that pharmaceuticals are a global challenge calling for multistakeholder approaches to prevent, reduce, and manage their entry into and presence in the environment. Studies developed in Brazil (Montagner et al., 2019) evaluate the occurrence of CEC in drinking, surface and groundwater and wastewater from São Paulo state. This study presents the results obtained in 10 years of analyses and reveals the presence of 58 CEC, including 14 pharmaceuticals and personal care products, 9 hormones, 17 pesticides, 8 industrial compounds and 10 illicit drugs, in a total of 708 samples. A preliminary risk assessment for aquatic life protection identified potential risks for caffeine, paracetamol, diclofenac, 17α-ethynylestradiol, 17β-estradiol, estriol, estrone, testosterone, triclosan, 4-n-nonylphenol, bisphenol A, atrazine, azoxystrobin, carbendazim, fipronil, imidacloprid, malathion and tebuconazole considering the frequency of detection and the number of positive samples but considering the drinking water criteria available in the literature no adverse effects to human health were expected to the studied scenario, except for 17β-estradiol. In the same year, Schulze et al. (2019) analysed the occurrence of emerging persistent and mobile organic contaminants in European water samples. These authors developed new analytical methods for CEC and tested the new methods in surface and groundwater samples. The new analytical methods are based on hydrophilic interaction liquid chromatography, reversed phase liquid chromatography, and supercritical fluid chromatography in combination with mass spectrometric detection. The methods were developed aiming to detect 57 target CEC and were applied to 14 samples from Germany, Spain and The Netherlands. The analyses reveal that 75% (43 CEC) of the analysed compounds were detected at least in one sample, among them 23 CEC that have not been reported before in environmental waters. These last compounds include methyl sulfate, 2-acrylamino-2-methylpropane sulfonate, benzyltrimethylammonium, benzyldimethylamine, trifluoromethanesulfonic acid, 6-methyl-1,3,5-triazinediamine, and 1,3-di-o-tolylguanidine and occur in 50% of the samples in concentrations ranging from ng/L to μg/L. These methods are very promising in the detection of a wide range of different CEC present in aquatic environments.

Last year, 2020, several works have been published relating to the occurrence of CEC in the environment, due to their relevance three works will be highlighted. Ngo et al. (2020) analysed the presence of pharmaceuticals and personal care products in Cau River (Vietnam). The samples were collected in surface water and sediments and the analysis showed the presence of 36 CEC with total concentrations ranging from 8.21 to 529 ng/L (water samples) and ranged from 17.4 to 172.8 μg/kg (sediments). The spatial distribution of CECs depends on population density and is evident that these compounds tended to concentrate downstream of the river in areas where population density is high. The most relevant and high concentrations found to correspond to caffeine, sulfamethoxazole, and lincomycin in water and triclocarban, levofloxacin, and griseofulvin in sediment showing that the antibiotic group was the main contributor to CEC contamination level. Kondor et al. (2020) investigated the presence of CECs in the Danube river and drinking water wells. The authors intended to examine the efficiency of riverbank filtration with regard to 111 CECs in a highly urbanised section of the river Danube. The comparison of 107 water samples from the Danube with 90 samples from relevant drinking water wells was performed in five sampling periods. The Danube samples showed the presence of 52 CECs and 10 of them were present in at least 80% of the samples. Higher CECs concentrations were found in winter which may be attributed to the role of the biological decomposition of CECs which is favoured by higher temperatures. The drinking water wells were detected and quantified 32 CECs. The results showed that riverbank filtration removes most CECs, around 95%, independently from the concentration in the river. The authors concluded that this type of water supply is safe for drinking water service. However, for four CECs, carbamazepine, lidocaine, tramadol, and lamotrigine, a low filtration efficiency was observed, less than 50%, however, no correlations were observed between the concentrations detected in the Danube and in the wells. Due to the changes in the efficiency of riverbank filtration in space and time, predicting the occurrence and concentrations of these four PhACs poses a further challenge to ensuring a safe drinking water supply. Ngqwala and Muchesa (2020) reviewed the occurrence and potential impacts of pharmaceuticals in aquatic environments, in South Africa. The authors discussed the pharmaceuticals commonly detected in SA waters: Cancer drugs (2), diuretics (1), anti-epileptics (1), antidepressants (8), anxiolytics (3), tranquillisers (1), steroids and related hormones (5), beta-blockers (6), analgesics and anti-inflammatories (13), antibiotics (12), lipid regulators (6), Anti-ulcer agents (3), proton pump inhibitors (1) and antiretrovirals (17). Some of these compounds have been found in extremely high concentrations, as for example, ibuprofen reached a maximum of 221 mg/L in a WWTP influent and

62 mg/L in surface water, and the antiretroviral drugs, emtricitabine and efavirenz, with concentrations of 172 and 140 mg/L, respectively (WWTP influent). The presence of so high concentrations is a good reason for intensive research into the fate of pharmaceuticals in the environment and in the development of efficient water treatment technologies and to improve national, provincial, regional and municipal regulations and legislation. There is a need for cooperation of stakeholders, manufacturers, regulators, veterinarians, pharmacists and consumers to agree on ways to decrease the harmful levels of pharmaceuticals in the environment.

11.3 BIOLOGICAL METHODS TO REMOVE CONTAMINANTS OF EMERGING CONCERN IN WASTEWATER

The extensive use of personal care products and pharmaceuticals, for human consumption and veterinary usage, led to the detection of several contaminants of emerging concern (CECs) in wastewater and aquatic compartments. In recent years, a large number of reports about this subject were published proving that increasing scientific attention is being focused on the occurrence of CECs in the environment (Beek et al., 2016; Geissen et al., 2015). This environmental problem is also an issue of special concern for the European Union (EU) and, in this regard, a framework for an EU action in the field of water policy was established by Directive 2000/60/EC (2000),[1] aiming at achieving good ecological and chemical status of surface water. This directive was updated by Directive 2013/39/EU (2013)[2] recommending the monitoring of 45 priority substances and highlighting the demand to develop new (waste)water treatment solutions. Conventional wastewater treatment plants (WWTPs) are inadequate for the removal of these pollutants, and alternative efficient and cost-effective methods to remove CECs are needed. The search for low-cost and environmentally-friendly technologies is strongly recommended and supported by the governments due to the high public acceptance and interest.

Several authors suggested methods to remove these compounds from aqueous systems including biological technologies such as anaerobic digestion, membrane bioreactors, biodegradation by pure cultures and activated sludge (Yang et al., 2016; Radjenovic et al., 2007; Marchlewicz et al., 2016; Kruglova et al., 2014) and physicochemical methods as activated carbon adsorption, clays adsorbents, ozonation and ozone/activated carbon coupling and chemical oxidation (García-Mateos et al., 2015; Khazri et al., 2017; Chedeville et al., 2016; Sabri et al., 2012). The advantages and disadvantages of all these methods are well known and it is generally accepted that the biological methods are more eco-friendly and presented fewer disadvantages. The use of bacteria and other microorganisms to remove contaminants is increasingly applied in wastewater treatment. Most of the microorganisms are present in the form of microbial aggregates, such as sludge flocs and granules. The use of granular aerobic sludge is regarded as a promising technology, to that effect, in wastewater treatment (Adav et al., 2008). Granular biomass has already been used to treat different troublesome wastewaters including petrochemical, pulp and paper industry, hypersaline and oily, sulfur-laden effluents, and applied in heavy metals removal, among others (Caluwé et al., 2017; Wei et al., 2017; Farooqi and Basheer, 2017).

Kent and Tay (2019) used a sequencing batch reactor inoculated with aerobic granular sludge for the treatment of 17α-ethinylestradiol, 4-nonylphenol, and carbamazepine in wastewater. The CECs concentration in the influent wastewater was maintained near 500 µg/L during the experiment. 17α-Ethinylestradiol and 4 nonylphenol were degraded by granular activated sludge at similar or slightly higher rates compared to activated sludge, without negative effect on the granule particle size or sludge bed height and carbamazepine had limited removal using granular activated sludge and appeared to interfere with the mechanism of granulation. The stabilised degradation for 17α-Ethinylestradiol and 4 nonylphenol in the sequential batch reactor, was 16.09 and 20.05

[1] https://eur-lex.europa.eu/legal-content/EN/TXT/?uri=CELEX%3A32000L0060
[2] https://eur-lex.europa.eu/legal-content/EN/ALL/?uri=CELEX%3A32013L0039

μg/g, respectively and there was limited degradation of carbamazepine during the anaerobic phase. The authors concluded that adsorption was initially the main removal mechanism but it decreased as active sites were saturated, while degradation rates improved as the sludge acclimatised to the contaminants. All three compounds had minimal effect on the chemical oxygen demand, ammonia-based nitrogen, or phosphate removal capabilities of the granular activated sludge.

Mery-Araya et al. (2019) studied the use of carbon substrate as a selection pressure to enhance the potential of aerobic granular sludge microbial communities for removing CECs. It is well known that carbon substrate influenced microbial community composition and abundance within aerobic granular sludge systems, and therefore, the potential for biotransforming CECs, giving an understanding of microbial communities and possible ways to manipulate them. For that, the authors proposed an aerobic granular sludge system supplemented with either acetate, 2-propanol, glycerol, or a 1:1:1 mixture of all three to investigate the relationships between carbon supplements and biotransformation of tylosin, sulfamethoxazole and trimethoprim (antibiotics), ibuprofen and naproxen (anti-inflammatory drugs) and carbamazepine (antiepileptic drug). The results show that the biotransformation of CECs is 45.9% greater in communities supplemented with glycerol (60.3±30.2 μg/LVSS) compared to acetate (20.9±29.7 μg/LVSS). These results are accordingly to the literature since several database surveys of metabolic reactions indicate that microbial communities supplemented with glycerol have the greatest capacity for the degradation of aromatic compounds, while those supplemented with acetate community have the lowest.

In 2020, Khan et al. (2020), aiming to increase the removal of CECs, tested the integration of aerobic granulation and UV/H_2O_2 processes in a continuous flow system for the degradation of sulfolane in contaminated water. Sulfolane is an organosulfur compound widely used in the chemical industry as a solvent for extractive distillation and chemical reactions. This novel approach was able to eliminate more than 99.99% of sulfolane in less than 6.3 hours of combined retention time. The authors evaluated the degradation kinetics, in batch and continuous flow operation, and the results showed zero and first order for aerobic granulation and UV/H_2O_2 processes, respectively. The results showed that the continuous flow-through system allowed a healthy granular system to flourish which was evidenced by the identification of multivalent elemental ions, such as Ca^{2+}, Fe^{3+} and P, and diverse bacterial communities that aided in stable granular structure. According to the authors, H_2O_2 is one of the key operational parameters and one of the major costs for UV/H_2O_2. The presence of hydrogen peroxide is necessary for fast degradation and high percentage degradation of sulfolane but the excess of H_2O_2 pose a potential threat to the discharge water receiving body. The selection of H_2O_2 concentration ultimately depends on the contaminant concentration, water matrix and retention time.

Ahmed et al. in 2017 reviewed the progress in biological and chemical treatment technologies for CEC removal from wastewater. The authors concluded that endocrine disrupting chemicals were better removed by activated sludge, membrane bioreactors and aeration processes among different biological processes, for surfactants and personal care products are efficiently removed by activated sludge processes while for pharmaceuticals and pesticides the best option seems to be the use of biological activated carbon. Other biological methods, such as the use of microalgae treatment processes, are able to remove a huge range of CECs to some extent. The authors also suggested several chemical methods with efficient removal rates as chemical oxidation processes such as ozonation/H_2O_2, UV photolysis/H_2O_2 and photo-Fenton processes. As a recommendation, the authors indicate hybrid systems such as MBR followed by reverse osmosis, nanofiltration/ultrafiltration, ozonation followed by biological activated carbon or activated sludge followed by ultrafiltration or by gamma radiation as the best option for the removal of endocrine disruptors, pesticides and pharmaceuticals. Del Álamo and his colleagues, analysed the removal of pharmaceutical compounds from urban wastewater by an advanced bio-oxidation process based on fungi *Trametes versicolor* immobilised in a continuous RBC system. The authors prepared solutions of synthetic and real wastewater spiked with 12 pharmaceutical compounds at a concentration of 50 μg/L and treated these solutions with the proposed method. The pharmaceuticals used, from different therapeutic groups, are antipyrine, clofibric acid, atenolol, caffeine, carbamazepine, diclofenac, gemfibrozil, hydrochlorothiazide, ibuprofen, ranitidine, sulfamethoxazole, and sulpiride. The advanced bio-oxidation process achieved a very good performance in terms of the removal of total organic carbon and reduction of $N\text{-}NH_4^+$ and $P\text{-}PO_4^{3-}$ nutrients. The pharmaceuticals sulfamethoxazole, atenolol, caffeine, ibuprofen and sulpiride were removed with removal efficiencies ranging from 80% to 95%, for antipyrine, clofibric acid, carbamazepine, diclofenac, gemfibrozil and ranitidine were eliminated with removal values ranging from 50% to 70% and hydrochlorothiazide was removed in around 40%. The authors also evaluated the anaerobic digestion of the fungal sludge generated upon the treatment and obtained a methane yield of 250 mL CH4 g_{VS}^{-1}. These results evidenced that the method proposed by del Álamo et al. is a promising alternative for the sustainable wastewater treatment of urban effluents, combining advanced oxidation with the biological operation for the removal of CECs.

In 2018, Londoño and Peñuela, tested, in batch assays, the anaerobic biodegradation of pharmaceuticals and personal care products. The compounds selected were methylparaben, ibuprofen and triclosan at a concentration of 300, 500, 1000, and 2000 μg/L. The study aimed to identify the mechanisms of biodegradation and sorption in the degradation of these compounds. The inoculum was granular sludge from a laboratory-scale anaerobic reactor. As for removal percentages, the authors found high degradation percentages for methylparaben, around 99%, with negligible sorption in the sludge. For ibuprofen, the removal was close to 0% and for triclosan the results have shown high biomass sorption and low biodegradation. The authors also highlighted that the concentrations tested, do not present negative or inhibitory effects on the microbial populations of the system. Arous and coauthors (2019), performed a mini-review were described the last innovative biological approaches for CECs removal from wastewater. The authors highlighted as good options for CECs removal: (a) phytoepuration or constructed wetlands, that are systems that use natural processes involving wetland vegetation, soils, and their associated microbial assemblages to improve water quality. These processes are well known for their effectiveness, low-cost and simple maintenance. The removal of CECs from wastewater by constructed wetlands takes place by the combination/integration of three main processes, biodegradation, adsorption and oxidation; (b) membrane bioreactors which integrate the activated sludge process and membrane separation as microfiltration or ultrafiltration, these processes are proven to be efficient in the treatment of wastewater containing micropollutants and pathogenic microorganisms due to the intense biological activity in the immediate vicinity of the membrane surface, (c) biofiltration, as soil filtration and biological filtration, have been investigated for the removal of CECs with interesting results; and (d) enzymatic treatment, enzymes as laccases or peroxidases can modify the chemical structures of several emerging contaminants. These enzymes are very promising biocatalysts for the transformation of toxic organic pollutants into less toxic or non-toxic end products, thereby mitigating or eliminating contamination from the environment. The authors concluded that even though biological processes have been proven to be suitable for the degradation of some organic micropollutants, some extreme recalcitrant contaminants are still undegraded or only partially removed and recommended the development of more efficient innovative processes with an emphasis in the use of coupled systems which can bridge the deficiencies in existing single biological technologies for the removal of these complex pollutants present in the water environment. Kanaujiya and coauthors (2019) performed a similar review and concluded that methods such as activated sludge process, biofilm-based reactor, and trickling bed bioreactor, two-phase partitioning bioreactor, membrane-based reactor, and cell-immobilised bioreactor systems have been used for the treatment of CECs contaminated wastewaters with good results. The authors affirm that these treatment systems are capable of tolerating high pollutant loads and are also able to treat highly water-insoluble pollutants. The authors also analysed the use of hybrid systems and concluded that these combined systems are not only capable of improving the treatment efficiency but also eliminating any accumulation of the toxic by-product produced during the treatment. Among the different hybrid systems, a combination of different biological systems is found to be highly efficient in treating micropollutant-containing wastewater.

In 2020, Ferreiro et al. (2020) tested the removal of CECs in the effluent of WWTP using biological and continuous-mode ultrafiltration processes. The WWTP is located in Biscay (Spain), and the authors analysed the removal of 39 high-occurrence CECs, including pharmaceutical products, industrial additives, food additives, herbicides and personal care products, by both methods. The removal of these CECs was analysed globally and after each operation. For evaluation of the efficiencies, the removal rates of the biological and of the continuous mode ultrafiltration treatments, including their variability over a day and a week in relation to the ammonium content, were examined in the influent of the WWTP. As conclusion, the results show that for the biological treatment, a wide range of different removal rates was obtained due to the different CEC's biodegradability and concentration, while for the continuous mode ultrafiltration treatment the removal rates were lower, but more constants between CECs. For biological treatment genistein, methylparaben, progesterone, testosterone, caffeine, and acetaminophen showed removal efficiencies above 99.5% but in contrast irbesartan, carbamazepine, diuron, and phenytoin showed average removal rates below 20.0%, and solutions with a higher concentration of CEC (above 1500 ng/L) presented high efficiencies above 80.0% in almost all cases. For the continuous mode ultrafiltration treatment, removal rates were not higher than 30.0% in most cases, except for the case of amitriptyline, which reached 63.0%. Low removal yields observed at low and/or high concentrations could be explained by fouling or irreversibly adsorbed material on the filtration membrane. The authors concluded that a biological treatment complemented with continuous mode ultrafiltration treatment allows a higher efficiency in removing CEC than the biological process alone. Nevertheless, ultrafiltration can be satisfactorily used as a tertiary treatment in order to help remove the small residual toxicity in the last WWTP stream. Luísa Angeles and his colleagues (2020) assessed the pharmaceutical removal and reduction in toxicity provided by advanced wastewater treatment systems. The study performed by these authors compares the removal efficiencies of 38 pharmaceuticals including antimicrobials, antidepressants, and other pharmaceuticals at different stages of the wastewater treatment process of seven WWTPs. These seven WWTPs differ from each other in the treatment technologies employed which include biological, physical, advanced chemical oxidation, and a combination of two or more of these technologies. In addition, to the removal percentages, the ecotoxicological effects of the WWTP effluents were also evaluated based on the behavioural effects in larval zebrafish. The authors found that the primary treatment did not provide significant reductions in the pharmaceutical concentrations, while secondary and tertiary processes resulted in the significant removal of several of the pharmaceuticals monitored. In the WWTPs using granular activated carbon or ozonation, as tertiary treatment, the removal of pharmaceuticals achieved 95% for several compounds, indicating the effectiveness of GAC and ozonation in removing pharmaceuticals from WWTPs. The authors highlighted that despite the efficiency of ozonation to remove the pharmaceuticals, removal may not necessarily mean complete degradation, since oxidation processes are known to form transformation products. One of the WWTPs analysed is a highly advanced treatment plant having four tertiary treatment stages. This WWTP has been found as the most efficient in removing pharmaceuticals, with all compounds having an overall % removal ranging from 97% to 100%. Notably, the final effluents of the seven WWTPs in this study did not show any significant behavioural alterations in zebrafish, indicating that despite differences in removal efficiencies all the treatment processes investigated are sufficient in preventing short-term biological effects. Quintelas et al. (2020), tested the removal of specific pharmaceutical compounds, ibuprofen and paracetamol, using activated sludge. For that, a bench reactor was inoculated with activated sludge and fed with a synthetic medium and spiked with ibuprofen and paracetamol solutions in concentrations ranging from 0.4 to 1 mg/L. The authors consider that, although the concentrations found in the environment are in the ng/L–µg/L range, higher concentrations are found in wastewater originating from hospitals or elderly houses (Langenhoff et al., 2013) and in what concerns to Portugal, ibuprofen consumption has been increasing during the last decade, moving from the 15th to 8th position in the sales ranking (Paíga et al., 2013). Considering the increase in the consumption of nonprescribed drugs, such as ibuprofen and paracetamol, and considering the presence of higher concentrations in specific locations the authors decided to test the reactor for higher concentrations than those found in the environment. The values of uptake found by the authors increased from 0.192 to 0.660 mg/g for ibuprofen and from 0.104 to 0.341 mg/g for paracetamol, with the increase in the initial concentration. The removal efficiency reached values from 99.1% to 99.5% and is independent of the initial ibuprofen concentration and for paracetamol ranged from 93.3 to 98.8 decreasing with the increase in the initial concentration. The mechanism of removal was also analysed and the results show that both, pseudo-first order and pseudo-second order kinetic models, can describe the mechanism. At same time, batch assays were performed in order to assess the toxicity of both pharmaceuticals into the activated sludge using quantitative image analysis (QIA). For ibuprofen experiments, QIA studies showed that this compound favours the growth of aggregated biomass rather than filamentous bacteria. The authors also performed a comparison between the use of activated sludge and adsorption by a commercial porous ceramic material and Pinus bark and can conclude that the biological technology allowed better results (99.5% against 19.4% and 9.3%, respectively for the ceramic material and Pinus bark) for ibuprofen, and for paracetamol, the results are comparable for biological technology and adsorption by the ceramic material. Finally, as conclusion, it is possible to affirm that activated sludge is a promising technology that could be applied with success to the removal of pharmaceutical contaminated effluents.

Maryjoseph and Ketheesan (2020), from Sri Lanka, reviewed microalgae-based wastewater treatments for the removal of emerging compounds. The authors found that the last decade has seen a renewed importance in microalgae for bioremediating emerging compounds from wastewater. Several reports have been published describing the use of microalgae to remove CECs at the laboratory scale but little attention has been paid to the application of microalgae to real contaminated effluents. The passage from laboratory tests to the real world presents a series of challenges. The laboratory tests are performed at controlled conditions and dynamics of operational conditions such as pH, temperature, dissolved oxygen, hydraulic retention time, bacteria-microalgae interactions, light limitation/inhibition and mixing conditions in the real-world reactor system may significantly deviate from the laboratory conditions. The effects of multiple CECs in wastewater in real world application are unpredictable and a pre-treatment unit prior to the microalgae-based system may be needed to selectively remove the contaminants in the conventional system followed by the microalgae-based system. A problem that can be attributed to the use of microalgae is the bioadsorption and biouptake of microalgae that do not provide a sustainable solution, both mechanisms facilitate the transformation of contaminant from one form to another form contributing to the toxicity by CECs in the final algal biomass and prevent the ultimate use of microalgae such as biofuel, biocrude, animal feed, biofertiliser and this fact could make the entire process unsustainable and/or uneconomical. The authors recommended future research for the identification and characterisation of the by-products formed during bioremediation process considering that the relative toxicity of such by-products and their presence might prevent the ability of reusing such wastewater.

Wu and Yin (2020) proposed a new concept of microbial niche nexus sustaining biological wastewater treatment which can achieve the efficient removal of known and unknown pollutants by tuning microbial niches to accommodate diverse microbial communities. These niches for specific microbes can enhance the removal of CECs and alleviate their toxicity. This is a very recent approach and future research should be performed. The type of pollutants in wastewater determines the enriched diverse functional microorganisms and to advance the idea of microbial niche nexus is essential to understand the microbial ecology, including the function, regulation, and interaction of known and unknown microorganisms. Exploring and re-constructing diverse microbial niches particularly based on time and space sequences in wastewater treatment systems is the crucial way to solve the increasing challenges and meet future demands.

Recently, in 2021, a series of interesting works have been published dedicated to the topic of CECs removal. For its relevance, three works will be highlighted. Alenzi and colleagues (2021) investigated the pharmaceuticals effect and removal, at environmentally relevant concentrations, from sewage sludge during anaerobic digestion. To perform the assays two glass reactors of 7 L were operated at 38°C with a hydraulic retention time of 43 days. Pharmaceuticals such as clarithromycin, clotrimazole, erythromycin, fluoxetine, ibuprofen, sertraline, simvastatin and tamoxifen were added to the reactors at concentrations varied from 19 μg/L, for tamoxifen, to 878 μg/L, for simvastatin. The results showed, in terms of effect in the biomass, that the presence of pharmaceuticals induce instability resulting in an accumulation of volatile fatty acids, increase the percentage of Firmicutes (>70%) and decreased the percentages of Bacteroidetes and Euryarchaeota (<5%), which seems to be the cause of volatile fatty acids accumulation and resultant decrease in the biogas production. Removal rates differed significantly depending on the pharmaceutical study. The concentration of some compounds like clotrimazole, fluoxetine, ibuprofen and tamoxifen increased during the cycles coinciding with the accumulated predicted values in the digester, suggesting that these compounds were very persistent and their removal was non-existent, but for compounds such as clarithromycin, erythromycin, simvastatin or sertraline, the concentrations remained lower than the predicted valued. The authors concluded that anaerobic microorganisms offer enhanced removal of the antibiotics clarithromycin and erythromycin over aerobic techniques. Dong et al. (2021) analysed the impact of triclosan on phosphate removal and microbial community in an activated sludge process. Triclosan is a broad-spectrum antimicrobial agent that is widely applied in pharmaceutical and personal care products, such as soaps, detergents, cosmetics, skin care lotions, disinfectants, and other sanitation goods. The authors found that a concentration of 100–150 μg/L of triclosan is able to inhibit the microbial growth and reduce the phosphorus removal efficiency of PO_4^{3-}–P and total phosphorus removal in an order of 16% and 8%, respectively. Moreover, the differences in the microbial community structures of the reactor-control and reactor-with triclosan groups gradually expanded, no obvious difference was observed in the final stage, and the interrelationships of microbes in the latter weakened. The long-term addition of TCS impairs the growth of polyphosphate-accumulating organisms. Hassan and his colleagues (2021) studied the removal of antibiotics by bioelectrochemical technology and analysed the problematic effects of this class of pharmaceuticals on microbial communities. Antibiotics presents potential role in exacerbating the emergence of antibiotics-resistance bacteria and antibiotics-resistance genes which is a very serious problem with an impact on human health. According to the authors, bioelectrochemical technologies combined with microbial fuel cells and microbial electrolytic cells, due to bacterial metabolism and the electrochemical-redox reactions, were considered promising alternatives to improve the removal of antibiotics.

All the technologies presented above have promising results in the removal of CECs; however, some challenges and recommendations should be considered regarding the use of these technologies.

11.4 CHALLENGES AND FUTURE PERSPECTIVES

The pollution caused by CECs is an important issue that deserves the attention of the scientific community. In recent years, a huge number of reports about this subject were published focusing on the use of biological treatments for the removal of these compounds from the environment. Although good results have been obtained with these technologies some work should be performed aiming to optimise the removal methods. According to several authors, there is a lack of detailed information on the degradation mechanisms involved, the influence of operational variables on CECs removal, reaction kinetics and reactor design for optimum performance and more work should be done to fill this gap. Ahmed et al. (2017) suggested the integration of existing treatment systems with nanoscale science and engineering, the development of more reliable methods for wastewater sample preparations, analytical techniques and validation protocols for the analysis of CECs especially in complex environmental samples, suitable sampling protocols should be defined to evaluate the removal performance of different WWTP processes at various operational conditions.

After analysing hundreds of works describing de use of different treatments for the removal of CECs we believe that the best systems capable of effectively removing CECs is the use of hybrid technologies based on combined chemical and biological treatment processes such as UV photolysis in the presence of H_2O_2 followed by MBR or biological activated carbon, ozonation in presence of H_2O_2 followed by MBR or biological activated carbon, photo-Fenton followed by MBR or biological activated carbon and we recommended further work to optimised these technologies. Ahmed and his colleagues (2017) recommended that new knowledge in genetic engineering should be introduced to select and amplify the most effective microbes for CECs degradation, which will reduce hydraulic retention time and save capital costs in reactor design and that the robustness and feasibility of full-scale chemical oxidation processes should be extensively investigated to ensure CECs removal efficiencies and minimise toxic by-products.

Another interesting research avenue in the degradation/removal of CECs is the search for novel biomaterials and bio-nano-catalysts following green agenda to establish highly effective and environmentally friendly treatment processes. Geissen et al. (2015) highlighted that the combination of regulations, use of monitoring data and application of tools for risk assessment will contribute to the reduction of the presence of CECs in the environment and consequently to a clean world.

ACKNOWLEDGEMENTS

The authors thank the Portuguese Foundation for Science and Technology (FCT) under the scope of the strategic funding of UIDB/ BIO/04469/2020unit. Cristina Quintelas and Daniela Mesquita thank FCT for funding through program "DL 57/2016 – Norma transitória".

BIBLIOGRAPHY

Adav, S.S., Lee, D.J., Show, K.Y., Tay, J.H., Aerobic granular sludge: Recent advances, *Biotechnology Advances*, 2008, 26: 411–423.

Ahmed, M., Zhou, J., Ngoa, H., Guo, W., Thomaidis, N., Xu, J., Progress in the biological and chemical treatment technologies for emerging contaminant removal from wastewater: A critical review, *Journal of Hazardous Materials*, 2017, 323: 274–298.

Alenzi, A., Hunter, C., Spencer, J., Roberts, J., Craft, J., Pahl, O., Escudero, A., Pharmaceuticals effect and removal, at environmentally relevant concentrations, from sewage sludge during anaerobic digestion, *Bioresource Technology*, 2021, 319: 124102.

Angeles, L., Mullen, R., Huang, I., Wilson, C., Khunjar, W., Sirotkin, H., McElroy, A., Aga, D., Assessing pharmaceutical removal and reduction in toxicity provided by advanced wastewater treatment systems, *Environmental Science: Water Research Technology*, 2020, 6: 62–77.

Arous, F., Hamdi, C., Bessadok, S., Jaouani, A., Innovative biological approaches for contaminants of emerging concern removal from wastewater: A mini-review, *Advances in Biotechnology and Microbiology*, 2019, 13(5): 114–120.

Beek, T., Weber, F.A., Bergmann, A., Hickmann, S., Ebert, I., Hein, A., Kuster, A., Pharmaceuticals in the environment–Global occurrences and perspectives, *Environmental Toxicology and Chemistry*, 2016, 35: 823–835.

Caluwé, M., Dobbeleers, T., Daes, J., Miele, S., Akkermans, V., Daens, D., Geuens, L., Kiekens, F., Blust, R., Dries, J., Formation of aerobic granular sludge during the treatment of petrochemical wastewater, *Bioresource Technology*, 2017, 238: 559–567.

Chedeville, O., Di Giusto, A., Delpeux, S., Cagnon, B., Oxidation of pharmaceutical compounds by ozonation and ozone/activated carbon coupling: A kinetic approach, *Desalination and Water Treatment*, 2016, 57: 18956–18963.

del Álamo, A., Pariente, M., Vasiliadou, I., Padrino, B., Puyol, D., Molina, R., Martínez, F., Removal of pharmaceutical compounds from urban wastewater by an advanced bio-oxidation process based on fungi Trametes versicolor immobilized in a continuous RBC system, *Environmental Science and Pollution Research*, 2018, 25: 34884–34892.

Dong, X., He, Y., Peng, X., Jia, X., Triclosan in contact with activated sludge and its impact on phosphate removal and microbial community, *Bioresource Technology*, 2021, 319: 124134.

Farooqi, I.H., Basheer, F., Treatment of adsorbable organic halide (AOX) from pulp and paper industry wastewater using aerobic granules in pilot scale SBR, *Journal of Water Process Engineering*, 2017, 19: 60–66.

Fernández, M., Fernández, M., Laca, A., Laca, A., Díaz, M., Seasonal occurrence and removal of pharmaceutical products in municipal wastewaters, *Journal of Environmental Chemical Engineering*, 2014, 2: 495–502.

Ferreiro, C., Gómez-Motos, I., Lombraña, J., de Luis, A., Villota, N., Ros, O., Etxebarria, N., Contaminants of emerging concern removal in an effluent of wastewater treatment plant under biological and continuous mode ultrafiltration treatment, *Sustainability*, 2020, 12: 725.

García-Mateos, F.J., Ruiz-Rosas, R., Marqués, M.D., Cotoruelo, L.M., Rodríguez-Mirasol, J., Cordero, T., Removal of paracetamol on biomass-derived activated carbon: Modeling the fixed bed breakthrough curves using batch adsorption experiments, *Chemical Engineering Journal*, 2015, 279: 18–30.

Geissen, V., Mol, H., Klumpp, E., Umlauf, G., Nadal, M., van der Ploeg, M., van de Zee, S., Ritsema, C.J., Emerging pollutants in the environment: A challenge for water resource management, *International Soil and Water Conservation Research*, 2015, 3: 57–65.

Golovko, O., Kumar, V., Fedorova, G., Randak, T., Grabic, R., Seasonal changes in antibiotics, antidepressants/psychiatric drugs, antihistamines and lipid regulators in a wastewater treatment plant, *Chemosphere*, 2014, 111: 418–426.

Hassan, M., Zhu, G., Lu, Y-Z., AL-Falahi, A., Lu, Y., Huang, S., Wan, Z., Removal of antibiotics from wastewater and its problematic effects on microbial communities by bioelectrochemical technology: Current knowledge and future perspectives, *Environmental Engineering Research*, 2021, 26: 190405.

Kanaujiya, D., Paul, T., Sinharoy, A., Pakshirajan, K., Biological treatment processes for the removal of organic micropollutants from wastewater: A Review, *Current Pollution Reports*, 2019, 5: 112–128.

Kent, J., Tay, J., Treatment of 17α-ethinylestradiol, 4-nonylphenol, and carbamazepine in wastewater using an aerobic granular sludge sequencing batch reactor, *Science of the Total Environment*, 2019, 652: 1270–1278.

Khan, M., Yu, L., Hollman, J., Tay, J., Achari, G., Integration of aerobic granulation and UV/H_2O_2 processes in a continuous flow system for the degradation of sulfolane in contaminated water, *Environmental Science: Water Research and Technology*, 2020, 6: 1711–1722.

Khazri, H., Ghorbel-Abid, I., Kalfat, R., Trabelsi-Ayadi, M., Removal of ibuprofen, naproxen and carbamazepine in aqueous solution onto natural clay: Equilibrium, kinetics, and thermodynamic study, *Applied Water Science*, 2017, 7: 3031–3040.

Kondor, A., Jakab, G., Vancsik, A., Filep, T., Szeberényi, J., Szabo, L., Maász, G., Ferincz, A., Dobosy, P., Szalai, Z., Occurrence of pharmaceuticals in the Danube and drinking water wells: Efficiency of riverbank filtration, *Environmental Pollution*, 2020, 265: 114893.

Kruglova, A., Ahlgren, P., Korhonen, N., Rantanen, P., Mikola, A., Vahala, R., Biodegradation of ibuprofen, diclofenac and carbamazepine in nitrifying activated sludge under 12°C temperature conditions, *Science of Total Environment*, 2014, 499: 394–401.

Langenhoff, A., Inderfurth, N., Veuskens, T., Schraa, G., Blokland, M., Kujawa-Roeleveld, K., Rijnaarts, H., Microbial removal of the pharmaceutical compounds ibuprofen and diclofenac from wastewater, *Biomed Research International*, 2013, 325806.

Lei, M., Zhang, L., Lei, J., Zong, L., Li, J., Wu, Z., Wang, Z., Overview of emerging contaminants and associated human health effects, *BioMed Research International*, 2015, Article ID 404796.

Londoño, Y.A., Peñuela, G.A., Study of anaerobic biodegradation of pharmaceuticals and personal care products: Application of batch tests, *International Journal of Environmental Science and Technology*, 2018, 15: 1887–1896.

Loos, R., Carvalho, R., António, D.C., Comero, S., Locoro, G., Tavazzi, S., Paracchini, B., Ghiani, M., Lettieri, T., Blaha, L., Jarosova, B., Voorspoels, S., Servaes, K., Haglund, P., Fick, J., Lindberg, R.H., Schwesig, D., Gawlik, B.M., EU-wide monitoring survey on emerging polar organic contaminants in wastewater treatment plant effluents, *Water Research*, 2013, 47: 6475–6487.

Marchlewicz, A., Domaradzka, D., Guzik, U., Wojcieszyńska, D., Bacillus thuringiensis B1(2015) is a Gram-positive bacteria able to degrade naproxen and ibuprofen, *Water, Air, & Soil Pollution*, 2016, 227: 197.

Maryjoseph, S., Ketheesan, B., Microalgae based wastewater treatment for the removal of emerging contaminants: A review of challenges and opportunities, *Case Studies in Chemical and Environmental Engineering*, 2020, 2: 100046.

Mery-Araya, C., Lear, G., Perez-Garcia, O., Astudillo-Garcia, C., Singhal, N., Using carbon substrate as a selection pressure to enhance the potential of aerobic granular sludge microbial communities for removing contaminants of emerging concern, *Bioresource Technology*, 2019, 290: 121705.

Montagner, C., Sodré, F., Acayaba, R., Vidal, C., Campestrini, I., Locatelli, M., Pescara, I., Albuquerque, A., Umbuzeiro, G., Jardim, W., Ten years-snapshot of the occurrence of emerging contaminants in drinking, surface and ground waters and wastewaters from São Paulo State, Brazil, *Journal of the Brazilian Chemical Society*, 2019, 30: 614–632.

Ngo, T., Van, D-A., Tran, H., Nakada, N., Tanaka, H., Huynh, T., Occurrence of pharmaceutical and personal care products in Cau River, Vietnam, *Environmental Science and Pollution Research*, 2020, in press.

Ngqwala, N.P., Muchesa, P., Occurrence of pharmaceuticals in aquatic environments: A review and potential impacts in South Africa, *South African Journal of Science*, 2020, 116: 5730.

Païga, P., Santos, L., Amorim, C.G., Araújo, A.N., Montenegro, M.C., Pena, A., Delerue-Matos, C., Pilot monitoring study of ibuprofen in surface waters of north of Portugal, *Environmental Science and Pollution Research*, 2013, 20: 2410–2420.

Quintelas, C., Mesquita, D., Torres, A., Costa, I., Ferreira, E., Degradation of widespread pharmaceuticals by activated sludge: Kinetic study, toxicity assessment, and comparison with adsorption processes, *Journal of Water Process Engineering*, 2020, 33: 101061.

Radjenovic, J., Petrovic, M., Barceló, D., Analysis of pharmaceuticals in wastewater and removal using a membrane bioreactor, *Analytical and Bioanalytical Chemistry*, 2007, 387: 1365–1377.

Sabri, N., Hanna, K., Yargeau, V., Chemical oxidation of ibuprofen in the presence of iron species at near neutral pH, *Science of Total Environment*, 2012, 427–428, 382–389.

Salthammer, T., Emerging indoor pollutants, *International Journal of Hygiene and Environmental Health*, 2020, 224: 113423.

Sauvé, S., Desrosiers, M., A review of what is an emerging contaminant, *Chemistry Central Journal*, 2014, 8: 15.

Schulze, S., Zahn, D., Montes, R., Rodil, R., Quintana, J., Knepper, T., Reemtsma, T., Berger, U., Occurrence of emerging persistent and mobile organic contaminants in European water samples, *Water Research*, 2019, 153: 80–90.

Shah, M.P. *Advanced Oxidation Processes for Effluent Treatment Plants*. Elsevier, 2020.

Shah, M.P. *Microbial Bioremediation and Biodegradation*. Springer, 2020.

Shah, M.P. *Removal of Emerging Contaminants through Microbial Processes*. Springer, 2021.

Wei, D., Zhang, K., Wang, S., Sun, B., Wu, N., Xu, W., Du, B., 2017. Characterization of dissolved organic matter released from activated sludge and aerobic granular sludge biosorption processes for heavy metal treatment via a fluorescence approach, *International Biodeterioration and Biodegradation*, 2017, 124: 326–333.

Wu, G., Yin, Q., Microbial niche nexus sustaining biological wastewater treatment, *npj Clean Water*, 2020, 3, Article 33.

Yang, S., Hai, F.I., Price, W.E., McDonald, J., Khan, S.J., Nghiem, L.D., Occurrence of trace organic contaminants in wastewater sludge and their removals by anaerobic digestion, *Bioresource Technology*, 2016, 210: 153–159.

12 Nanotechnology-Based Remediation Techniques to Eliminate Heavy Metal Pollutants from Wastewater

Rimmy Singh and Rachna Bhateria
Maharshi Dayanand University

Sharma Mona
Guru Jambheshwar University of Science & Technology

CONTENTS

12.1 INTRODUCTION

For the well-being and sustainability of humanity, adequate amount of clean and safe water is a cornerstone for the current water system. Pathogen-free water does not only reduce the disease burden but also improves the quality of life. Our natural capital of aquatic biodiversity and healthy ecosystems also reaches to next level with improvement in water systems (Costanza and Daly, 1992). With growing populations, adequate supply of water supports urban settlements, essential goods manufacturing, energy, and food production. Investment in current water technology development is highly needed for sustainable and universal delivery and access to critical water services. Adaptation of rigid infrastructure systems, proper planning for the countermeasures of changing climate and most importantly protection of water resources (Mauter et al., 2018). For the sustenance of life on this planet, water is one of the most necessities. With rapidly flourishing industrialization and urbanization, water utilization is observed to have doubled every 15 years. According to "World Health Organization (WHO)" reports, 40% of the planet's population is facing challenges of water scarcity or approximately 2 million or more people are devoid of access to

DOI: 10.1201/9781003164982-12

enough or clean water. Nearby 1.6 million population end up to death from avoidable water-borne diseases such as diarrhoea and 90% of them are children under 5 years of age according to WHO estimates (Pandit and Kumar, 2015). On the other hand, industrial and related anthropogenic activities have contributed to increased concentration of pollutants discharged in wastewater and groundwater directly or indirectly. Thereof, the health of millions of people is at the risk around the globe (Fu et al., 2014). Lack of improved sanitation has excruciating and undeviating impact on human health and hence, concern towards the availability of clean and potable water arises. Exposure to various pathogens and chemical toxicants creates a threat of entering the living system via food chain, for instance, irrigating crops end up in contaminated water (due to leaching of pesticides and fertilizers et.) and leads to bioaccumulation of toxicants in aquatic organisms (Schwarzenbach et al., 2010). Anthropogenic industries including mining, electroplating, agricultural plants, etc. lack eco-friendly technologies or natural activities like storms, volcanoes, earthquakes, etc., which contribute to lethal pollutants in wastewater discharged by them (Hokkanen et al., 2016; Ipek, 2005). These pollutants can be categorized into organic, inorganic, and biological (Gupta and Ali, 2013).

Nanotechnology has been one of the most emerging technologies leading towards achieving the goal of sustainable water treatment methods. Nanomaterials are undoubtedly efficient structures that have highly advanced properties. "World Health Organization (WHO) discovered that around 1 billion individuals on the planet have no access to consumable water and that 2.6 billion individuals lack access to sanitation" (WHO, 2004). The undeniable fact is that the quality of recycled water should be improved to fill this gap of adjusting conventional treatment techniques to more prohibitive necessities. Nanotechnology applications could address the most valuable alternatives for the accomplishment of this objective. "National Nanotechnology Initiative" (NNI) characterized nanotechnology as "the arrangement and control of particle dimensions within 1 and 100 nanometres (nm), where interesting properties empower novel application that are not possible when working with bulk materials or even single molecules or atoms". The understanding of basic structures and properties offers "the chance of utilizing novel material in the treatment of surface water, groundwater, and wastewater particularly for the expulsion of heavy metals, natural and inorganic solutes, and microorganism" (De Luca and Ferrer, 2017).

12.2 WATER POLLUTION: NEEDS A RELIABLE SOLUTION

Water contamination is one of the major concerns because of its harmfulness to plants, humans, and animals. Higher concentrations of water contaminants rigidly affect climate and plant metabolic exercises hence affecting food production. Water pollutants can enter human tissues through different assimilation pathways, for example, soil-food chain, direct ingestion, inward breath, dermal contact and oral admissions (Singh and Prasad, 2015). Water pollutants, for example, substantial metals in slanted levels can act as "ecological toxins" under marine and earthbound environments (Nazemi, 2012; Veschasit et al., 2012). Emerging and priority contaminants are recurrently detected in natural waters, sewage effluents and drinking water sources at treacherous points that may endanger communal and environmental health. This fact emphasizes the prerequisite of capital vitalization and technological modernization of marginally operative water treatment infrastructure. Pollution control and improved water treatment systems not only ease productivity loss and health care costs but also accompanies socioeconomic capacity and economic development (Hutton et al., 2004; Van Minh and Hung, 2011). For boosting up the financial burden of maintenance and operation of ageing wastewater and drinking water treatment plants, technological innovation has become critical.

The predicted global population is 7.9 billion by 2020 which may encounter the planet with extreme water scarcity. At both small and large scales, water quality monitoring has become challenging due to low concentration detection of contaminants, complexity, and variability of the quality matrix (Vikesland, 2018). Therefore, in recognition of these issues, it is of utmost consideration to eliminate these contaminants to provide better public health (Dyson, 1996). Different methods of water purification are receiving attention from researchers worldwide (Gupta and Nayak, 2012; Saleh et al., 2011; Saleh and Gupta, 2012; Gupta et al., 2011).

In the course of time, heavy metal pollution is a matter of overall concern. Heavy metal pollutants, likewise, called arising impurities, comprise an immense number of substances of human-centred or common source, including industrial waste, chemicals, and agrochemicals. In water assets, these materials are ordinarily present at low concentrations and every substance has a structure and activity, that not only detects and investigates as well as expulse from potable water and effluent treatment plants (Bolong et al., 2009; Luo et al., 2014). Another concern is the absence of the greatest allowable concentrations of these pollutants, and thus, many precautionary measures and checking moves are made to guarantee that these pollutants do not pave their path in surface waters (Bolong et al., 2009). Arising contaminants are consumed in enormous amounts, and they cause negative effects on the environment (Petrović et al., 2003). A significant certainty is that because of their essence in water assets and subsequently in drinking water, they can go about as human and amphibian living beings' endocrine disruptors (Bolong et al., 2009). Likewise, it ought to be considered that heavy metals when combined with other substances can cause synergistic impacts, making it more troublesome for their detection, evaluation, and expulsion (Luo et al., 2014). In the last century, the water interest for anthropogenic exercises has extended sevenfold because of the quadrupled of overall population (Pendergast and Hoek, 2011). A report on "Progress on Sanitation and Drinking Water-2015 Update and MDG Assessment" by the World Health Organization (WHO) shows that around 2.6 billion people groups have gotten to upgraded drinking water source in the year 1990 may increment to 663 million individuals getting to improved drinking water source continuously 2015. The number of people living in water alarm regions will increase to around 3.9 billion by 2030, as evaluated by the World Water Council (Xu et al., 2018).

12.3 SCIENCE OF NANOMATERIALS: NANOTECHNOLOGY

Nanotechnology is a novel and emerging technology dominating various fields of research. The term 'nano' was derived from a Greek word for 'dwarf'. During the 19th century, the concept of nanotechnology fangled its roots towards technological advancement. Richard Feynman in his lecture, "There is a plenty of room at the bottom" explained how things behave on a small scale. According to him, "atoms on small scale behave like nothing on large scale, for they satisfy the laws of quantum mechanics. As we go down and fiddle around with the atoms, we are working with different laws and we can expect to do different things" (Feynman, 1959). A nanometre (nm) is "one billionth of a metre or about one hundred thousandth of the width equal to a human hair". Therefore, 21st century has come up with leapfrogging novel opportunities associated with nanotechnology for the current water sector. Modifications and control over chemical structure, material size and morphology results in magnificent transitions in particle properties. Nanotechnology has offered advanced materials which endow water treatment technologies with highly exceptional antimicrobial, quantum, catalytic, electrical, adsorptive and optical properties (Alvarez et al., 2018). For water purification, nanotechnology can be a panacea as multifunctional distributed treatment plants can be empowered by engineered nanomaterials. These treatment systems are easily deployable and small moreover they can reduce water loss, contamination problem and energy requirements for the supply of water through large and leaky distribution systems over long distance (Zodrow et al., 2017). High flexibility and capacity are offered by distributed nanotechnology-enabled treatment systems to remediate complex matrix which otherwise requires complex and large treatment infrastructure. In multifunctional advanced materials, multiple treatment functions can also be facilitated with engineered nanomaterials (for instance, adsorption, disinfection, and catalytic degradation) that can target different contaminants with removal processes as they possess different physicochemical properties. This can be done to reduce their footprint, to simplify treatment methods, and to enhance their efficiency. Nanotechnology can be a significant alternative approach to develop infrastructure-intensive technologies that can treat wastewaters to marginally safe levels

(Alvarez et al., 2018). Engineered nanomaterials (ENMs) empower next generation multifunctional materials for the heavy metal extraction from industrial effluents. Fabricated nanomaterials have elevated surface-volume relationship, hydrophilic and hydrophobic interactions, high reactivity, tunable pore size etc. which are substantially important in the process of adsorption and catalysis (Das et al., 2014). Nanoadsorbent with highly efficient adsorption properties has dominated various sectors of research and application. Nanoparticles are materials having minimum of one dimension within the 100 nm of diameter (Laurent et al., 2008). These materials can be categorised as "zero dimensional materials (0D), one dimensional material (1D), two dimensional materials (2D) and three-dimensional materials (3D) (Tiwari et al., 2012)" based on their overall shape. The significance of nanoparticles was noted when investigators revealed that the physiochemical properties (e.g., optical) of a substance are influenced by its size. Small, 20-nm palladium (Pd), gold (Au), silver (Ag) and platinum (Pt), nanoparticle have distinctive dark black, wine red colour, black and yellowish grey colour, respectively. When each dimension in a particle is in the nanometric range then it is called as iso-dimensional nanoparticle for instance spherical nanoparticles of silica. For different materials and fields, the definition of nanoparticles may vary. Nanoparticles are often called as "nanoclusters or clusters" which can be defined as association of million molecules or atoms (Ferrando et al., 2008).

It is currently believed that nanoparticles (1–100 nm) show one-of-a-kind chemical and physical properties that contrast from those of the relating mass materials (Hadjipanayis and Siegel, 2012; Chow and Gonsalves, 1996; Edelstein and Cammaratra, 1998; Liz-Marzan and Kamat, 2004; Schmid, 2011; Ozin and Arsenault, 2005; Nagarajan and Hatton, 2008). The reliance of the properties of nanoscale materials on the size, shape, and composition of the nanomaterials is a wonder of both central logical interest and numerous reasonable and technological applications (Hadjipanayis and Siegel, 2012; Chow and Gonsalves, 1996; Edelstein and Cammaratra, 1998; Liz-Marzan and Kamat, 2004; Schmid, 2011). These properties are frequently extraordinary, and at times unrivalled, to those of the relating mass materials.

From the beginning of time till now, carbon-based water filtration and refinement parts have been used. For water refinement, utilizing carbon as wood charcoal was drilled by "Egyptians and Sumerians in 3750 BC" (Patrick, 1995). During the 1940s, activated carbon (Pollard et al., 1992) and its different forms were discovered, including nanostructured forms such as "carbon nanofibers (CNFs), carbon nanotubes (CNTs), and graphene-based materials" that were used for the treatment via disinfection, catalytic wet air oxidation, layer-based separation, and adsorption, and for detecting and sensing (Das et al., 2017). Metal and its oxide nanoparticles have additionally been accounted for in the process of adsorption, oxidation, photocatalysis, sanitization, and detection. Notwithstanding being a functioning area for quite a long time, "nanomaterials-based treatment" and detecting innovations are yet to possess an enormous portion of the market because of instability regarding affectability and selectivity, greater expense, and field-level issues during activity. Since the end of the 19th century, nanostructured materials have been blended and employed in experimental and innovative fields. Back in 1989, nanomaterial and nanotechnology investigations had fangled their roots when new instrumentations, such as Scanning Tunnelling microscopy, led to profound information on "nano dimensions" of the materials. Consequently, a few nanomaterials were presented in medication, biotechnology, software engineering, and space investigation. Furthermore, inventive nanostructured forms of materials and cycles have been added to the advancement of synthetic manufacturing through new nanocatalysts and items (Sannino et al., 2017).

By means of the description given by the "National Nanotechnology Initiative (NNI), the U.S. Government research and development (R&D) program", nanotechnology is recognizable as "the understanding and control of matter at dimension between approximately 1 and 100 nanometres (nm), where unique phenomena enable novel application not feasible when working with bulk materials or even single atoms or molecules". Moreover, as indicated by NNI "nanomaterials are all nanoscale materials or materials that contain nanoscale structures inside or on their surfaces.

These can incorporate designed nano-objects, for example, nanoparticles, nanotubes, nanoplates, and naturally occurring nanoparticles, for example, volcanic debris, ocean shower, and smoke". At long last, it becomes imperative to remember that nanostructures are none of the unfamiliar materials but rather "nanoscale form" of familiar materials. From the time, when the historical backdrop of "nanotechnology" theoretically began, with the discourse that "Richard Feynman gave on December 29th, 1959 at the yearly gathering of American Physical Society SWCNT (Feynman, 1959)", there have been a great deal of advances. Unexpectedly, his stunning vision zeroed in consideration on the chance of working with little materials. Subsequently, nanotechnology got perhaps the most recent test for people, the control of materials at the nuclear level (De Luca and Ferrer, 2017).

12.3.1 Nanomaterials: Types

Iron-based nanomaterials for example, magnetite and nano zerovalent iron have been explored for wastewater treatment. Likewise, at mechanical scale, the creation of novel nanomaterials at a mechanical scale has energized the advancement in innovations that are dependent on them. Specifically, carbon-based nanomaterials are accessible at lower costs than they were before, and they are under concentrated investigations for their applications in water decontamination and natural remediation (Sannino et al., 2017). Engineered nanomaterials include nanoparticles, nanospheres, nanorods, dendrimers, carbon nanotubes (CNTs), etc. Metal-based nanoadsorbents, mainly titanium oxide, alumina, zinc oxide, and iron oxides, have been anticipatedly used for the decontamination of metals from water and effluents as they are effective and economical.

However, to improve the quality of nZVI nanoparticles qualities, focus on the limitation of their conglomeration and adjustment, moreover, to balance their reactivity and treatment time during application is crucial. Because of solid appealing interparticle powers, nZVI will expand in size, agglomerate while reaching micron size, and bring about restricted versatility in permeable media (O'Carroll et al., 2013). Coatings with natural mixtures or different polymers can settle their size and scattering. Carbon nanotubes (CNTs) are alluring because of their critical underlying, real substance features (Popov, 2004). These CNTs possess round, and hollow constructions inside which carbon-particle hexagons are organized in a helical design about a needle pivot (Iijima, 1991; Shah, 2020, 2021). As per a few examinations, CNTs present more noteworthy advantages than financially accessible granular enacted carbon (GAC) (Mukhopadhyay, 2011). While utilizing CNTs to eliminate "natural organic matter (NOM)" from wastewater, elevated adsorption and desorption rates were shown by CNTs as compared to GAC filters (Su and Lu, 2007). However, minimal reactivation temperature is required and the lesser weight reduction. It is also demonstrated that CNTs are more economical than GAC for the expulsion of NOM from water sources (Mukhopadhyay, 2011). By concerning other impurity elimination, for example, cyanobacterial poison (e.g., microcystins MCs), CNTs additionally indicated adsorption limits better than GAC. Specifically, Yan et al. (2006) showed that the adsorption of MCs on the CNTs is dictated by the surface zone and the outer distance across the CNTs (Yan et al., 2006; Shah, 2020). Based on hybridized carbon molecule layers, CNTs are partitioned into single-walled carbon nanotubes (SWCNTs) and multi-walled carbon nanotubes (MWCNTs) (Theron et al., 2008). However, SWCNTs comprise a solitary graphite sheet wrapped into a round and hollow cylinder while MWCNTs are rather made from a variety of concentrically settled nanotubes (Baughman et al., 2002).

The beginnings of the distinctions can be clarified by the high surface-to-volume proportion, scattering factors and the quantum size impact one of a kind to a particular length scale. For instance, semiconductor nanoparticles in the size scope of w1–w20 nm in measurement (frequently called quantum spots) have short-range structures that are equivalent to the mass semiconductors yet have optical and additionally electronic properties which are significantly unique in relation to the mass. The imprisonment of electrons inside a semiconductor nanocrystal brings about the move of the band hole to higher energy with more modest translucent size. This impact is known as the "quantum size impact" (Hadjipanayis and Siegel, 2012; Chow and Gonsalves, 1996; Edelstein and Cammaratra, 1998; Liz-Marzan and Kamat, 2004; Schmid, 2011; Ozin and Arsenault, 2005; Nagarajan and Hatton, 2008).

12.4 SCOPE OF NANOTECHNOLOGY ENABLED NANOMATERIALS IN WATER DECONTAMINATION

12.4.1 Nanosorbents

Till date, several adsorbents have dominated the market with characteristics like large surface area, easy to manufacture and regenerate, apt pore size and structure and with excellent mechanical properties (Figure 12.1). Natural clay, silica gel, biomaterials, carbon materials, molecular sieves, etc. are some of the commonly used adsorbents (Liu et al., 2014; Abdolali et al., 2016; Habiba et al., 2017; Cai et al., 2017). In the last few years, research has been conducted at high levels to discover low-cost and efficient adsorbents for water decontamination. The most extensively explored materials for heavy metal removal are metal oxides including nanometal oxides which have characteristic features of high reactivity and large surface-to-volume ratio. The robust efforts to discover

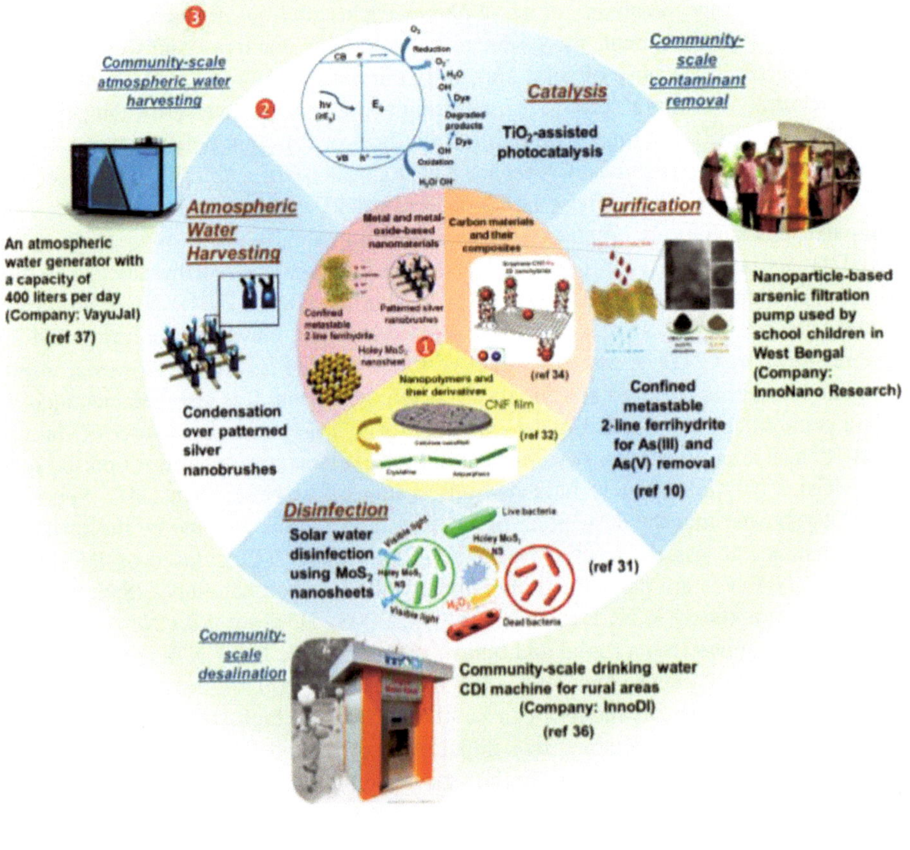

FIGURE 12.1 Representation of translation of materials from lab to market (Nagar and Pradeep, 2020; Kumar et al., 2017; Sarkar et al., 2018; Vadahanambi et al., 2013).

cost-effective alternatives and the advancement of nano-techniques have increased with the expansion of novel class of nanoparticles for wastewater remediation. Nanomaterials that are often characterized by their high surface area have gained attention due to their excellent properties and wide range of applications.

Remediation of water frameworks polluted with heavy metals is getting expanded consideration as more data on their harmful impacts on people and the climate gets known. Likewise, expanded population growth is putting tension on water assets and boundless worldwide industrialization is delivering a huge amount of toxins contaminating a greater amount of the world's water supplies. Ceramic metal oxides are generally utilized for the remediation of substantial metals, yet these materials bind ambiguously and reversibly and can without much of a stretch become soaked with pervasive species, like Ca, Mg, and Zn (Fryxell, 2007). Furthermore, division and restoration of these sorbent materials from the cleaned water can be highly challenging (Warner et al., 2010).

12.4.2 Nanotubes

Due to their uncommon chemical, physical, and electrical properties, CNTs are utilized in numerous applications including clinical science, natural designing, electrical designing, and material science (Shirvanimoghaddam et al., 2018). The primary chemically dynamic sites in carbon nanotubes are situated around imperfect portions like pentagons, arranged against a cylinder body which largely comprises hexagons; and it can be this that gives CNTs their incredible capacity to interface with different mixtures (Abbasi, 2017). Functionalizing CNTs is the key step to improve their productivity. CNTs have high efficacy due to specifically their molecule size and the nature of their surface compound organization. From aqueous solution, carbon nanotubes have been extensively used to eliminate pollutants due to light mass density, strong interaction among contaminants and CNTs, large surface area and high porous and hollow structure (Kabbashi et al., 2009; Di et al., 2004; Gu et al., 2013; Wang et al., 2007; Chen et al., 2009; Shao et al., 2010).

12.4.3 Nano Catalysis

Heterogenous photocatalysts have been advanced by using metal NPs for contaminant degradation including dyes also. Amalgamation of noble metals with TiO_2 is used as cutting edge in the oxidation process which improves the functioning of TiO_2 as wide gap semiconductor in water treatment. The beneficial outcomes of metal NPs can be perceived as TiO_2 can be energized by "photons of ultra-bandgap energy, ordinarily energy higher than 3.2eV belongs to UV range (365nm), structures electron-hole pairs". With strong oxidation power, hydroxyl radicals can be generated when electron-hole pairs intermingle with "adsorbed species" which leads to "oxidation or/reduction reactions" (Sannino et al., 2017).

Nano zerovalent copper particles have represented an applicable movement for joined interaction, and "photoreduction from copper salt" prompts dynamic "photoreforming photocatalysts", extraordinarily impacted by structural and morphological constraints. Cu NPs during H_2 photoproduction go through oxidation-reduction process in association with the photooxidation of oxygenated compounds, playing an important part in the overall reaction. However, the lowest size Cu NPs (2nm) display high activity as for nanocopper (3.7nm) whereas diverse sound structure shows a great impact in the Cu–TiO_2 collaboration, improving the charge separation by developing more contact among them (Montini et al., 2011).

12.4.4 Membrane Separation

Membrane filtration methods are extremely competent in sequestering heavy metal ions from water and effluents. Large pores and high permeability is the feature of ultra-filtration membrane; thus, owing to their smaller diameter issues, heavy metal ions do not get trapped in these membranes. Large complexes or metal ions associated with water-soluble "macromolecular polymers" can be easily trapped in ultra-membranes. Membrane filtration is sustainable separation technology for metal and organic pollutants. This wastewater treatment method is highly efficient and selective, needs low energy, facile in operation and can be applied to a wide range of metals. Membrane technology is an advanced water treatment technology and is used widely, however, its high operational cost, ease of membrane blocking, high requirement of additional complex agents and low efficiency have restricted its use to broad areas.

For TDS under 45,000 g/L, membrane separation technology is efficient for treating water. Such innovations "(Reverse Osmosis (RO), electrodialysis (ED), forward Osmosis (FO), nanofiltration (NF))" have acquired tremendous success because of cost adequacy (0.26–0.54 US $/m^3), low energy prerequisite (3–8 kWhelec/m^3), and little CO_2 discharge "(1.7–2.8 kg/m^3 for seawater RO)" into the air (Shahzad et al., 2017, 2018). Now in the desalination market, Reverse Osmosis has outperformed thermal technologies by supplanting them as conventional. In the desalination interaction, "membrane module is the most energy-escalated part that comprised of almost 71% (2.5–4 kWh/m^3 for seawater RO)" of the Specific energy consumption (SEC) (Qasim et al., 2019). While a greater part of RO frameworks has accomplished "a SEC of 2.3 kWh/m^3, thermodynamic breaking point remains at 0.76 kWh/m^3, for a feed having a TDS of 35,000, showing degree for development". Though, considering recovering of 50% energy, the practical attainable SEC limit ascends to 1.06 kWh/m^3 (Elimelech and Philip, 2011). Decrease in retrieval rate is able to enhance SEC yet drive upgrade functioning and investment expenses. The cutting-edge coatings ought to consequently conquer the compromise among permeability, rejection, and improved selectivity. Between the membrane materials, "nanocomposite polymeric membranes" are industrially fruitful because of their minimal effort and achievability for large-scale production. Polyamide films are generally utilized because of their high selectivity and porousness contrasted with ordinary cellulose acetic acid derivation-based membranes. Polyamide layers are made from a wide cross-connected "nonporous polyamide layer" upheld by a permeable "polysulphone layer" at the base (Tan et al., 2018).

12.4.5 Microbial Disinfection

The occurrence of microorganisms in drinking water is the main source of intense waterborne ailments around the planet, particularly in developing nations. More than 2,000,000 deaths per year, especially of children, are attributed to biologically hazardous water and preventable waterborne diarrheal infections (WHO and UNICEF, 2017).

Designed nanomaterials give novel functionalities to microbe inactivation, like enormous surface area and explicit reactivity, through components that have not been treated with conventional disinfectants (Li et al., 2008). Engineered nanomaterial-empowered inactivation mechanisms such as cell wall disruption by nanoscale structures (Bhadra et al., 2015; Ivanova et al., 2013; Gazit, 2007; Kang et al., 2007), surface-dynamic phases stimulated by electrostatic interaction (Wang et al., 2015), photochemical generation of responsive oxygen species (Cho et al., 2005; Liu et al., 2016; Zhang et al., 2010), and target delivery of disinfecting agents (Xiu et al., 2012). The special surface properties and reactivity of ENMs prompting biofilm hindrance (Bhadra et al., 2015) give another undiscovered chance to lessen microorganisms in consumable water, as microbial regrowth in treated water during storage and transport is a common concern (Propato and Uber, 2004). Drawn-out utilization of commercial disinfectants has prompted the improvement of resistance by certain microbes like "*Giardia* and *Cryptosporidia*". Thus, it becomes mandatory to rethink over traditional disinfection strategies and opt for novel technologies (Li et al., 2008). The significant "antimicrobial mechanisms of nanoparticles are by interfering with transmembrane electron movement, upsetting electron movement, oxidizing cell components, creating secondary products like Reactive Oxygen Species (ROS) which can additionally cause harm. In contrast to commercial disinfectants, nanoparticles do not create toxic disinfecting by-products since they are not strong oxidants and are generally idle" (Li et al., 2008).

"Hydrophilic metal oxide nanoparticles" are normally inserted into membranes to diminish fouling by expanding "hydrophilicity" (Qu et al., 2013). For instance, "alumina, titania, silica and zeolite nanoparticles have been added to polymeric ultrafiltration membranes" (Bae and Tak, 2005). Laboratory studies have been performed to decide the capability of CNT films in the treatment of water. These nanotube films have been concentrated to be compelling against microorganisms, infection, turbidity and furthermore towards the evacuation of natural pollutants. The films have been distinguished as encouraging materials for desalination and as a choice to turn around "reverse osmosis membranes" while these channels require less energy because of their high penetrability. CNT layers what is more are relied upon to be more strong and simpler to clean and reuse (Srivastava et al., 2004). The expense of carbon nanotube adsorbents as of now runs between US$ 25 and 38/kg. Huge scope business creation might cut down the expense significantly to US$ 10 (Dalton et al., 2004).

12.4.6 Nanosensor

New sensor technologies have been created by consolidating synthesis approaches at microscale and nanoscale to make little, convenient, and profoundly precise sensors that can identify the trace amount of synthetic or biochemical water substances (Ahmed et al., 2014). One can identify microbes in considerably less time, as nanosensors illuminate the presence of microorganisms in the samples. One can set aside time and money by utilizing nanosensors for the quantification of microorganisms in water samples. Diverse nanomaterials, for example, quantum specks, nanoshells, and CNTs are utilized in the nanosensors for water quality examination (Ahmed et al., 2014). "Nanosensors with gold (Au)/PANI nanocomposites were utilized to identify Hg^{2+} particles in water tests dependent on Raman intensity response" (Wang et al., 2011). Similarly, Gupta et al. (2013) built up a nanosensor dependent on glassy carbon cathode supported with amino phenyl and graphene oxide for synchronous detection of cadmium and chromium in real-time samples. "Colorimetric and fluorescence-based nanosensors" were modified with gold nanoparticles for Al(III) discovery and exhibited high affectability, selectivity, ease, and continuous estimations in real water tests without pre-treatment (Abubaker et al., 2019). The future of water treatment technologies will be smart-based nanosensors and IoT (Figure 12.2).

12.5 CONCLUSION

Some nanotechnology-based water treatment solutions are effectively accessible and available; however, a lot more are in the pipeline. Nonetheless, prolonged haul assessment of these solutions is needed as the greater part of them have been attempted and promoted based on laboratory research. Nanotechnology applications in effluent treatment are demonstrating different ways to deal with conventional treatment techniques, which have innovation constraints. The more persuasive methods for water decontamination ought to encounter cost-saving standards particularly for the elimination of heavy metals that become obligatory when present in trace levels. These extra rigid water specification cut-off points become difficult to be achieved by ordinary advancements. In addition, the sludge and toxic releases associated with conventional treatment brings about additional treatment requiring high energy or potentially expensive arrangements. Nano-based innovations appear to perform well by expanding evacuation productivity of a few toxins in comparison with traditional cycles and advancements.

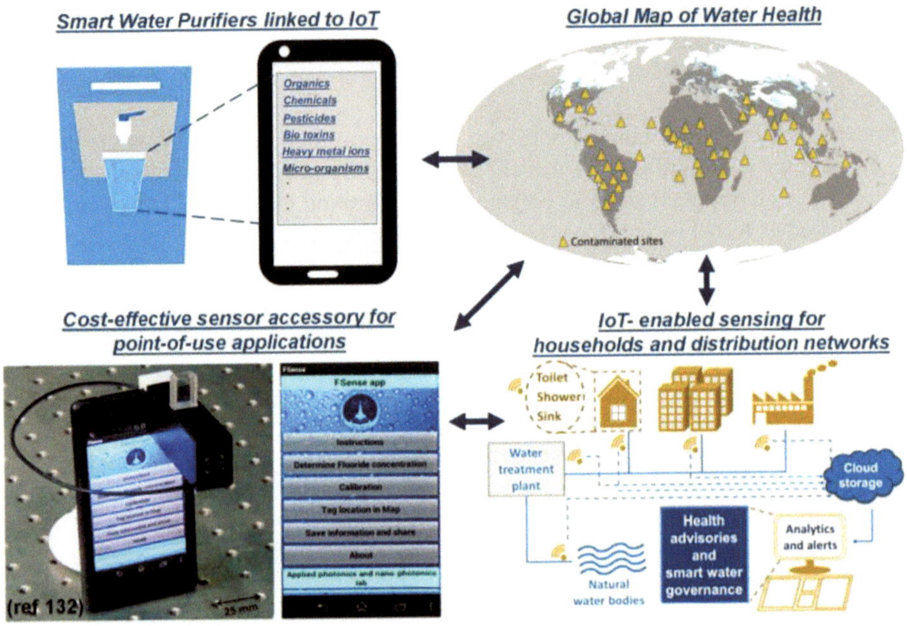

FIGURE 12.2 Smartphone with nanosensors and IoT-based domestic water-purifying systems (Hussain et al., 2017).

REFERENCES

Abbasi, M. (2017). Synthesis and characterization of magnetic nanocomposite of chitosan/SiO_2/carbon nanotubes and its application for dyes removal. *Journal of Cleaner Production*, *145*, 105–113.

Abdolali, A. et al. (2016). A breakthrough biosorbent in removing heavy metals: Equilibrium, kinetic, thermodynamic and mechanism analyses in a lab-scale study. *Science of the Total Environment*, *542*, 603–611.

Abubaker, M., Ngah, C. W. Z. C. W., Ahmad, M., & Kuswandi, B. (2019). Functionalized gold nanoparticles as optical nanosensors for determination of aluminum (III) ions in water samples. In *Green Synthesis, Characterization and Applications of Nanoparticles* (pp. 459–484). Elsevier.

Ahmed, T., Imdad, S., Yaldram, K., Butt, N. M., & Pervez, A. (2014). Emerging nanotechnology-based methods for water purification: A review. *Desalination and Water Treatment*, *52*(22–24), 4089–4101.

Alvarez, P. J., Chan, C. K., Elimelech, M., Halas, N. J., & Villagrán, D. (2018). Emerging opportunities for nanotechnology to enhance water security. *Nature Nanotechnology*, *13*(8), 634.

Bae, T. H., & Tak, T. M. (2005). Effect of TiO_2 nanoparticles on fouling mitigation of ultrafiltration membranes for activated sludge filtration. *Journal of Membrane Science*, *249*(1–2), 1–8.

Baughman, R. H., Zakhidov, A. A., & de Heer, W. A. (2002). Carbon nanotubes—The route toward applications. *Science*, *297*, 787–792.

Bhadra, C. M. et al. (2015). Antibacterial titanium nano-patterned arrays inspired by dragonfy wings. *Scientific Reports*, *5*, 16817.

Bolong, N., Ismail, A. F., Salim, M. R., & Matsuura, T. (2009). A review of the effects of emerging contaminants in wastewater and options for their removal. *Desalination*, *239*(1–3), 229–246.

Cai, J. et al. (2017). Electrospun composite nanofiber mats of Cellulose@ Organically modified montmorillonite for heavy metal ion removal: Design, characterization, evaluation of absorption performance. *Composites Part A: Applied Science and Manufacturing*, *92*, 10–16.

Chen, C., Hu, J., Shao, D., Li, J., & Wang, X. (2009). Adsorption behavior of multiwall carbon nanotube/iron oxide magnetic composites for Ni (II) and Sr (II). *Journal of Hazardous Materials*, *164*(2–3), 923–928.

Cho, M., Chung, H., Choi, W. & Yoon, J. (2005). Diferent inactivation behaviors of MS-2 phage and Escherichia coli in TiO_2 photocatalytic disinfection. *Applied and Environmental Microbiology*, *71*, 270–275.

Chow, G. M., Gonsalves, K. E. (1996). *Nanotechnology: Molecularly Designed Materials*. American Chemical Society, Washington, DC.

Costanza, R., & Daly, H. E. (1992). Natural capital and sustainable development. *Conservation Biology*, *6*(1), 37–46.

Dalton, A. B. et al. (2004). Continuous carbon nanotube composite fibers: Properties, potential applications, and problems. *Journal of Materials Chemistry*, *14*(1), 1–3.

Das, R., Ali, M. E., Abd Hamid, S. B., Ramakrishna, S., & Chowdhury, Z. Z. (2014). Carbon nanotube membranes for water purification: A bright future in water desalination. *Desalination*, *336*, 97–109.

Das, R., Vecitis, C. D., Schulze, A., Cao, B., Ismail, A. F., Lu, X., Chen, J., Ramakrishna, S. (2017). Recent advances in nanomaterials for water protection and monitoring. *Chemical Society Reviews*, *46*, 6946–7020.

De Luca, A., & Ferrer, B. B. (2017). Nanomaterials for water remediation: Synthesis, application and environmental fate. In *Nanotechnologies for Environmental Remediation* (pp. 25–60). Springer, Cham.

Di, Z. C., Li, Y. H., Luan, Z. K., & Liang, J. (2004). Adsorption of chromium (VI) ions from water by carbon nanotubes. *Adsorption Science & Technology*, *22*(6), 467–474.

Dyson, T. (1996). *Population and Food: Global Trends and Future Prospects*. Psychology Press.

Edelstein, A. S., & Cammaratra, R. C. (Eds.). (1998). *Nanomaterials: Synthesis, Properties and Applications*. CRC Press.

Elimelech, M., & Phillip, W. A. (2011). The future of seawater desalination: Energy, technology, and the environment. *Science*, *333*, 712–717.

Ferrando, R., Jellinek, J., & Johnston, R. L. (2008). Nanoalloys: From theory to applications of alloy clusters and nanoparticles. *Chemical Reviews*, *108*(3), 845–910.

Feynman, R. P. (1959, December). Plenty of room at the bottom. In *APS Annual Meeting*.

Fryxell, G. E. (2007). *Synthesis of Nanostructured Hybrid Sorbent Materials Using Organosilane Self-assembly on Mesoporous Ceramic Oxides* (p. 159). Imperial College Press, London, UK.

Fu, F., Dionysiou, D. D., & Liu, H. (2014). The use of zero-valent iron for groundwater remediation and wastewater treatment: A review. *Journal of Hazardous Materials*, *267*, 194–205.

Gazit, E. (2007). Self-assembled peptide nanostructures: The design of molecular building blocks and their technological utilization. *Chemical Society Reviews*, *36*, 1263–1269.

Gu, H., Rapole, S. B., Huang, Y., Cao, D., Luo, Z., Wei, S., & Guo, Z. (2013). Synergistic interactions between multi-walled carbon nanotubes and toxic hexavalent chromium. *Journal of Materials Chemistry A*, *1*(6), 2011–2021.

Gupta, V. K., Agarwal, S., & Saleh, T. A. (2011). Chromium removal by combining the magnetic properties of iron oxide with adsorption properties of carbon nanotubes. *Water Research*, *45*(6), 2207–2212.

Gupta, V. K., & Ali, I. (2013). Environmental water: advances in treatment, remediation and recycling. *Newnes*.

Gupta, V. K., & Nayak, A. (2012). Cadmium removal and recovery from aqueous solutions by novel adsorbents prepared from orange peel and Fe_2O_3 nanoparticles. *Chemical Engineering Journal*, *180*, 81–90.

Gupta, V. K., Yola, M. L., Atar, N., Ustundağ, Z., & Solak, A. O. (2013). A novel sensitive Cu (II) and Cd (II) nanosensor platform: Graphene oxide terminated p-aminophenyl modified glassy carbon surface. *Electrochimica Acta*, *112*, 541–548.

Habiba, U., Afifi, A. M., Salleh, A., & Ang, B. C. (2017). Chitosan/(polyvinyl alcohol)/zeolite electrospun composite nanofibrous membrane for adsorption of Cr^{6+}, Fe^{3+} and Ni^{2+}. *Journal of Hazardous Materials*, *322*, 182–194.

Hadjipanayis, G. C., & Siegel, R. W. (Eds.). (2012). *Nanophase Materials: Synthesis-Properties-Applications* (Vol. 260). Springer Science & Business Media.

Hokkanen, S., Bhatnagar, A., & Sillanpää, M. (2016). A review on modification methods to cellulose-based adsorbents to improve adsorption capacity. *Water Research*, *91*, 156–173.

Hussain, I., Ahamad, K. U., & Nath, P. (2017). Low-cost, robust, and field portable smartphone platform photometric sensor for fluoride level detection in drinking water. *Analytical Chemistry*, *89*, 767–775.

Hutton, G., Haller, L., Water, S., & World Health Organization. (2004). Evaluation of the costs and benefits of water and sanitation improvements at the global level (No. WHO/SDE/WSH/04.04). World Health Organization.

Iijima, S. (1991). Helical microtubules of graphitic carbon. *Nature*, *354*, 56–58.

Ipek, U. (2005). Removal of Ni (II) and Zn (II) from an aqueous solution by reverse osmosis. *Desalination*, *174*(2), 161–169.

Ivanova, E. P. et al. (2013). Bactericidal activity of black silicon. *Nature Communications*, *4*, 2838.

Kabbashi, N. A., Atieh, M. A., Al-Mamun, A., Mirghami, M. E., Alam, M. D. Z., & Yahya, N. (2009). Kinetic adsorption of application of carbon nanotubes for Pb (II) removal from aqueous solution. *Journal of Environmental Sciences*, *21*(4), 539–544.

Kang, S., Pinault, M., Pfeferle, L. D. & Elimelech, M. (2007). Single-walled carbon nanotubes exhibit strong antimicrobial activity. *Langmuir, 23*, 8670–8673.

Kumar, A. A., Som, A., Longo, P., Sudhakar, C., Bhuin, R. G., Gupta, S. S., Anshup, Sankar, M. U., Chaudhary, A., Kumar, R., & Pradeep, T. (2017). Confined metastable 2-line ferrihydrite for affordable point-of-use arsenic-free drinking water. *Advanced Materials, 29*, 1604260

Laurent, S., Forge, D., Port, M., Roch, A., Robic, C., Vander Elst, L., & Muller, R. N. (2008). Magnetic iron oxide nanoparticles: Synthesis, stabilization, vectorization, physicochemical characterizations, and biological applications. *Chemical Reviews, 108*(6), 2064–2110.

Li, Q., Mahendra, S., Lyon, D. Y., Brunet, L., Liga, M. V., Li, D., & Alvarez, P. J. (2008). Antimicrobial nanomaterials for water disinfection and microbial control: Potential applications and implications. *Water Research, 42*(18), 4591–4602.

Liu, C. et al. (2016). Rapid water disinfection using vertically aligned MoS_2 nanoflms and visible light. *Nature Nanotechnology, 11*, 1098–1104.

Liu, D., Li, Z., Zhu, Y., Li, Z., & Kumar, R. (2014). Recycled chitosan nanofibril as an effective Cu (II), Pb (II) and Cd (II) ionic chelating agent: Adsorption and desorption performance. *Carbohydrate Polymers, 111*, 469–476.

Liz-Marzan, L. M., & Kamat, P. V. (2004). Nanoscale materials. In *Nanoscale Materials* (pp. 1–3). Springer, Boston, MA.

Luo, Y. et al. (2014). A review on the occurrence of micropollutants in the aquatic environment and their fate and removal during wastewater treatment. *Science of the Total Environment, 473*, 619–641.

Mauter, M. S., Zucker, I., Perreault, F., Werber, J. R., Kim, J. H., & Elimelech, M. (2018). The role of nanotechnology in tackling global water challenges. *Nature Sustainability, 1*(4), 166–175.

Montini, T., Gombac, V., Sordelli, L., Delgado, J. J., Chen, X., Adami, G., & Fornasiero, P. (2011). Nanostructured Cu/TiO_2 photocatalysts for H_2 production from ethanol and glycerol aqueous solutions. *ChemCatChem, 3*, 574.

Mukhopadhyay, S. (2011). *Nanoscale Multifunctional Materials: Science and Applications*. Wiley, New York.

Nagar, A., & Pradeep, T. (2020). Clean Water through Nanotechnology: Needs, gaps, and fulfillment. *ACS Nano, 14*(6), 6420–6435.

Nagarajan, R., & Hatton, T. A. (Eds.). (2008). *Nanoparticles: Synthesis, Stabilization, Passivation, and Functionalization*. American Chemical Society.

Nazemi, S. (2012). Concentration of heavy metal in edible vegetables widely consumed in Shahroud, the North East of Iran. *Journal of Applied Environmental and Biological Sciences, 2*(8), 386–391.

O'Carroll, D., Sleep, B., Krol, M., Boparai, H., & Kocur, C. (2013). Nanoscale zero valent iron and bimetallic particles for contaminated site remediation. *Advances in Water Resources, 51*, 104–122.

Ozin, G. A., & Arsenault, A. C. (2005). *Nanochemistry: A Chemical Approach to Nanomaterials*. RSC Publishing.

Pandit, A. B., & Kumar, J. K. (2015). Clean water for developing countries. *Annual Review of Chemical and Biomolecular Engineering, 6*, 217–246.

Patrick, J. W. (Ed.). (1995). *Porosity in Carbons: Characterization and Applications*. Wiley.

Pendergast, M. M., & Hoek, E. M. (2011). A review of water treatment membrane nanotechnologies. *Energy & Environmental Science, 4*(6), 1946–1971.

Petrović, M., Gonzalez, S., & Barceló, D. (2003). Analysis and removal of emerging contaminants in wastewater and drinking water. *TrAC – Trends in Analytical Chemistry, 22*(10), 685–696.

Pollard, S. J. T., Fowler, G. D., Sollars, C. J., & Perry, R. (1992). Low-cost adsorbents for waste and wastewater treatment: A review. *Science of the Total Environment, 116*(1–2), 31–52.

Popov, V. (2004). Carbon nanotubes: Properties and application. *Materials Science and Engineering R: Reports, 43*, 61–102.

Progress on Drinking Water, Sanitation and Hygiene: 2017 Update and SDG Baselines (World Health Organization (WHO) and United Nations Children's Fund (UNICEF), 2017).

Propato, M. & Uber, J. G. (2004). Vulnerability of water distribution systems to pathogen intrusion: How effective is a disinfectant residual? *Environmental Science & Technology, 38*, 3713–3722.

Qasim, M., Badrelzaman, M., Darwish, N. N., Darwish, N. A., & Hilal, N. (2019). Reverse osmosis desalination: A state-of-the-art review. *Desalination, 459*, 59–104.

Qu, X., Alvarez, P. J., & Li, Q. (2013). Applications of nanotechnology in water and wastewater treatment. *Water Research, 47*(12), 3931–3946.

Saleh, T. A., Agarwal, S., & Gupta, V. K. (2011). Synthesis of $MWCNT/MnO_2$ and their application for simultaneous oxidation of arsenite and sorption of arsenate. *Applied Catalysis B: Environmental, 106*(1–2), 46–53.

Saleh, T. A., & Gupta, V. K. (2012). Column with CNT/magnesium oxide composite for lead (II) removal from water. *Environmental Science and Pollution Research, 19*(4), 1224–1228.

Sannino, D., Rizzo, L., & Vaiano, V. (2017). Progress in nanomaterials applications for water purification. In *Nanotechnologies for Environmental Remediation* (pp. 1–24). Springer, Cham.

Sarkar, D., Mondal, B., Som, A., Ravindran, S. J., Jana, S. K., Manju, C. K., Pradeep, T. Holey. (2018). MoS 2 nanosheets with photocatalytic metal rich edges by ambient electrospray deposition for solar water disinfection. *Global Challenges, 2*, 1800052.

Schmid, G. (Ed.). (2011). *Nanoparticles: From Theory to Application.* John Wiley & Sons.

Schwarzenbach, R. P., Egli, T., Hofstetter, T. B., Von Gunten, U., & Wehrli, B. (2010). Global water pollution and human health. *Annual Review of Environment and Resources, 35*, 109–136.

Shah, M. P. (2020a). *Advanced Oxidation Processes for Effluent Treatment Plants.* Elsevier.

Shah, M. P. (2020b). *Microbial Bioremediation and Biodegradation.* Springer.

Shah, M. P. (2021). *Removal of Emerging Contaminants through Microbial Processes.* Springer.

Shahzad, M. W., Burhan, M., Ang, L., & Ng, K. C. (2017). Energy water-environment nexus underpinning future desalination sustainability. *Desalination, 413*, 52–64.

Shahzad, M. W., Burhan, M., Ang, L., & Ng, K. C. (2018). Adsorption desalination: Principles, process design, and its hybrids for future sustainable desalination. *Emerging Technologies for Sustainable Desalination Handbook*, 3–34.

Shao, D., Hu, J., & Wang, X. (2010). Plasma induced grafting multiwalled carbon nanotube with chitosan and its application for removal of UO, Cu^{2+}, and Pb^{2+} from aqueous solutions. *Plasma Processes and Polymers, 7*(12), 977–985.

Shirvanimoghaddam, K., Czech, B., Abolhasani, M. M., & Naebe, M. (2018). Sustainable periodically patterned carbon nanotube for environmental application: Introducing the cheetah skin structure. *Journal of Cleaner Production, 179*, 429–440.

Singh, A., & Prasad, S. M. (2015). Remediation of heavy metal contaminated ecosystem: An overview on technology advancement. *International Journal of Environmental Science and Technology, 12*(1), 353–366.

Srivastava, A., Srivastava, O. N., Talapatra, S., Vajtai, R., & Ajayan, P. M. (2004). Carbon nanotube filters. *Nature Materials, 3*(9), 610–614.

Su, F., & Lu, C. (2007). Adsorption kinetics, thermodynamics and desorption of natural dissolved organic matter by multiwalled carbon nanotubes. *Journal of Environmental Science and Health, Part A, Toxic/Hazardous Substances and Environmental Engineering, 42*, 1543–1552.

Tan, Z., Chen, S., Peng, X., Zhang, L., & Gao, C. (2018). Polyamide membranes with nanoscale turing structures for water purification. *Science, 360*, 518–521.

Theron, J., Walker, J. A., & Cloete, T. E. (2008). Nanotechnology and water treatment: Applications and emerging opportunities.

Tiwari, J. N., Tiwari, R. N., & Kim, K. S. (2012). Zero-dimensional, one-dimensional, two-dimensional and three-dimensional nanostructured materials for advanced electrochemical energy devices. *Progress in Materials Science, 57*(4), 724–803.

Vadahanambi, S., Lee, S. H., Kim, W. J., & Oh, I. K. (2013). Arsenic removal from contaminated water using three-dimensional graphene carbon nanotube-iron oxide nanostructures. *Environmental Science & Technology, 47*, 10510–10517.

Van Minh, H., & Hung, N. V. (2011). Economic aspects of sanitation in developing countries. *Environmental Health Insights, 5*, EHI-S8199.

Veschasit, O., Meksumpun, S., & Meksumpun, C. (2012). Heavy metals contamination in water and aquatic plants in the Tha Chin River, Thailand. *Agriculture and Natural Resources, 46*(6), 931–934.

Vikesland, P. J. (2018). Nanosensors for water quality monitoring. *Nature Nanotechnology, 13*(8), 651–660.

Wang, Y. et al. (2015). High-performance capacitive deionization disinfection of water with graphene oxide-graf-quaternized chitosan nanohybrid electrode coating. *ACS Nano, 9*, 10142–10157.

Wang, X., Shen, Y., Xie, A., Li, S., Cai, Y., Wang, Y., & Shu, H. (2011). Assembly of dandelion-like Au/PANI nanocomposites and their application as SERS nanosensors. *Biosensors and Bioelectronics, 26*(6), 3063–3067.

Wang, H., Zhou, A., Peng, F., Yu, H., & Yang, J. (2007). Mechanism study on adsorption of acidified multi-walled carbon nanotubes to Pb (II). *Journal of Colloid and Interface Science, 316*(2), 277–283.

Warner, C. L. et al. (2010). High-performance, superparamagnetic, nanoparticle-based heavy metal sorbents for removal of contaminants from natural waters. *ChemSusChem, 3*(6), 749–757.

Water, S. & World Health Organization. (2004). Meeting the MDG drinking water and sanitation target: a mid-term assessment of progress/WHO/UNICEF Joint Monitoring Programme. In *Meeting the MDG Drinking Water and Sanitation Target: A Mid-Term Assessment of Progress/WHO/UNICEF Joint Monitoring Programme.*

Xiu, Z. M., Zhang, Q. B., Puppala, H. L., Colvin, V. L. & Alvarez, P. J. J. (2012). Negligible particle-specifc antibacterial activity of silver nanoparticles. *Nano Letters, 12*, 4271–4275.

Xu, J., Cao, Z., Zhang, Y., Yuan, Z., Lou, Z., Xu, X., & Wang, X. (2018). A review of functionalized carbon nanotubes and graphene for heavy metal adsorption from water: Preparation, application, and mechanism. *Chemosphere, 195*, 351–364.

Yan, H. et al. (2006). Adsorption of microcystins by carbon nanotubes. *Chemosphere, 62*, 142–148.

Zhang, D. Q., Li, G. S. & Yu, J. C. (2010). Inorganic materials for photocatalytic water disinfection. *Journal of Materials Chemistry, 20*, 4529–4536.

Zodrow, K. R. et al. (2017). Advanced materials, technologies, and complex systems analyses: Emerging opportunities to enhance urban water security.

13 Microplastics in Wastewater
A Review of the Current Knowledge on Detection, Occurrence, and Removal

B. Yamini Lakshmi and Baranidharan Sundaram
National Institute of Technology

S. Muthukumar
BIT Mesra

CONTENTS

13.1 INTRODUCTION

Since 1950s plastic production increased enormously, present worldwide plastic production exceeds 348 million tonnes and it was estimated by the year 2035 the quantity will be doubled due to its wide usage primarily because of its durability and multipurpose utility. Plastics a huge category of synthetic materials, in which 5000 grades of plastic materials are produced from various combinations of polymers and other chemicals. Continuous increase in plastic production and poor management of the plastic waste management has led to dumping it into water bodies across the world.

Microplastics (MPs) are the category of plastics having size smaller than 5 mm, which cannot be directly visible to the naked eye. Because of the negative effect on the ecosystem, MPs are considered as emerging pollutants (Bouwmeester et al., 2015). MPs can be classified as primary and secondary

DOI: 10.1201/9781003164982-13

MPs. Primary MPs which intentionally manufactured to the size less than 5 mm such as beads, pellets, fibres, fragments etc., and secondary MPs are formed due to weathering, fragmentation or decomposition from larger plastics debris (Dowarah and Devipriya, 2019; Shah, 2020, 2021). Primary MPs enter directly into the marine environment through surface runoff, streams, rivers and final effluent from wastewater treatment plant, secondary MPs are due to weathering, chemical and biological breakdown of meso and macro plastics waste existing in the environment (Naidu, 2019). Further, the ageing of the MPs in the environment also release chemical additives. MPs act as on adsorbent and transportable pollutant due to its small size, high surface to volume ratio and hydrophobic nature. Moreover, MPs also act as appropriate route for absorbable and movable pollutants such as heavy metals, persistent organic pollutants (POP), microorganisms, and virus. It is also given in few review paper, MPs are also ingested by plants and animals. Therefore, the accumulation enter into the food chain of the plants and animals and then to the human organisms, which finally risk to the human healthiness (Peng et al., 2017).

Wastewater treatment plants (WWTPs) has been increased globally due to increase in population growth. Wastewater treatment should be considered necessary to treat influent wastewater arising from households, sewers, and industry by removing organic loads and pathogenic microorganisms to save and protect receiving water bodies such as rivers, lakes, and seas, which are used for drinking water, fishing, and other water activities. The main question is whether WWTP can also remove MPs from influents and stop their entry into surface waters. From the past few years, study on fate of MPs in wastewater has started to gain more attention studies shown that removal of the MPs from WWTPs are examined and concluded that WWTPs do not remove plastics materials completely from wastewater (Leslie et al., 2017). Subsequently, high quantities of MPs are released into the environment on daily basis via treated effluent and sludge products (Ziajahromi et al., 2017). Global studies have demonstrated that, removal of MPs from wastewater is >95% from current wastewater systems (Ziajahromi et al., 2017; Mintenig et al., 2016; Shah, 2021). This leads to enormous volume of the final treated effluent is discharged into the environment from the conventional WWTP. Thus, after the treatment millions of MPs are discharged into the aquatic bodies (Gies et al., 2018). In addition to the contamination of the water bodies, MPs from the sludge which uses as a fertilizer, results soil contamination in the agricultural fields (Murphy et al., 2016; Talvitie et al., 2017).

There is a huge necessity to investigate the study of occurrence of MPs in wastewater regarding the MPs load and to what extent they enter into the streams via effluent from the WWTPs. Multiple studies around the world have shown and confirmed the presence of MPs in WWTPs and their escape into receiving waters (Gündoğdu et al., 2018). Both primary and secondary MPs are released on a daily basis to aquatic environments and the quantities of escaped particles depend not only on treatment stages of the WWTPs but also on the size of the WWTP filters, as well as size and range of MPs. It is shown that removal of MPs from tertiary treatment plant is higher in quantity than MPs from secondary treatment plant (Ziajahromi et al., 2017; Sun et al., 2018; Lares et al., 2018). Perhaps, major focus was made on inlet and outlet samples concentration, only a few authors have started studying MPs at different stages within the WWTP (Murphy et al., 2016), studied MPs abundance at different stages of WWTP and found majority of MPs are removed from the Grease/oil chamber. While Carr et al. (2016) observed contradictory results that 1 MP/L from the effluent of tertiary wastewater suggests which not a source of MPs into the environment (Mintenig et al., 2016) studied and sampled 12 WWTPs in Germany and estimated yearly discharges ranging between 9×10^7 and 4×10^9 MP particles. Simon et al. (2018) estimated from ten largest Danish WWTPs discharge around 3 tonnes of MP/year in a size range of 10–500 μm.

Murphy et al. (2016) collected bulk samples of wastewater before a sieving step (65 μm mesh size), Carr et al. (2016) used fixed sieves of sizes (400–445 μm) in WWTPs (Mintenig et al., 2016), applied enzymatic – oxidative purification in combination with FTIR. Different authors in their studies used various techniques for sampling, separation, treatment, and detection of MPs in WWTPs. This chapter aims to determine comparison on detection, occurrence, analysis, and removal of MPs at different stages of WWTP. The crucial characteristics such as shape, size, concentration, polymer identification, removal efficiency is also presented and compared globally from various papers.

13.2 SOURCES AND TRANSFER OF MPs INTO WWTPs

WWTPs are one of the key sources of MPs in the environment, and several studies explained that MPs are present even after the secondary treatment. Although it was very limited knowledge regarding the various input sources of MPs in WWTPs, but primary sources are generally considered as prominent source of plastics to WWTPs. Even though studies had improved regarding understanding plastic pollution in WWTPs, an appropriate source for plastics entering the WWTPs are very limited. Based on operational conditions, population growth and treatment process applied, concentrations of MPs are facial cleanser – 4.1–1240.0 μm, facial scrubs – 8–2000 μm, toothpaste – 90–300 μm, shower gel – 110–970 μm, body scrubs - <200 μm, facial scrubs <100 μm (Fendall and Sewell, 2009; Kalčíková et al., 2017; Napper et al., 2015; Carr et al., 2016).

As per the literature, removal of MPs from the primary treatment is about 78%–98% and from the secondary treatment 7%–20% of the smaller MPs (Murphy et al., 2016; Talvitie et al., 2017) study showed 98% removal of MPs at a WWTP in Scotland, which is serving population of 6,50,000 citizens and after the secondary treatment about 6.5×10^7 MPs are released into the aquatic environment.

Various personal care and cosmetic are also sources of MPs through which they enter into Domestic WWTP. Major portion of these products consists of polyethylene. Fibre is the other commonly found MP discharged from the WWTPs. According to Talvitie et al. (2017), fibres are present for about 70% from the influent of the WWTP. During the laundry of the clothing material, more than 1900 PE are released per wash into the wastewater.

Besides MPs released into the waterbodies, MPs are also found in the sludge from WWTPs. This sludge is used as a soil fertilizer in the agricultural fields, which leads to contamination of MPs in the soil (Mohapatra et al., 2016). Carr et al. (2016) estimated that 1.09×10^9 MP/day in the form of sludge enters the environment. Nizzetto et al. (2016) estimated about 63,000 tons of MPs are entered in Europe and North America farmlands. In Australia about greater than 2800 tons of the MPs are applied into the agroecosystems in form of biosolids.

In Italy, WWTP receives approximately 1×10^9 Mps/day in the influent and removal efficiency of 84% after the tertiary treatment discharges 1.6×10^8 MPs/day into the aquatic environment. In that study size of MPs are in range of 0.5–.01 mm in the final effluent and sludge after the secondary treatment. Focusing on shape of the MPs, lines of 41% (according to author lines means fibres) from the effluent (Magni et al., 2019; Mason et al., 2016) 17 WWTPS in US discharging 0.05 MP/L from the final effluent, which is more than 4 million of MPs/day.

13.3 SAMPLING AND SEPARATION TECHNIQUES

13.3.1 Sampling

Sampling plays an important role for identification of MPs. MPs sampling in open water bodies like oceans and river are carried using different sized mesh nets are used and dragged at certain speeds by boat. In case of WWTPs which is having high organic content, no specific standardized and universal method is determined for MPs sampling. With different physical conditions there is a different sampling methods mentioned in the current studies. MPs sampling are categorized as auto sampler collection, container collection and filtration collection. Table 13.1 explains the various sampling methods from various studies. Equipment required during the sampling methods are 10L steel bucket and glass bottle, vacuum system, RPS 20 model sampler, sieves, filter device with an electric pump and self-constructed sequential filtering setup etc.,

Xu (2019) explained wastewater is collected in pump or steel bucket called as container collection and stored in a glass bottle for no change in wastewater volume reduction. Another study by Gündoğdu et al. (2018) to collect the 24 hours composite samples for a volume of 5L from the influent and effluent of the WWTP called as auto sampling. Talvitie et al. (2017) reported discrete sampling in which sampling is collected for every 15 minutes interval in a whole day.

TABLE 13.1
Sampling Method and Equipment Used for Waste Water and Sludge Sampling

Method	Equipment	Sample Location	Reference
Container	Steel buckets and glass bottle	Inlet and outlet	Xu (2019)
Auto samplers	Vaccum system RPS 20 model sampler	Inlet and outlet	Gündoğdu et al. (2018)
Container and filtration	Steel bucket and glass bottle	After grit chamber and final outlet	Park et al. (2020)
Filtration	Steel bucket and steel sieve	Inlet, after chlorination, final outlet point	Ruan et al. (2019)
Filtration	Filter device with an electric pump	Inlet, outlet, SST	Talvitie et al. (2015)
Container	Glass jar	Inlet, PST, sludge	Gies et al. (2018)
Container	Steel buckets	Inlet, outlet and SST	Murphy et al. (2016)
Container	Glass jar	Inlet, PST, and SST, sludge	Leslie et al. (2017)
Container	Glass beaker and stainless steel bucket	Sludge, influent, and SST	Lares et al. (2018)
Container	Steel bucket	Inlet, SST, and final outlet point.	Magni et al. (2019)
Surface filtration	Plastic	Influent, SST, and tertiary effluent.	Carr et al. (2016)
Pumping and sieving	Plastic	Primary and SST	Mason et al. (2016)
Pumping and filtration	Plastic	Influent and SST	
Pumping and sieving	Plastic	PST, SST and tertiary effluent	Ziajahromi et al. (2017)

Note: PST, primary sedimentation tank; SST, secondary sedimentation tank.

Filtration is the most commonly used technique to take the samples from the wastewater in the practical conditions. The separation step is used carried by the series of sieves in which continuous effluent is passed. The collected MPs are immediately transferred into the container cleaned by pure water for laboratory analysis. In general, mesh sizes are in a range of 38–4750 μm and chosen mesh size are used based on the MPs size to be collected (Wang and Wang, 2018). And 47 μm sieve is used for the filtration of 10L wastewater by Liu et al. (2019). It is also observed while using single sieve there will be clogging of the particles in a short duration of time. Lv et al. (2019) used series of sieves having mesh size of 500, 250, 125, 62.5 and 25 μm for filtration of MPs from wastewater.

Since effluents consists of high percentage of organic load which tends to blocking of the sieves. Not only sieve blocking, the method also influences the changes of microparticle morphology during separation process. According to Michielssen et al. (2016), Ziajahromi et al. (2017), microfibers have high length to thickness ratio which allows MPs to be retained longitudinally in the sieve. Sampling step should be carried out in a well-performed way in order to get the identical results with less separation error.

Sampling technique should be selected based on the aim of the study. During the sampling stage, adequate quantity should be collected for accurate results. If more samples have to be collected, then filtration sampling has to be adopted. However, for accurate daily estimation of MP concentration, automated sampling is a better option to process the work simple way.

13.3.2 Density Separation

Separation technique will be applied to separate MPs from wastewater after the application of sampling and sieving. Each polymer behaves in a peculiar way with different chemical compositions. During the separation of MPs from sample, the concentration of the polymers is the main characteristic. In general, polymers are such as high-density polyethylene, low-density polyethylene, polyethylene, polyethylene terephthalate, polystyrene, polyamide, polyurethane, and polyvinylchloride having density in varying from 0.88 to 1.65 g/cm^3. Commonly these polymers are used in milk bottle, caps, pipes, plastic bags, trays, films, foam, fishing nets, gaskets, wire insulations, shutter-proof windows, containers etc.

To separate the MPs from wastewater samples, high-density solutions like NaCl, (NaI), and $ZnCl_2$, are mixed with the samples and stirred for fixed time (Lusher et al., 2017). Further, the formed supernatant consisting of the plastic particles is separated by filtration under pressure for further process. Various authors studied and performed density separation and filtration for wastewater, sludge sample by using NaI, $ZnCl_2$, and $NaCl_2$ and sieves like filter paper of 11 and1.2 μm glass fibre filters. Because of no standard method for separation of polymers, comparing the number of MPs reported in different sources is difficult. In this way, the establishment of standard protocol is primarily important for data assessments.

13.3.3 Filtration

In general, the supernatant obtained from the separation technique is allowed to pass through the filter papers of size 1–2 μm. Tweezers are used for the separation of the particles retained on the filter papers for identification (Rocha-Santos, 2018). Different filter media are used namely glass fibre, nitrocellulose and polycarbonate filters. Though filtration is a simple technique, blocking of pores of the filter due to particulate matter leads to the reduction in efficiency (Wang and Wang, 2018). To overcome this problem, reducing the volume of the filter sample or by adding the coagulants solid particles can be formed.

13.4 SAMPLE PROCESSING – DIGESTION

The main aim of the digestion is to remove the biological matter present in the samples obtained from separation technique, as it leads to confusion and overestimates during the identification of polymers. In the following subsections, different types of digestion methods are explained in brief.

13.4.1 Digestion with Acid Substances

From various studies, acids like HNO_3, H_2SO_4, and HCL are used for chemical digestion, which makes the samples to destroy or damage polymers present in the MPs (He et al., 2018). To reduce this, low concentration of acids are taken during the process of removal of biological matter (Cole et al., 2014; Naidu, 2019) studies have shown that by using acid digestion some polymers may destroy, which may result in the breakdown of some plastics, leading to mistake during the quantity estimation. Due to chemical digestion, polymers that have a high level of tendency for damage are polyamide (PA), polyurethane (PU), and black tyre rubber elastomer. Polymers like PET, PP, PC, PTFE, HDPE, LDPE, and EVA will give less damage like loss of colour. From the research, it will show the resistance of polymers to chemical digestion, so that precautions can be taken for the original sample to be digested. The use of acid digestion leads to the disintegration of some plastics, which leads to the mistake in the identification of MPs.

13.4.2 Digestion with Alkaline Substances

Unlike acid digestion, alkaline digestion also causes morphology damage to the MPs. In alkaline digestion solutions like 10 M NaOH, 10% KOH are used at a temperature 60°C which causes damage to the nylon fibres and fragments (Cole et al., 2014). From Hurley et al. (2018) studies, using of 1 M and 10 M NaoH and 10% KOH at 60°C is not appropriate for the removal of biological material from sludge and soil matrixes. In addition to the acid and alkaline digestion, the other method called as digestion by oxidation gives higher results for the removal of organic matter. Unlike acid and alkaline solutions, which cause damage to the polymers and subsequently affect the estimation of the polymer count.

13.4.3 Digestion by Oxidation

Currently, oxidation with H_2O_2 is being used with different contact time and temperature depending on organic matter present in the effluent. Digestion will be performed under heat until H_2O_2 evaporates. Ziajahromi et al. (2017) studied effluents from primary, secondary and tertiary WWTP, sample is passed through the sieves and material retained on the mesh is removed by distilled water and chemical digestion performed for 1 week with 2 mL hydrogen peroxide. Gies et al. (2018) for the digestion of primary, secondary, and influent samples used 30% of H_2O_2 at room temperature. performed digestion with 15% of H_2O_2 for supernatant of the wastewater and sludge obtained from the density separation with NaCl and filtered with cellulose nitrate membrane filter of 8 μm (Li et al., 2018) in which sample after the density separation with NaCl and supernatant passed through the 30 μm sieve and digested with 100 mL of 30% H_2O_2.

For making the process quick, catalyst such as Fenton's reagent (Fe^{2+}) is used during oxidation process. Table 13.2 shows the application of hydrogen peroxide with Fenton's reagent for oxidation. pH is the important factor to be controlled when Fenton reagent is added to the sample. Ideal condition for performing the reaction with Fetaon's reagent is pH-3. Many studies are carried out with different combinations of hydrogen peroxide, catalyst, time and temperature for better results (Gies et al., 2018; Magni et al., 2019). Using Fenton's reagent as a catalyst during the separation of MPs achieved 86% removal of the organic matter and by KOH and NaOH application it was observed only 67% removal of organic matter (Tagg et al., 2017), also described that the spectra generated from the ATR-FTIR after digestion with Fenton's reagent have no major changes in the polymers like PP, PE, PVC, and nylon.

From the present literature, it was pointed out that using high temperature for oxidation leads to negative results (Carr et al., 2016). Improper estimation for quantity analysis is demonstrated by a study (Gies et al. 2018) utilizing temperature greater than 60°C heat melted microbead quantity. Hurley et al. (2018) proposed that for positive results of MPs identification, oxidation based on Fenton's reagent should be maintained at a temperature less than 40°C.

13.4.4 Enzymatic Digestion

Enzymatic digestion is another method for the extraction of biological matter from the samples of wastewater. Cole et al. (2014) by using proteinase – K and achieved digestion efficacy of 88%. By increasing the time for incubation during digestion efficacy can be improved by 97%.

TABLE 13.2
Oxidation with Fenton's Reagent

Sample	Oxidation	Duration	Temp	Reference
Ww	H_2O_2 with 30%	1 week	Room temp	Gies et al. (2018)
Ww	H_2O_2 of 15%	3 days	Room temp	Magni et al. (2019)
Ww	H_2O_2 of 30%	Until H_2O_2 evaporated	60°C	Ziajahromi et al. (2017)
Sludge	30% of H_2O_2	24 hours		Li et al. (2018)
Sludge	30% of H_2O_2	5 hours	70°C	Dyachenko et al. (2017)
Ww	H_2O_2, 30% + $FeSO_4$ $7H_2O$	60 minutes	75°C	Lares et al. (2018)
Sludge	15% H_2O_2	3 days	Room temp	Magni et al. (2019)
Soil and sludge	30% H_2O_2	6 hours	60°C	Lusher et al. (2017)
Sludge	30% H_2O_2	10 days	Room temp	Gies et al. (2018)
Ww	H_2O_2, 30% + $FeSO_4 7H_2O$	30 minutes	70°C	Dyachenko et al. (2017)
Soil and sludge	H_2O_2 30% + $FeSO_4$ $7H_2O$	2 hours	Room temp	Lusher et al. (2017)

Note: Ww, wastewater.

No damage was observed to the samples. However, it is a very expensive and complex procedure. To make it cost-effective, an enzyme trypsin, collagenase, and papain were used which is less cost to proteinase –K. Mintenig et al. (2016) used protease, lipase, and cellulose enzymes in the enzymatic digestion. Simon et al. (2018) used cellulose enzyme for the removal of biological matter from wastewater samples. Despite the acceptable results for the removal of the organic matter and MPs separation, the procedure of this method involved greater than 10 days and several steps, which may lead to the contamination of the sample loss as per the authors.

13.5 OCCURRENCE OF MPS/IDENTIFICATION TECHNIQUES

For identification of MPs after pretreatment, commonly used techniques are visual counting, stereomicroscope, Fourier Transform infrared spectroscopy (FT-IR), Raman Spectroscopy, and Scanning Electron microscopy.

13.5.1 Physical Method

Visual counting is used to count the number of plastics into various categories like size, shape, and colour. The naked eye can separate plastics of size less than 2 mm, while smaller to this microscope, such as stereomicroscope, dissection microscope, and photograph microscope, is required. The most commonly used microscope to examine the MPs in the WWTPs is stereomicroscope after visual counting method, which can identify particles up to microns (Li et al., 2018). SEM method is applied for the identification of MPs by its high-resolution images. After the chemical digestion changes in the MPs, surface can also be detected using the SEM method. In addition to above, polarized light optical microscopy (PLOM) method is a primary technique for its quick and cost-effective for the identification of MPs in wastewater. It is mainly used for the identification of PP, PET, and PE. Particles identified by PLOM were confirmed approximately 81% as MPs.

13.5.2 Chemical Method

Apart from the physical characteristics of the MPs, chemical characteristics are also needed to be identified for reduction in error during the identification of MPs. FTIR-ATR, Raman Spectroscopy are used for the polymer identification of the MPs. FTIR and Raman are limited for particles >1 μm for identification. FTIR consists of reflectance, transmission, and attenuated total reflectance (ATR), which will be adopted based on the identification of shape, size, and thickness of the sample. ATR-FTIR is highly efficient for the identification of particles greater than 500 μm. μ-ATR-FTIR is used the analyse smaller particles identification. MPs were identified by the ATR-FTIR technique for secondary effluent samples of WWTP located in Valencia.

13.5.3 Thermo – Analytical Method

Apart from common visual and Raman analytical technique, thermo analytical method is used on MPs characterization. Thermal analysis methods are namely pyrolysis gas chromatograph- mass spectrometry (Pyro- GC-MS), thermogravimetric analysis coupled with differential scanning calorimetry (TGA-DSC) are the time-saving methods for treating complex MPs in the wastewater samples. From this method, mass of MPs can also be obtained which will be helpful to quantity MPs pollution from WWTPs. However, this method is rarely used for MPs analysis if no information is provided. Table shows the comparison of MPs identification method.

However, these techniques should be used accordingly since they yield significant facts for plastics research in regard to the sources, types, and size of plastics in samples. Table 13.3 gives a summary on the advantages and disadvantages of each digestion and identification technique.

TABLE 13.3
Summary on Identification Techniques of MPS in WWTPs

Technique	Advantages	Disadvantages
Oxidation with hydrogen peroxide at ambient temperature.	Removal efficiency of organic matter is greater than 80% from wastewater samples. Samples are not much affected when exposed to temperature of 48 hours.	It takes longer period for removal of organic matter in large volumes of samples. Based on the composition of the biological content in the sample sometime organic matter is not completely removed. It also causes discolouration of PET and PP.
Oxidation with H_2O_2 above 60°C	Decrease the reaction time from weeks to days. Around 80-87% of the organic matter is removed from the samples.	Complex procedure with time-consuming. Degradation of the polymers takes place such as PA, Nylon, PMMA and PS. Decolourisation, reduction in size of plastic particles are also observed. A new method with lower temperature may overcome this matter although increase in the reaction time.
Fenton's reagent with hydrogen peroxide at ambient temperature.	Without impact on plastics, complex biological matter is breakdown within short period of time. Biological material removal efficiency of 85%–100% is obtained from soil and sludge by using this method. It is time saving and cost effective method. It also gives better results for larger volume of samples.	It requires high temperature for completing the reaction, which may result degradation of the polymers. Ice baths are recommended in the laboratories to reduce the temperature to optimum level for controlling destruction of the polymers.
Visual counting	It is very easy and inexpensive method. It is used to measure size, shape, colour and quantity directly. Larger size samples are quickly identified.	It is size dependent technique and has low magnification factor of microscopes and hence not used if particles are <300 µm. 70% of error will be observed in samples and underestimation of plastic particles take place. It is a time-consuming process
Spectroscopy (Raman)	Accuracy of the polymer identification is increased by this method. It is highly sensitive to non-polar functional groups and insensitive for disturbing of signals of water and atmospheric CO. it is suitable for plastic particles in a range of 1 to 20 µm. The technique has provision to produce visual image, from which dimension of the particles can be determined.	It is hard to identify particles of size less than 1 µm. Samples require detailed purification to avoid the undesirable changes prior to analysis so that interruption will not take place during identification. It requires skilled labour and time-consuming procedure.

(Continued)

TABLE 13.3 (*Continued*)
Summary on Identification Techniques of MPS in WWTPs

Technique	Advantages	Disadvantages
μ-FTIR and ATR-FTIR	Polymer types of plastics are identified quickly and directly by comparing the resulting spectra with the known spectra. It is fast, quite reliable and well-established technique. ATR-FTIR is used to analyse larger particles of size >500 μm and down to that size μ-ATR-FTIR is practiced. μ-ATR-FTIR gives accurate results.	It is time-consuming and high skilled labour is required for analysis of the particles. Contaminants and additives overlap and create difficulty in data interpretation. Information in respect to organic additives is not provided. For comparison of the results a known spectra should be created for easy identification. FTIR is expensive and requires highly skilled labour for its operation and data processing.
Thermo analytical methods	It gives faster identification of particles with their characteristics. Irrespective of the particle size and shape, it gives better results during quantitative estimation. Information regarding the chemical structure, polymer and additives can also be achieved.	For this technique sample of weight of 0.1 to 0.5 mg should be taken for better results. It is very time-consuming method compared to above all methods. Particles shall be handled very carefully with tweezers. Operating cost of this technique is very high if the blockage of the contaminants occurs.

TABLE 13.4
Globally Occurrence and Removal of MPs from Different STPs

Country	Shape and Type	MPs at Inlet, Oulet and Sludge (MP/L)	Removal Rate	Reference
Finland	Synthetic and textile particles	Inlet – 430 and 180 particles/L Outlet – 8.6 and 4.9 particles/L. Sludge – nil	96%–98%	Talvitie et al. (2015)
Scotland	Flakes, film, fibres, foam and bead	Inlet – 15.7/L Outlet – 0.25/L	98.4%	Murphy et al. (2016)
USA	Fibre, fragment, paint, chip, microbead and nurdle	Inlet – 80–140/L Outlet – 0.5–5.9/L	95.6%–99.4%	Michielssen et al. (2016)
USA	Irregular polyethylene	Inlet – 1/L Outlet – 0.0007/L	99.9%	Carr et al. (2016)
USA	Fibres, fragments, pellets, foam and films	Outlet – 0.05/L	99.9%	Mason et al. (2016)
Russia	Synthetic, black fibres and textiles	Inlet – 467 textile/L, 160 synthetic/L and 3160 black fibres/L Outlet – 16/L, 7/L and 1.0/L	96%	Lares et al. (2018)
Netherlands	Foils, spheres and fibres	Inlet – 68–910/L Outlet – 51–81/L Sludge – 510–760/kg	72%	Leslie et al. (2017)
Finland	Fibres, fragments, films, flakes and spherical	Inlet – 57.6/L Outlet – 1/L Sludge – 170.9×10^3/L	99%	Talvitie et al. (2017)
Canada	Fibres, foam, granule, sheet, pellet and fragment	Inlet – 31.1/L Outlet – 2.6/L Sludge – 14.9×10^3/L	97%–99.9%	Gies et al. (2018)

13.6 REMOVAL OF MPs FROM WWTPs

In general, WWTP consists of primary, secondary, and rarely tertiary treatment units for the removal of organic matter by chemical and biological processes. While in connection to the removal of MPs from these treatment units, most of the studies review that the pretreatment unit itself removes the majority of the MPs in a range of 72%–98%. In the case of secondary treatment unit, an additional 7%–20% will be removed and whereas removal of MPs from tertiary units depends on the method of technology adopted. A total of 34 STPs in New York were studied, and it was concluded that tertiary treatment such as microfiltration, and rapid sand filters, did not guarantee the absence of microbeads from the effluents. Talvitie et al. (2017) compared step-wise treatment units of STP and declared that the highest removal of MPs is from MBR (99.9%), RSF (97%), dissolved air flotation (95%), disc filter (40%–98.5%). Michielssen et al. (2016) also reported that MBR removes MPs with the efficiency of 97.2%, respectively. Murphy et al. (2016) observed removal of MPs from the grit chamber is 44.6%, primary sedimentation tank – 33.8%, and biological treatment tank – 20.1%.

Regarding the removal of MPs from the different units of STP is given in Table 13.4 from various global monitoring studies.

13.7 CONCLUSION AND PERSPECTIVES

MPs concentration has become a serious issue and has attracted global attention. However, the point source of MPs is from final discharge after treatment and sludge from the WWTPs into the environment, which should get more analysis and well controlled. This chapter explains the various methodologies for sampling, separation, filtration, sample processing, occurrence and removal of these emerging contaminants. The total process is dependent and should be selected according to the objective and treatment necessity as well as receiving aquatic body quality. Based on the literature study, attention for future research should be given to characterization of tiny MPs (size less than 20 μm) and uniformity in the MPs concentration per unit. Some digestion methods are aggressive to the MPs and a basic method with reagent should be developed for better results. For the concentration of MPs in WWTPs, long-term data is required for scientific assessment.

REFERENCES

Bouwmeester, H., Hollman, P.C.H., Peters, R.J.B., 2015. Potential health impact of environmentally released micro- and nanoplastics in the human food production chain: Experiences from nanotoxicology. *Environ. Sci. Technol.* 49, 8932–8947.

Carr, S.A., Liu, J., Tesoro, A.G., 2016. Transport and fate of microplastic particles in wastewater treatment plants. *Water Res.* 91, 174–182. https://doi.org/10.1016/j.watres.2016.01.002.

Chang, M., 2015. Reducing microplastics from facial exfoliating cleansers in wastewater through treatment versus consumer product decisions. *Mar. Pollut. Bull.* 101, 330–333. https://doi.org/10.1016/j.marpolbul.2015.10.074.

Cole, M., Webb, H., Lindeque, P.K., Fileman, E.S., Halsband, C., Galloway, T.S., 2014. Isolation of microplastics in biota-rich seawater samples and marine organisms. *Sci. Rep.* 4, 1–8. https://doi.org/10.1038/srep04528.

Dowarah, K., Devipriya, S.P., 2019. Microplastic prevalence in the beaches of Puducherry, India and its correlation with fishing and tourism/recreational activities. *Mar. Pollut. Bull.* 148, 123–133. https://doi.org/10.1016/j.marpolbul.2019.07.066.

Dyachenko, A., Mitchell, J., Arsem, N., 2017. Extraction and identification of microplastic particles from secondary wastewater treatment plant (WWTP) effluent. *Anal. Methods.* 9, 1412–1418. https://doi.org/10.1039/c6ay02397e.

Fendall, L.S., Sewell, M.A., 2009. Contributing to marine pollution by washing your face: Microplastics in facial cleansers. *Mar. Pollut. Bull.* 58, 1225–1228. https://doi.org/10.1016/j.marpolbul.2009.04.025.

Gies, E.A., LeNoble, J.L., Noël, M., Etemadifar, A., Bishay, F., Hall, E.R., Ross, P.S., 2018. Retention of microplastics in a major secondary wastewater treatment plant in Vancouver, Canada. *Mar. Pollut. Bull.* 133, 553–561. https://doi.org/10.1016/j.marpolbul.2018.06.006.

Gündoğdu, S., Çevik, C., Güzel, E., Kilercioğlu, S., 2018. Microplastics in municipal wastewater treatment plants in Turkey: A comparison of the influent and secondary effluent concentrations. *Environ. Monit. Assess.* 190. https://doi.org/10.1007/s10661-018-7010–y.

He, D., Luo, Y., Lu, S., Liu, M., Song, Y., Lei, L., 2018. Microplastics in soils: Analytical methods, pollution characteristics and ecological risks. *TrAC – Trends Anal. Chem.* 109, 163–172. https://doi.org/10.1016/j.trac.2018.10.006.

Hurley, R.R., Lusher, A.L., Olsen, M., Nizzetto, L., 2018. Validation of a method for extracting microplastics from complex, organic-rich, environmental matrices. *Environ. Sci. Technol.* 52, 7409–7417. https://doi.org/10.1021/acs.est.8b01517.

Kalčíková, G., Alič, B., Skalar, T., Bundschuh, M., Gotvajn, A.Ž., 2017. Wastewater treatment plant effluents as source of cosmetic polyethylene microbeads to freshwater. *Chemosphere.* 188, 25–31. https://doi.org/10.1016/j.chemosphere.2017.08.131.

Lares, M., Ncibi, M.C., Sillanpää, M., Sillanpää, M., 2018. Occurrence, identification and removal of microplastic particles and fibers in conventional activated sludge process and advanced MBR technology. *Water Res.* 133, 236–246. https://doi.org/10.1016/j.watres.2018.01.049.

Lei, K., Qiao, F., Liu, Q., Wei, Z., Qi, H., Cui, S., Yue, X., Deng, Y., An, L., 2017. Microplastics releasing from personal care and cosmetic products in China. *Mar. Pollut. Bull.* 123, 122–126. https://doi.org/10.1016/j.marpolbul.2017.09.016.

Leslie, H.A., Brandsma, S.H., van Velzen, M.J.M., Vethaak, A.D., 2017. Microplastics en route: Field measurements in the Dutch river delta and Amsterdam canals, wastewater treatment plants, North Sea sediments and biota. *Environ. Int.* 101, 133–142. https://doi.org/10.1016/j.envint.2017.01.018.

Li, X., Chen, L., Mei, Q., Dong, B., Dai, X., Ding, G., Zeng, E.Y., 2018. Microplastics in sewage sludge from the wastewater treatment plants in China. *Water Res.* 142, 75–85. https://doi.org/10.1016/j.watres.2018.05.034.

Liu, X., Yuan, W., Di, M., Li, Z., Wang, J., 2019. Transfer and fate of microplastics during the conventional activated sludge process in one wastewater treatment plant of China. *Chem. Eng. J.* 362, 176–182. https://doi.org/10.1016/j.cej.2019.01.033.

Lusher, A.L., Hurley, R.R., Vogelsang, C., Nizzetto, L., Olsen, M., 2017. Mapping microplastics in sludge. https://doi.org/10.13140/RG.2.2.25277.56804.

Lv, X., Dong, Q., Zuo, Z., Liu, Y., Huang, X., Wu, W., 2019. Microplastics in a municipal wastewater treatment plant: Fate, dynamic distribution, removal efficiencies, and control strategies. *J. Clean. Prod.* https://doi.org/10.1016/j.jclepro.2019.03.321.

Magni, S., Binelli, A., Pittura, L., Giacomo, C., Della, C., Carla, C., Gorbi, S., Regoli, F., 2019. Science of the total environment the fate of microplastics in an Italian wastewater treatment plant. *Sci. Total Environ.* 652, 602–610. https://doi.org/10.1016/j.scitotenv.2018.10.269.

Mahon, A.M., O'Connell, B., Healy, M.G., O'Connor, I., Officer, R., Nash, R., Morrison, L., 2017. Microplastics in sewage sludge: Effects of treatment. *Environ. Sci. Technol.* 51, 810–818. https://doi.org/10.1021/acs.est.6b04048.

Mason, S.A., Garneau, D., Sutton, R., Chu, Y., Ehmann, K., Barnes, J., Fink, P., Papazissimos, D., Rogers, D.L., 2016. Microplastic pollution is widely detected in US municipal wastewater treatment plant effluent. *Environ. Pollut.* https://doi.org/10.1016/j.envpol.2016.08.056.

Michielssen, M.R., Michielssen, E.R., Ni, J., Duhaime, M.B., 2016. Fate of microplastics and other small anthropogenic litter (SAL) in wastewater treatment plants depends on unit processes employed. *Environ. Sci. Water Res. Technol.* 2, 1064–1073. https://doi.org/10.1039/c6ew00207b.

Mintenig, S.M., Löder, M.G.J., Primpke, S., Gerdts, G., 2016. Identification of microplastic in effluents of waste water treatment plants using focal plane array-based micro-Fourier-transform infrared imaging. *Water Res.* https://doi.org/10.1016/j.watres.2016.11.015.

Mohapatra, D.P., Cledón, M., Brar, S.K., Surampalli, R.Y., 2016. Application of wastewater and biosolids in soil: Occurrence and fate of emerging contaminants. *Water. Air. Soil Pollut.* 227. https://doi.org/10.1007/s11270-016-2768-4.

Murphy, F., Ewins, C., Carbonnier, F., Quinn, B., 2016. Wastewater treatment works (WwTW) as a source of microplastics in the aquatic environment. *Environ. Sci. Technol.* 50, 5800–5808. https://doi.org/10.1021/acs.est.5b05416.

Naidu, S.A., 2019. Preliminary study and first evidence of presence of microplastics and colorants in green mussel, Perna viridis (Linnaeus, 1758), from southeast coast of India. *Mar. Pollut. Bull.* 140, 416–422. https://doi.org/10.1016/j.marpolbul.2019.01.024.

Napper, I.E., Bakir, A., Rowland, S.J., Thompson, R.C., 2015. Characterisation, quantity and sorptive properties of microplastics extracted from cosmetics. *Mar. Pollut. Bull.* 99, 178–185. https://doi.org/10.1016/j.marpolbul.2015.07.029.

Nizzetto, L., Futter, M., Langaas, S., 2016. Are agricultural soils dumps for microplastics of urban origin? *Environ. Sci. Technol.* 50, 10777–10779. https://doi.org/10.1021/acs.est.6b04140.

Park, H.J., Oh, M.J., Kim, P.G., Kim, G., Jeong, D.H., Ju, B.K., Lee, W.S., Chung, H.M., Kang, H.J., Kwon, J.H., 2020. National reconnaissance survey of microplastics in municipal wastewater treatment plants in Korea. *Environ. Sci. Technol.* 54, 1503–1512. https://doi.org/10.1021/acs.est.9b04929.

Peng, G., Zhu, B., Yang, D., Su, L., Shi, H., Li, D., 2017. Microplastics in sediments of the Changjiang Estuary, China. *Environ. Pollut.* 225, 283–290. https://doi.org/10.1016/j.envpol.2016.12.064.

Rocha-Santos, T.A.P., 2018. Editorial overview: Micro and nano-plastics. *Curr. Opin. Environ. Sci. Heal.* 1, 52–54. https://doi.org/10.1016/j.coesh.2018.01.003.

Ruan, Y., Zhang, K., Wu, C., Wu, R., Lam, P.K.S., 2019. A preliminary screening of HBCD enantiomers transported by microplastics in wastewater treatment plants. *Sci. Total Environ.* 674, 171–178. https://doi.org/10.1016/j.scitotenv.2019.04.007.

Shah, M.P., 2020. *Advanced Oxidation Processes for Effluent Treatment Plants.* Elsevier.

Shah, M.P., 2020. *Microbial Bioremediation and Biodegradation.* Springer.

Shah, M.P., 2021. *Removal of Emerging Contaminants through Microbial Processes.* Springer.

Simon, M., van Alst, N., Vollertsen, J., 2018. Quantification of microplastic mass and removal rates at wastewater treatment plants applying Focal Plane Array (FPA)-based Fourier Transform Infrared (FT-IR) imaging. *Water Res.* 142, 1–9. https://doi.org/10.1016/j.watres.2018.05.019.

Sun, J., Dai, X., Wang, Q., Van Loosdrecht, M.C.M., Ni, B., 2018. Microplastics in wastewater treatment plants: Detection, occurrence and removal. *Water Res.* https://doi.org/10.1016/j.watres.2018.12.050.

Tagg, A.S., Harrison, J.P., Ju-Nam, Y., Sapp, M., Bradley, E.L., Sinclair, C.J., Ojeda, J.J., 2017. Fenton's reagent for the rapid and efficient isolation of microplastics from wastewater. *Chem. Commun.* 53, 372–375. https://doi.org/10.1039/c6cc08798a.

Talvitie, J., Heinonen, M., Pääkkönen, J.P., Vahtera, E., Mikola, A., Setälä, O., Vahala, R., 2015. Do wastewater treatment plants act as a potential point source of microplastics? Preliminary study in the coastal Gulf of Finland, Baltic Sea. *Water Sci. Technol.* 72, 1495–1504. https://doi.org/10.2166/wst.2015.360.

Talvitie, J., Mikola, A., Koistinen, A., Setälä, O., 2017. Solutions to microplastic pollution – Removal of microplastics from wastewater effluent with advanced wastewater treatment technologies. *Water Res.* 123, 401–407. https://doi.org/10.1016/j.watres.2017.07.005.

Wang, W., Wang, J., 2018. Investigation of microplastics in aquatic environments: An overview of the methods used, from field sampling to laboratory analysis. *TrAC – Trends Anal. Chem.* 108, 195–202. https://doi.org/10.1016/j.trac.2018.08.026.

Xu, X., 2019. Chemosphere microplastics in the wastewater treatment plants (WWTPs): Occurrence and removal. *Chemosphere.* 235, 1089–1096. https://doi.org/10.1016/j.chemosphere.2019.06.197.

Ziajahromi, S., Neale, P.A., Rintoul, L., Leusch, F.D.L., 2017. Wastewater treatment plants as a pathway for microplastics: Development of a new approach to sample wastewater-based microplastics. *Water Res.* https://doi.org/10.1016/j.watres.2017.01.042.

14 Emerging Nanofiber Technology for the Removal of Metal Ions in Wastewater Treatment Plants

Njabulo S. Mdluli
University of South Africa

Nilesh S. Wagh and Jaya Lakkakula
Amity University

Philiswa N. Nomngongo
University of Johannesburg

Nomvano Mketo
University of South Africa

CONTENTS

14.1 INTRODUCTION

Over the years, wastewater has been reported as one of the major sources of pollution globally. This wastewater results from industrial and anthropogenic activities [1]. However, the most commonly reported threat to fresh water is wastewater from industrial activities. This is because various industrial processes consume a huge amount of water and, at later stages, this water is released as wastewater effluent containing various pollutants [2,3]. Therefore, if the wastewater effluent is not effectively treated to remove present pollutants, nearby rivers, dams, and lakes will be polluted [4]. Furthermore, the pollution of surface water caused by domestic and industrial wastewater effluents has also been reported to cause disturbance in aquatic ecosystem [5]. Therefore, wastewater effluent is a major contributor to freshwater scarcity, as it plays a vital role in surface water pollution [6].

DOI: 10.1201/9781003164982-14

The wastewater effluents are primarily polluted with non-degradable heavy metal ions, which pose a significant environmental risk due to their high toxicity and ability to accumulate in living organisms and ecosystems [7]. Developing countries like South Africa rely on surface water for domestic, agricultural, and industrial purposes [8]. Therefore, if the surface water is contaminated with metals, consumers will be at higher risks, since metal ions have long-lasting health effects, thereby causing death [9]. Ebenrte et al. performed a report on heavy metal emissions and their effect on the environment in South African mine effluents in the Limpopo province. This study reported that mining was a major cause of heavy metals in all the river basins. In Limpopo, 22 sub-basins were chosen for the analysis, and all of the sub-basins suggested that metals had an effect on water quality (As, Co, Cu, Fe, Hg, Mo, and Pb) [10]. The evaluation of heavy metals in municipal sewage sludge in Limpopo province was stated by Kudakwashe et al., (South Africa). In this study, the concentration levels of Ag, Pb, and Cd were studied in different towns (Louis Trichardt, Thohoyandou, Tzaneen, Polokwane) of the province [8]. The sewage sludge samples were acid-digested and then analyzed using the ICP-OES for the determination of Ag, Cd, and Pb. These results indicated that the concentration of metals differed much between the different towns. For example, Ag concentration levels were 21.93±0.38, 6.13±0.12, 13.40±0.46, and 6.89±0.1 mg/kg in Louis Trichardt, Thohoyandou, Tzaneen, Polokwane, respectively [8]. A related research on heavy metals in wastewater and sewage sludge was conducted by Agoro et al. This research was carried out in a number of municipal treatment plants in the Eastern Cape province (South Africa). The concentrations of metals reported in this study were 1.17, 0.14, 69.789, and 0.099 mg/kg for Cu, Cd, Fe, and Pb, respectively, using atomic absorption spectrometry. The high levels of heavy metals reported from these wastewater samples give evidence for the need to develop cost-effective and efficient strategies for metal removal in aqueous solutions.

Therefore, several wastewater purification strategies have been put to play in ensuring that there is minimal harm to living organisms and the environment. Adsorption, membrane process, biological process, chemical reactions, coagulation-flocculation, flotation, electrochemical procedure, ion exchange, sedimentation, and precipitation are some of the wastewater purification methods [11]. Some of the reported purification methods like ozonation and chlorinated reverse osmosis are less favored due to their harmful nature and high operational cost [12]. However, among the recorded pollutant removal techniques, sorption is regarded as the most cost-effective, versatile, low-production secondary emission, high-efficiency, and relatively simple process [13]. In addition, nanotechnology has played a huge role in the development of effective and novel adsorptive removal methods [5]. Nanofiber technology is an emerging tool under nanotechnology, and it is well-reported for wastewater treatment for various types of pollutants [14]. Other than their material size, these nanomaterials are typically classified by differences in biological, chemical, and physical properties [13]. This is because nanofibers have a very small diameter in nanometre scale with a very high surface area and tuneable porosity [15]. The advantages that come with the use of nanofibers are numerous, and one of them is the fact that various cost-effective materials can be utilized in the synthesis or fabrication of nanofibers [16]. These materials include natural and synthetic polymers, metals, metal oxides, carbon-based and composite nanomaterials, etc. [2]. In addition, nanofibers can be easily modified, thus changing the physiochemical properties, making it possible to remove a wide range of pollutants [17]. Metal-based (inorganic-based), carbon-based, composite, rigid, porous, non-porous, hollow, and core–shell nanofibers are among the various types of nanofibers mentioned in the literature [18]. Different fabrication techniques, such as bottom-up, top-down, physical, chemical, biological, non-spinning, and electrospinning, can be used to create these various forms of nanofibers. However, the mostly reported fabrication technique is electrospinning. The latter is highly favored as it is simple to operate, low cost, and normally yields very long fiber [19,20].

14.2 ELECTROSPINNING TECHNIQUE

Electrospinning is a technique for producing nanofibers with diameters ranging from 40 to 200 nm in which fiber solution is spined using electrostatic forces. A high-voltage power supply, a spinneret,

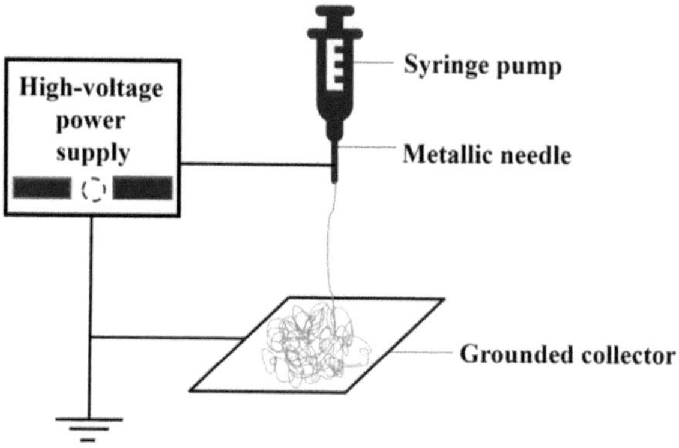

FIGURE 14.1 An outline of the electrospinning technology. (Reproduced with permission from A. M. Nasir et al., "Recent progress on fabrication and application of electrospun nanofibrous photocatalytic membranes for wastewater treatment: A review," J. Water Process Eng., vol. 40, no. November 2020, p. 101878, 2021 [4].)

and a collector are the key components of this technique (see Figure 14.1) [21]. Polymer droplets overcome surface tension to form a jet, which evaporates the solvent and solidifies to form ultrafine fiber during the electrospinning process. Several factors influence this newly formed electrospun nanofiber which include factors such as processing parameters, solution parameters, and ambient parameters. The most basic processing parameter for electrospinning is spinning voltage [21]. When the voltage is increased such that a critical voltage is reached, this result to the formation of "Taylor cone". Then, a further increase of the voltage will result in the formation of the jet which eventually forms a nanofiber. It is worthy to indicate that, the high voltage helps stretching the fiber, thus reducing the fiber diameter [11]. The other significant parameter for electrospinning processing is the spinning solution injection speed. This parameter can be controlled by adjusting the speed of the syringe pump such that the slower the syringe pump, the longer the electrospinning time. Various researchers have reported that, the faster the speed, the larger the fiber diameter and the lesser the electrospinning time. For the above-mentioned reasons, the electrospinning parameters need to be properly controlled for the formation of the most ideal nanofibers [22]. Environmental factors such as relative humidity and temperature, on the other hand, may influence the nanofiber's morphology. At relative low humidity, the evaporation rate is increased which is key in the formation of thinner fibers. On the other hand, the effect of temperature has opposite effects on the diameter of the nanofiber, with low temperatures promoting thinner nanofiber while high temperatures promote thicker nanofibers [22].

In summary, this electrospinning process can be categorized into three: the processing conditions, the intrinsic properties, and the ambient parameters. These parameters help to generate fibers with controllable diameters, bead-free, and single nanofibers. In addition, the main factor that affects the final fiber diameter is polymer solution viscosity. Therefore, various researchers have highlighted on how to fabricate different nanofibers (metal-based, carbon-based, membrane, and polymer-based). Pereao et al. published a study on the relationship between electrospinning parameters and nanofiber diameter and volume. Continuous individual polycaprolactone (PCL) nanofibers with diameters ranging from 350 nm to 1 m have the capacity to be collected at lengths of 35–50 cm, according to the study. The maximum fiber length, diameter, and uniformity were all affected by the polymer concentration, voltage, and plate size [16]. In hematite-based polymer nanofiber composites, the electrospinning technique is widely used. The first approach involves combining varying quantities of hematite nanoparticles into the polymer precursor solutions for electrospinning. This produces several polyacrolonitrile/hematite composite using a simple one-step fabrication route [23]. These more conventional nanoparticle-embedded composites are then further processed into polyacrylonitrile/hematite core–shell nanofibers using a hydrothermal treatment.

Another fabrication method reported is the single-pot nanofiber fabrication method. This method is less favored as it is likely to result in loss of the hematite surface area, and this happens as a result of embedding of the nanoparticle in the polymer matrix [24]. Therefore, the multi-step, which is electrospinning coupled with hydrothermal nanofiber fabrication, is highly favored as it promotes the external hematite surface area in the core–shell material [25].

The nanofiber properties and surface area correlate with the metal uptake (concentrations). For example, when comparing the core–shell, hallow, porous and nanoparticle-embedded nanofibers, the benefits of core–shell nanofibers were reported to out class the others [26]. Different nanofibers (core–shell, porous, and aligned nanofiber) normally differ in how they are fabricated. Core–shell nanofibers usually combine the benefits of core and shell materials, resulting in enhanced physical and chemical properties. These nanofibers are produced using a single needle and an electrospinning system with a single spinneret to prepare the core-sheath nanofiber directly from a homogeneous polymer solution. It is worth noting that, hallow nanofibers can be made from core–shell nanofibers by simply removing the core shell [27]. It's worth remembering that during phase separation, changing the humidity and temperature of the electrospinning results in the formation of porous nanofiber. The higher the humidity (<30%), the higher the rate of pores formation but the problem is that the pores are not uniform, and this negatively affects the physical properties of the nanofiber. Using a high-speed spinning drum as a collector often results in aligned nanofibers, which are parallel nanofibers. A well-designed collector can also help in the generation of more appropriate nanofiber [14]. For example, a collector with gaps in between generates parallel nanofibers with gaps in between. As a result, these nanofibers have been tested for the removal of metal ions, pharmaceuticals, and dyes, among other things. The following sub-sections discuss in greater depth the removal of metal ions using different types of nanofibers.

14.3 REMOVAL OF METAL ION BY USING NANOFIBERS

The use of nanofibers in wastewater treatment has become a subject of study in the last few decades [28]. The structural properties of nanofibers make them ideal for removing metal ions from wastewater. These properties include a high surface area-to-volume ratio, high porosity, and interconnectivity, all of which are ideal for heavy metal adsorption [16]. Furthermore, nanofibers are one of the most applied adsorbents due to their easy desorption (especially metal ions), high selectivity, low cost, high adsorption capacity of metal ions, reusability, and high mechanical strength [28]. There are two main processes that are a driving force toward adsorption which are physical and chemical adsorption. In physical adsorption, the electrostatic forces or intermolecular forces between adsorbent and metal ions facilitate adsorption. In physical adsorption, the presence of weak van de Wall forces brings about adsorption. As a result, the enthalpy of physisorption is very poor in this form of adsorption. The specific surface area of the adsorbent is the key determinant of adsorption efficiency in this method of adsorption [29]. Chemical adsorption, on the other hand, is based on the formation of a stable chemical bond between adsorption materials and adsorbent (metal ions). As a result, in chemical adsorption, the adsorbents' surface chemical properties are the primary determinant of adsorption performance.

Several different nanofibers have been reported for metal removal, and these nanofibers showed various adsorption capabilities, depending on the targeted metal ion. The type of nanofibers reported in the literature for metal ion removal is polymer-based, carbon-based, metal-decorated, and membrane-based nanofibers (see Table 14.1) [30].

14.3.1 POLYMER-BASED NANOFIBERS

Nanofibers made of polymers, particularly biopolymers, have recently gotten a lot of attention in wastewater treatment. The latter is also composed of different biomaterials which include cellulosic biopolymer, cellulose, cyclodextrin, alginic acid, chitosan, and proteins [31]. The most studied

TABLE 14.1
A Summary of the Different Types of Nanofiber Used in Metal Removal in Waste Water Treatment

Type of Nanofiber	pH	Temperature (°C)	Metal	Contact Time (hours)	Reusability	Adsorption Capacity (mg/g)	Adsorption Isotherm Mechanism	Adsorption Kinetics	Adsorption Efficiency (%)	Ref
Membrane	4	25	Ag^{2+}, Cu^{2+} & Pb^{2+}	48	3	5.1, 8.1 & 10.1	Langmuir and Freundlich	Pseudo-first order and second order equations	80	[20]
Membrane	7	25	Cu^{2+}	24	6	163.3	Langmuir and Freundlich	Pseudo-first-order and second-order kinetic models	90	[1]
Membrane	4.5	25	As (V)	24	3	30.8	Langmuir	Second-order kinetic model	86.2	[47]
Membrane	2	30	Cr (VI) & Pb (II)	0.5	5	509.7	Freundlich, Langmuir and Dubinin-Radushevich isotherm models	Pseudo-first-order, pseudo-second-order and intra-particle diffusion models	89	[51]
Membrane	5.5	25	Pb	12	N/A	469	Langmuir and Freundlich isotherms	First order, second order and intra-particle diffusion model	90	[46]
Carbon nanofiber	4.5 & 2.9	43	U(VI) & Eu(III) Pb(II) & Cr (VI)	40 min	3	125 & 91 423.7 & 221.3	Nitrogen sorption/desorption isotherm kinetics	N/A	N/A	[17]
Carbon nanofiber	5.5	25±1	Pb^{2+}	1	N/A	30	N/A	N/A	98	[2]

(*Continued*)

TABLE 14.1 (*Continued*)
A Summary of the Different Types of Nanofiber Used in Metal Removal in Waste Water Treatment

Type of Nanofiber	pH	Temperature (°C)	Metal	Contact Time (hours)	Reusability	Adsorption Capacity (mg/g)	Adsorption Isotherm Mechanism	Adsorption Kinetics	Adsorption Efficiency (%)	Ref
Carbon nanofiber	10	25	Cu^{2+}	0.5	N/A	8.88	Langmuir	Pseudo-second-order models	96	[3]
Carbon nanofiber	4.5	44	Pb	1 min	4	7.1	Langmuir	Pseudo-second-order kinetics	79	[6]
Carbon nanofiber	7	25	Cd(II)	2	N/A	1.22	Langmuir and Freundlich	Pseudo-second-order	98	[40]
Carbon nanofiber	3–6	25	Pb^{2+}	3	N/A	47	Langmuir	Pseudo-second-order Lagergren equation	N/A	[41]
Polymer nanofiber	6	25	Cu^{2+} and Pb^{2+}	15	N/A	62.3 & 168.3	Langmuir	Pseudo-first-order, second-order and double-exponential models	N/A	[33]
Polymer nanofiber	7–7.5	44	As and Cr	10 minutes	N/A	42.37 & 434.78	Langmuir and Freundlich	N/A	N/A	[52]
Polymer nanofiber	6	N/A	Pb^{2+} & Fe^{2+}	35 & 40 minutes	5	49.75 & 45.45	Langmuir and Freundlich	N/A	99.7 & 95.46	[53]
Metal-decorated	5.6–6.8	25	As	24	N/A	9.3	Langmuir and Freundlich	N/A	N/A	[27]
Metal-decorated	4.5 & 6	25	Pb^{2+}, Zn^{2+} & Cd^{2+}	15	5	95.25–125.75	Langmuir	Pseudo-second-order	93	[42]
Metal-decorated	N/A	N/A	Hg	0.5	N/A	57.4	N/A	N/A	N/A	[43]
Metal-decorated	N/A	25	Fe(III) & Zn(IV)	4	4	360 & 230	Langmuir, Freundlich and Temkin models	Pseudo-first order, second order and intra-particle diffusion models	90	[28]

N/A, not applicable.

biopolymers in water treatment are chitosan and cellulose, according to research. These biopolymers have a lot of functional groups including sulfhydryl, hydroxyl, amine, carboxylic, and phenolic groups [11]. It's worth noting that heavy metal removal is highly dependent on the adsorbent's functional groups. In addition, soy proteins also contain nearly all the amino acids and these amino acids contain all functional groups that are crucial in the adsorption of metals [21]. The adsorption efficiency for the different biomaterial differs because of the different functional groups present in the different biomaterials. For example, oatmeal has β-glucan and a lot of hydroxyl groups which play a major role in metal adsorption.

Different techniques have been used in polymer nanofibers. These techniques included physical, chemical, thermal and electrostatic fabrication techniques, and electrospinning. Aijaz et al., for example, published a paper on the synthesis of polymer nanofibers for the removal of nickel from wastewater. For the production of polymer nanofiber, Polyetherimide-Acrylonitrile (PEI-AN) and a solution of Fe/PEI-AN were mixed in an electrospinning machine of 30kV maximum capacity [32]. In a 10mL syringe with a syringe needle of 0.6mm, a solution of PEI -AN and Fe/PEI-AN was inserted. The syringe was then attached to the pump assembly and needle connected to high voltage (19kV) supply. The flow rate was set to 0.6 mL/h with a 15cm gap between the needle and the collector, and the humidity was set to 20%. The produced nanofiber membrane from the drum collector was placed in an oven for 24hours at 60°C [32].

Moreover, polymer-based fibers have been used in wastewater treatment in many tests. Abbasizadeh et al. announced the development of a novel electrospun polyvinyl alcohol/titanium oxide nanofiber adsorbent with mercapto groups for the removal of uranium (VI) and thorium (IV) from wastewater. At 45°C and pH levels of 4.5 and 5.0, the sorption capacities of U(VI) and Th(IV) were about 196.1 and 238.1 mg/g, respectively [33]. The sorption of the metal ions was endothermic and that was proven using calculations of Gibbs free energy, heat of the sorption, and entropy change. The use of nanofiber mats decorated with CeO_2 nanoparticles by the marcapoto group for efficient adsorption of Pb(II) and Cu(II) from wastewater was stated by Yari et al. The double-exponential model was used to describe the kinetic data for Pb(II) and Cu(II) removal, with maximum adsorption capacities of 90.9 and 88.3 mg/g, respectively. Adsorption was also found to favor high temperatures in this sample [34].

Metal-decorated polymer-based nanofibers are mostly iron oxides such as hematite. The use of hematite nanofibers is rapidly increasing, due to the easy separation of magnetic nanofibers, thereby causing cost-effective wastewater treatment practices [35]. Furthermore, iron oxide-based nanofibers are mostly reported as environmentally friendly adsorbents [14]. The chemical properties of iron oxide, like the zero charge near pH 7 ensure that both cationic and anionic targets analytes can adsorb onto the sorbent [14].

14.3.2 Carbon Based Nanofibers

Carbon-based nanofibers have also been widely used in wastewater treatment as they pose features that promote the efficient removal of water contaminants. These carbon-based nanofibers (CNFs) are cylindrical and made of graphene layers estimated to be of 100 μm length and a diameter ranging from 200 to 600nm [36,48–50]. The small size of the nanofibers promotes large surface areas of about 1877 m^2/g, and this feature favors CNFs over carbon nanotubes (CNTs). This high surface area also enhances high retention factor of the target analyte [37]. Because of the various functional groups (COOH, OH, and CO) that can be added to the surface of CNFs to enhance sorption capability, CNFs are favored for the adsorptive removal of various pollutants [38]. In general, CNFs are synthesized through chemical vapor deposition (CVD), hot filament-assisted sputtering, and template-assisted methods [39].

The adsorption capabilities of CNF on metals cannot be overlooked. Sun et al., for example, found that radionuclides with high adsorption power, such as U (VI) and Eu (III), had adsorption capacities of 125 and 91 mg/g, respectively. The optimum pH for adsorption was 4.5 and 2.9. Similar results with high adsorption capacities were reported for heavy metals including Pb II (423.7 mg/g) and Cr VI (221.3 mg/g) [17]. Ahmed et al. reported the efficient lead sorption from wastewater by CNFs. It was reported that a dose of 250 mg/L of the nanocomposite sample was added to a 20 mL lead solution with an initial concentration of 20 mg/L [2]and adsorption efficiencies were achieved at pH 5.5 and at temperature of 25 ± 1°C. This study reported high removal efficiencies of 98% and adsorption capacity of 30 mg/g was achieved in about 90–120 minutes. CNFs are used to strip copper from wastewater by Garcia-Diaz et al. This was accomplished by batch adsorption experiments in which the pH of the Cu^{2+} solution was modified with HCl and NaOH [3]. It was reported that at very low pH values (below 2), the rate of adsorption was very low, and this was due to the surface charge of the CNFs being positive causing repulsion to the Cu^{2+}. When the pH was increased from 4 to 10, there was an increase in the rate of adsorption as the surface was becoming more negative promoting the adsorption of Cu^{2+}. The maximum adsorption of 8.88 mg/g was achieved at pH 10 with a removal efficiency of 96% [3]. This experiment achieved its equilibrium in 30 minutes indicating that it was one of the best reported studies on metal removal using CNFs. Mahar et al. also published on the use of CNFs in water treatment for lead removal. This study proved to be less time-consuming as it took 1 minute to reach equilibrium and the adsorption capacity of 7.1 mg/g with adsorption efficiency (79%) was obtained at pH 4.5 and 44°C. These nanofibers proved to be reusable for four cycles [6].

In addition, Ihsanullah et al. investigated how to remove cadmium ions from liquid phase using acid-modified CNFs. The adsorption behavior of Cd(II) was evaluated using the Freudlich and Langmuir isotherm models in this analysis, which revealed that pH 7 was the most optimum for Cd(II) removal. The reported maximum adsorption capacity was 1.22 mg/g, indicating that the CNFs were appropriate adsorbents for water purification [40]. The adsorption efficiency of 98% was reported after a 2 hour contact time. Mishra and Verma have identified the use of carbon nanofibers mediated by surface ion imprinting for the removal of aqueous Pb (II). The optimum pH 3–5 for the adsorption of Pb ions was reported with adsorption capacity of 47 mg/g. In this study, the synthesized CNFs were first put in the FTIR to investigate the functional groups present [41]. The FTIR spectra showed several peaks which indicated the presence of different functional groups. Among the reported functional groups were the carboxylic and hydroxyl groups which promoted the binding of metal ions to these negatively charged binding sites. This, therefore, promoted the adsorption of metals as the surface of the adsorbent is negatively charged and the target analyte is positively charged.

14.3.3 Metal Based Nanofibers

There are several metal oxide nanomaterials (Fe_3O_4, CuO, and ZnO) which have been used in the synthesis of nanofiber. Metal oxide nanomaterials just like the other nanosorbent have proven to be even the most efficient in the removal of heavy metals in wastewater. Cupric oxide (CuO) has been found to be more effective than other metal oxides at removing arsenic. ZnO, on the other hand, is a highly stable, nontoxic substance with high adsorption efficiencies at pH levels ranging from 5.8 to 6.8. As a result, this nanomaterial is considered environmentally friendly [27]. The adsorption capacity for As was 9.3 mg/g at room temperature. It is worth noting that for ZnO to be more efficient in metal removal it needs to be coated with metal composites most favorable Fe-Ti, Fe-Zr, Ti-Ce, Fe-Cu, and Fe-Cr [27].

A variety of metal-decorated nanofibers have been used in wastewater treatment to remove metals and metalloids. Zhao et al. performed a study using sodium titanate nanofibers to extract heavy metal ions from wastewater [42]. The adsorption of heavy metals pursued pseudo-second-order kinetics and Langmuir isotherm models, according to this analysis. Furthermore, at pH 4.5–6, the maximal adsorption capacities for the metal ions studied (Cu^{2+}, Pb^{2+}, Zn^{2+}, and Cd^{2+}) ranged from 95.25 to 125.75 mg/g. The use of zonal-thiol functionalized silica nanofibers for the removal of heavy metal ions from wastewater was documented by Shouzhu and colleagues. With a recorded adsorptive capacity of 57.4 mg/g, these zonal silica nanofibers were effective in removing mercury (Hg) from wastewaters [43]. The increase in adsorptive capacity was due to the introduction of the thiol groups on the surface of the nanofibers which further increased the surface area. Efome et al. also looked at using Fe(III) and Zn(IV) metal-organic frameworks (MOF) sponsored on nanofiber to remove heavy metals from wastewater. These MOFs showed high adsorption efficiency above 90% for Pb and Hg removal [28].

14.3.4 Membrane Based Nanofibers

The main challenge in adsorption methods is that adsorbent can agglomerate together and easily loose adsorption characteristics [21]. To solve this problem, membrane-based nanofibers evolved [25]. Layered double hydroxides, a class of two-dimensional nanostructured compounds with positively and negatively charged ions to balance the charges, are used to make these nanofibers. From these layers, several functional groups which promote ion exchange are found, thereby enhancing the removal of metal pollutants from wastewaters [44]. Furthermore, the membrane nanofibers have a high porosity, which means they have higher permeability. This high porosity of the membrane plays a key role in the reduction of energy consumption. Other properties of the nanofibrous membranes, such as small pore size, large surface area, versatility in surface functionalities, and nature, make them ideal for selective pollutant adsorption. Different polymers can be electrospun together with particulate nanofillers and biological agents to form nanocomposites or hybrid nanofibrous membranes.

For explaining the adsorption mechanisms in wastewater treatment, the Langmuir and Freundlich isotherms were mostly reported. Chen et al. conducted research on the synthesis of nanofiber membranes using MgAl-ED-LDH and the adsorption properties of the nanofiber membranes synthesized were tested in the removal of Cu (II) from wastewater. The method showed that Cu (II) adsorption onto MgAl-EDTA-LDH was helped by modification of the chemical properties of the substrate forming some hydroxyl groups [45]. The Cu (II) reacted with the hydroxyl groups to form complexes thereby making the Cu (II) to be attached onto the adsorbent. Nie et al. reported the Kevlar-based nanoparticles as an efficient way for the removal of Ag^{+}, Cu^{2+} and Pb^{2+} in wastewater. The Kevlar nanofiber proved to be slightly negatively charged due to the ionization of the Kevlar surface making it to have high affinity for cationic species thus promoting the adsorption of Ag^{+}, Cu^{2+}, and Pb^{2+}. It was also reported that the very large surface area of the particles further increased the binding sites for the adsorption of cationic ions [20]. According to the findings, the metals' adsorption capacities were 5.1, 8.1, and 10.1 mg/g for Ag, Cu, and Pb ions, respectively, with a high adsorption potential of more than 80% for all metals. Qiu et al. examined the adsorption activity of Cu^{2+} in wastewater by fabricating porous membranes made up of nanofibers. The batch adsorption method was used to investigate Cu^{2+} adsorption on chitosan membrane fibers in this analysis [1]. The maximum adsorption efficiency of 163.3 mg/g was reported when the pH was 7, and the temperature was 25°C over a 24 hours period contact time. In addition, a higher adsorption efficiency of 90% was reported making this method to be most efficient in future for solving the problems of water contamination due to heavy metals. The chitosan membrane nanofibers showed to be even more cost-efficient as they have the ability to be reused six times. Huang et al. published another study in which polymer nanofiber membranes were used to remove heavy metals. It was reported that the chitosan-rectorite nanospheres embedded in aminated polyacrylonitrile nanofibers membranes showed a huge adsorption efficiency of 469 mg/g at an optimum pH of 5.5 [46]. This experiment was conducted for 12 hours and very high removal rate (90%) making this method of using membranes to be one of the best in the removal of Pb in wastewater.

Lin Mil et al. also mentioned using chitosan nanofiber membranes to remove As(V) from aqueous solutions. This method was tested using the batch process for 24 hours of contact time. It was discovered that the maximum adsorption of 30.8 mg/g was achieved at a pH of 4.5 and a temperature of 25°C [47]. This study also reported the adsorption rate to be 86.2% and the specific surface area was found to be $13.2\,m^2/g$. In an adsorption and membrane method, Koushkbaghi et al. investigated the use of aminated -Fe_3O_4 nanoparticles filled chitosan/PVA/PES dual layers nanofibrous membrane for the removal of Cr (VI) and Pb (II) ions from aqueous solutions. This study favored very low pH as the most efficient adsorption was achieved at pH 2 and the maximum adsorption was 509.7 mg/g with 89% removal rate [48]. These adsorption experiments were conducted at 30°C for 30 minutes. When looking at the removal rate, time, and temperature at which the experiment is conducted, this method is very good and can be used in future for the removal of metal from wastewater treatment plants.

14.4 AN OVERVIEW SUMMARY

When looking at Table 14.1, it can be deduced that there is no biasness in the number of reported studies between the different types of nanofibers as there is no huge difference in the number of publications. In terms of metals studied, Pb (II), Cu (II), and As have been well studied and this might be due to the hazardous nature of these metals to human health. The Langmuir and Freundlich isotherm mechanisms were mostly favored, and they both reported high adsorption efficiencies. The adsorption kinetics was often described using pseudo-first-order and second-order equations, whereas the intra-particle diffusion model and double-exponential models were rarely used. In addition, when looking at the trend in the number of publications from 2011 to 2020, the number of publications is fluctuating. However, there was only one publication in 2011 and non was reported in 2012 in the current topic. Then, an increase was observed in 2017 and 2018 whereby a total of eight publications were reported in the years combined (see Figure 14.2). From 2019 to 2020 a decrease in the number of publications was also reported with 2 and 1 publications reported, respectively.

14.5 CONCLUSION AND FUTURE RECOMMENDATIONS

The use of nanofiber technology looks to be the global problem solver in wastewater treatment. Nanofiber technology showed high efficiencies in the removal of analytes as the adsorption surface area is much increased when compared to volume ratio. With the use of nanofibers, very small

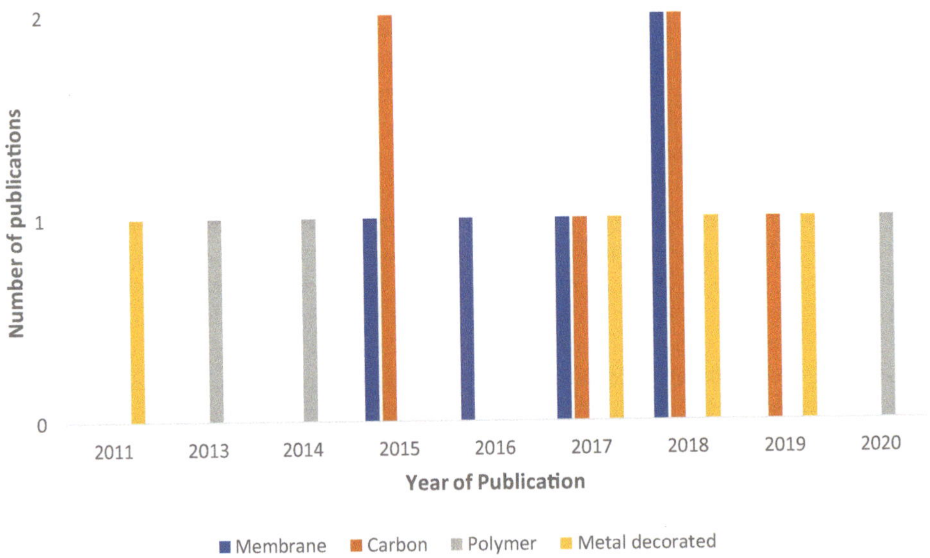

FIGURE 14.2 Number of publications on metal removal on wastewater using nanofiber technology.

quantities and reusability of the nanomaterial make this method to be highly favored in wastewater treatment as it is more efficient with low operational cost. Literature has also reported that the use of nanofiber material in wastewater treatment is environmentally-friendly, and no toxic secondary waste has been reported with this adsorption method. However, literature has indicated that adsorption using nanofiber material is still open to be further explored as very little has been reported in other metals. There are several toxic metals (Al, Co, Ni, Sb, etc.), that have not been studied yet, and their presence in wastewater is highly possible. It is therefore vital to study metals adsorption capacities on the different nanofibers. It is worth noting to state that further studies on metal removal need to be studied as to check which metal has high affinity for membrane, carbon, polymer and metal-decorated based nanofiber. In addition, the use of polymer nanofiber (specifically biopolymer) if well studied can be highly favored as it uses readily available plant waste materials in its synthesis which further cuts the operational costs.

ACKNOWLEDGMENT

The authors would like to acknowledge the University of South Africa and NRF THUTHUKA (113951) for their fanatical assistance.

REFERENCES

1. W. Qin, J. Li, J. Tu, H. Yang, Q. Chen, and H. Liu, Fabrication of porous chitosan membranes composed of nanofibers by low temperature thermally induced phase separation, and their adsorption behavior for Cu^{2+}, *Carbohydr. Polym.*, vol. 178, no. June, pp. 338–346, 2017.
2. Y. M. Ahmed, A. Al-Mamun, M. F. R. Al Khatib, A. T. Jameel, and M. A. H. A. R. AlSaadi, Efficient lead sorption from wastewater by carbon nanofibers, *Environ. Chem. Lett.*, vol. 13, no. 3, pp. 341–346, 2015.
3. F. Jos, Carbon nanofibers: A new adsorbent for copper removal from wastewater, *Metals (Basel)*, pp. 1–13, 2018.
4. A. M. Nasir et al., Recent progress on fabrication and application of electrospun nanofibrous photocatalytic membranes for wastewater treatment: A review, *J. Water Process Eng.*, vol. 40, no. November 2020, p. 101878, 2021.
5. Z. Cai, X. Song, Q. Zhang, and T. Zhai, Electrospun polyindole nanofibers as a nano-adsorbent for heavy metal ions adsorption for wastewater treatment, *Fiber Polym.*, vol. 18, no. 3, pp. 502–513, 2017.
6. F. K. Mahar et al., Rapid adsorption of lead ions using porous carbon nanofibers, *Chemosphere*, vol. 225, pp. 360–367, 2019.
7. A. Hołda, E. Kisielowska, and T. Niedoba, Bioaccumulation of Cr (VI) ions from aqueos solutions by Aspergillus niger, *Polish J. Environ.stud.*, vol. 20, no. 2, pp. 345–349, 2011.
8. K. K. Shamuyarira, and J. R. Gumbo, Assessment of heavy metals in municipal sewage sludge: A case study of Limpopo Province, South Africa, *Int. J. Environ. Res. Public Health*, vol. 11, no. June, pp. 2569–2579, 2014.
9. D. Gandavadi, S. Sundarrajan, and S. Ramakrishna, Bio-based nanofibers involved in wastewater treatment, *Macromol. Mater. Eng.*, vol. 1900345, pp. 1–15, 2019.
10. E. Pc, K. Shale, M. Sedibe, P. Tikilili, and A. Mc, South African mine effluents: Heavy metal pollution and impact on the ecosystem, *Int. J. Chem. Sci.*, vol. 15, no. 4, pp. 1–12, 2017.
11. O. K. Pereao, C. Bode-Aluko, G. Ndayambaje, O. Fatoba, and L. F. Petrik, Electrospinning: Polymer nanofibre adsorbent applications for metal ion removal, *J. Polym. Environ.*, vol. 25, no. 4, pp. 1175–1189, 2017.
12. M. A. Nunes, S. Pawlowski, A. S. G. Costa, R. C. Alves, M. B. P. P. Oliveira, and S. Velizarov, Science of the total environment valorization of olive pomace by a green integrated approach applying sustainable extraction and membrane-assisted concentration, *Sci. Total Environ.*, vol. 652, pp. 40–47, 2019.
13. R. Rashid, I. Shafiq, P. Akhter, M. J. Iqbal, and M. Hussain, A state-of-the-art review on wastewater treatment techniques: The effectiveness of adsorption method, *Environ. Sci. Pollut. Res.*, 2021.
14. Y. Liao, C. H. Loh, M. Tian, R. Wang, and A. G. Fane, Progress in electrospun polymeric nanofibrous membranes for water treatment: Fabrication, modification and applications, *Prog. Polym. Sci.*, vol. 77, pp. 69–94, 2018.

15. D. Wang, A critical review of cellulose-based nanomaterials for water purification in industrial processes, *Cellulose*, vol. 26, no. 2, pp. 687–701, 2019.
16. O. Pereao, C. Bode-Aluko, K. Laatikainen, A. Nechaev, and L. Petrik, Morphology, modification and characterisation of electrospun polymer nanofiber adsorbent material used in metal ion removal, *J. Polym. Environ.*, vol. 27, no. 9, pp. 1843–1860, 2019.
17. M. Ding, K. K. R. Bannuru, Y. Wang, L. Guo, A. Baji, and H. Y. Yang, Free-standing electrodes derived from metal–organic frameworks/nanofibers hybrids for membrane capacitive deionization, *Adv. Mater. Technol.*, vol. 3, no. 11, pp. 1–9, 2018.
18. L. N. Nthunya et al., A review of nanoparticle-enhanced membrane distillation membranes: Membrane synthesis and applications in water treatment, *J. Chem. Technol. Biotechnol.*, vol. 94, no. 9, pp. 2757–2771, 2019.
19. G. E. Romanos, C. P. Athanasekou, V. Likodimos, P. Aloupogiannis, and P. Falaras, Hybrid ultrafiltration/photocatalytic membranes for efficient water treatment, *Ind. Eng. Chem. Res.*, vol. 52, no. 39, pp. 13938–13947, 2013.
20. C. Nie, Z. Peng, Y. Yang, C. Cheng, and L. Ma, Kevlar based nanofibrous particles as robust, effective and recyclable absorbents for water purification, *J. Hazard. Mater.*, vol. 318, pp. 255–265, 2016.
21. J. Cui, F. Li, Y. Wang, Q. Zhang, W. Ma, and C. Huang, Separation and purification technology electrospun nano fiber membranes for wastewater treatment applications, *Sep. Purif. Technol.*, vol. 250, no. May, p. 117116, 2020.
22. S. Ilyas, R. Ranjan, H. Kim, N. Ilyas, and R. Sattar, Separation and purification technology extraction of nickel and cobalt from a laterite ore using the carbothermic reduction roasting-ammoniacal leaching process, *Sep. Purif. Technol.*, vol. 232, no. August 2019, p. 115971, 2020.
23. J. Płotka-wasylka, and N. Szczepan, Trends in analytical chemistry modern trends in solid phase extraction: New sorbent media, *Trends Anal. Chem.*, vol. 77, pp. 23–43, 2016.
24. A. D. G.-D. A. Herrera, C. Tejada-Tovar, and A. Villabona-Ortiz, Cd (II) and Ni (II) uptake by novel biosorbent prepared from oil palm residual biomass and Al_2O_3 nanoparticles, *Sustain. Chem. Pharm.*, vol. 15, no. June 2019, 2020.
25. A. Barhoum, K. Pal, H. Rahier, H. Uludag, I. S. Kim, and M. Bechelany, Nanofibers as new-generation materials: From spinning and nano-spinning fabrication techniques to emerging applications, *Appl. Mater. Today*, vol. 17, pp. 1–35, 2019.
26. F. Luo, J. Wang, Z. Yao, L. Zhang, and H. Chen, Polydopamine nanoparticles modified nanofiber supported thin film composite membrane with enhanced adhesion strength for forward osmosis, *J. Memb. Sci.*, vol. 618, no. April 2020, p. 118673, 2021.
27. K. T. Peter, A. J. Johns, N. V. Myung, and D. M. Cwiertny, Functionalized polymer-iron oxide hybrid nano fibers: Electrospun filtration devices for metal oxyanion removal, *Water Res.*, vol. 117, pp. 207–217, 2017.
28. J. E. Efome, D. Rana, T. Matsuura, and C. Q. Lan, Metal-organic frameworks supported on nanofibers to remove heavy metals, *J. Mater. Chem. A*, vol. 6, no. 10, pp. 4550–4555, 2018.
29. L. Peng, X. Zhang, Y. Sun, Y. Xing, and C. Li, Heavy metal elimination based on metal organic framework highly loaded on fl exible nano fibers, *Environ. Res.*, vol. 188, no. May, p. 109742, 2020.
30. R. Selvam, Synthesis and fabrication of nanomaterials, *KSR Gr. Inst.*, no. January 2015, pp. 311–314, 2018.
31. P. Wang et al., Adsorption of hexavalent chromium by novel chitosan/poly(ethylene oxide)/permutit electrospun nanofibers, *New J. Chem.*, vol. 42, no. 21, pp. 17740–17749, 2018.
32. M. O. Aijaz, M. R. Karim, H. F. Alharbi, N. H. Alharthi, F. S. Al-mubaddel, and H. S. Abdo, Magnetic/polyetherimide-acrylonitrile composite nanofibers for nickel ion removal from aqueous solution, pp. 1–17, 2021.
33. S. Abbasizadeh, A. R. Keshtkar, and M. A. Mousavian, Preparation of a novel electrospun polyvinyl alcohol/titanium oxide nanofiber adsorbent modified with mercapto groups for uranium(VI) and thorium(IV) removal from aqueous solution, *Chem. Eng. J.*, vol. 220, pp. 161–171, 2013.
34. P. Kampalanonwat, and P. Supaphol, The study of competitive adsorption of heavy metal ions from aqueous solution by aminated polyacrylonitrile nanofiber mats, *Energy Procedia*, vol. 56, no. C, pp. 142–151, 2014.
35. H. N. M. Ekramul Mahmud, A. K. Obidul Huq, and R. B. Yahya, The removal of heavy metal ions from wastewater/aqueous solution using polypyrrole-based adsorbents: A review, *RSC Adv.*, vol. 6, no. 18, pp. 14778–14791, 2016.

36. Y. Liu et al., Recent advances on electroactive CNT-based membranes for environmental applications: The perfect match of electrochemistry and membrane separation, *Chinese Chem. Lett.*, vol. 31, no. 10, pp. 2539–2548, 2020.
37. M. Anjum, R. Miandad, M. Waqas, F. Gehany, and M. A. Barakat, Remediation of wastewater using various nano-materials, *Arab. J. Chem.*, vol. 12, no. 8, pp. 4897–4919, 2019.
38. B. Hashemi, Carbon-based sorbents and their nanocomposites for the enrichment of heavy metal ions: A review, *Microchim. Acta*, 2019.
39. G. Zou et al., Carbon nanofibers: Synthesis, characterization, and electrochemical properties, *Carbon N. Y.*, vol. 44, no. 5, pp. 828–832, 2006.
40. Ihsanullah et al., Adsorptive removal of cadmium(II) ions from liquid phase using acid modified carbon-based adsorbents, *J. Mol. Liq.*, vol. 204, pp. 255–263, 2015.
41. S. Mishra, and N. Verma, Surface ion imprinting-mediated carbon nanofiber-grafted highly porous polymeric beads: Synthesis and application towards selective removal of aqueous Pb (II), *Chem. Eng. J.*, vol. 313, pp. 1142–1151, 2017.
42. M. Zhao et al., Application of sodium titanate nano fibers as constructed wetland fillers for efficient removal of heavy metal ions from wastewater, *Environ. Pollut.*, vol. 248, pp. 938–946, 2019.
43. S. Li, X. Yue, Y. Jing, S. Bai, and Z. Dai, Colloids and Surfaces A: Physicochemical and engineering aspects fabrication of zonal thiol-functionalized silica nanofibers for removal of heavy metal ions from wastewater, *Colloids Surfaces A Physicochem. Eng. Asp.*, vol. 380, no. 1–3, pp. 229–233, 2011.
44. R. R. Choudhury, S. K. Sahoo, and J. M. Gohil, Potential of bioinspired cellulose nanomaterials and nanocomposite membranes thereof for water treatment and fuel cell applications, *Cellulose*, vol. 27, no. 12, pp. 6719–6746, 2020.
45. H. Chen et al., Preparation of MgAl-EDTA-LDH based electrospun nanofiber membrane and its adsorption properties of copper(II) from wastewater, *J. Hazard. Mater.*, vol. 345, pp. 1–9, 2018.
46. M. Huang et al., Chitosan-rectorite nanospheres embedded aminated polyacrylonitrile nanofibers via shoulder-to-shoulder electrospinning and electrospraying for enhanced heavy metal removal, *Appl. Surf. Sci.*, vol. 437, pp. 294–303, 2018.
47. L. L. Min, Z. H. Yuan, L. Bin Zhong, Q. Liu, R. X. Wu, and Y. M. Zheng, Preparation of chitosan based electrospun nanofiber membrane and its adsorptive removal of arsenate from aqueous solution, *Chem. Eng. J.*, vol. 267, pp. 132–141, 2015.
48. M. P. Shah. *Removal of Emerging Contaminants through Microbial Processes*. Springer, 2021.
49. M. P. Shah. *Advanced Oxidation Processes for Effluent Treatment Plants*. Elsevier, 2020.
50. M. P. Shah. *Microbial Bioremediation and Biodegradation*. Springer, 2020.
51. S. Koushkbaghi, A. Zakialamdari, M. Pishnamazi, H. F. Ramandi, M. Aliabadi, and M. Irani, Aminated-Fe_3O_4 nanoparticles filled chitosan/PVA/PES dual layers nanofibrous membrane for the removal of Cr(VI) and Pb(II) ions from aqueous solutions in adsorption and membrane processes, *Chem. Eng. J.*, vol. 337, no. October 2017, pp. 169–182, 2018.
52. M. Bhaumik, H. J. Choi, R. I. McCrindle, and A. Maity, Composite nanofibers prepared from metallic iron nanoparticles and polyaniline: High performance for water treatment applications, *J. Colloid Interface Sci.*, vol. 425, pp. 75–82, 2014.
53. A. R. Esfahani, Z. Zhang, Y. Y. L. Sip, L. Zhai, and A. H. M. A. Sadmani, Removal of heavy metals from water using electrospun polyelectrolyte complex fiber mats, *J. Water Process Eng.*, vol. 37, no. March, p. 101438, 2020.

15 Quantitative Image Analysis as a Valuable Tool to Assess Aerobic Wastewater Treatment Systems

Daniela P. Mesquita and Cristina Quintelas
University of Minho

A. Luís Amaral
University of Minho
Instituto Superior de Engenharia de Coimbra
Instituto de Investigação Aplicada

Eugénio C. Ferreira
University of Minho

CONTENTS

DOI: 10.1201/9781003164982-15

15.1 INTRODUCTION

Clean water is essential for a healthy life, with human demand exerting pressure on the available supply of freshwater. Thus, it is now clear that in many parts of the world, there is a growing scarcity of water. In the case of industrialized countries, the lack of water is mainly caused by the contamination of ground and surface waters due to intensive industrial and/or agricultural activities. On the other hand, in developing countries, the main cause of serious water pollution problems is untreated wastewaters, most severe near large urban centers and leading to low water availability. Thus, growing water scarcity necessitates the reuse of treated wastewater as well as energy, cost-efficient and lower footprint treatment systems. For that purpose, biological WWT plays a fundamental role in improving water quality as well as promoting water recycling.

Wastewater discharges from domestic, industrial, and commercial practices may contain pollutant levels affecting the quality of receiving waters (e.g., ground waters, surface waters, watercourses) and soils. With the awareness of a global environmentally sustainable development, Agenda 21 and the Rio Declaration on Environment and Development, recognized the protection and management of freshwater resources from contamination as one of the priorities in the world. However, much remains to be done, since, in most cases, effluent discharge does not always comply with pre-treatment requirements and with the binding pollutant levels. Also, in the past few years, the increase of the world population, growth of human activities, industrialization and production of novel products have increased the pollutants abundance and complexity in wastewaters and led to a high degree of environmental pollution problems, serious health hazards, and threats to entire ecosystems.

A set of established biological processes are nowadays available, with novel still emerging, for the treatment of highly complex wastewaters and improvement of wastewater treatment plants (WWTP) efficiency all over the world. Indeed, biological WWT processes (aerobic and/or anaerobic) are widely employed to efficiently reduce contaminants (organic matter and nutrients – nitrogen and phosphorus) discharge levels in watercourses, both from domestic and industrial wastewaters. Concomitantly, the assessment of wastewater and sludge quality needs to be routinely evaluated by a number of physical, chemical, and operational parameters, paramount to meet the legally binding discharge values (imposed by environmental policies). However, most of the employed traditional methodologies are costly, labor-intensive, and time-consuming, and a few may even present environmental hazards due to the use of harmful chemicals and perilous end products.

Recently, quantitative image analysis (QIA) techniques have been shown to be valid alternatives for monitoring key operational parameters, with the collected data paramount for the timely assessment of WWT malfunctions and the establishment of appropriate corrective actions, thus increasing the treated wastewater quality. The continuous improvement, and affordability, of microscopes and computers led to more widespread use of QIA techniques, largely to assess the biomass morphology and structure properties, mainly at the laboratory scale. However, QIA is not yet regularly implemented in real WWTP and future implementation still needs to be put into practice more widely.

This chapter gathers the most relevant published work and discusses the recent advances, in the application of QIA techniques to assess biological WWT processes. Section 15.1 presents an introduction to the WWT problematic. Section 15.2 presents an overview of the set of biological WWT processes currently available. Section 15.3 analyzes the biomass and microbiota present in WWT processes. Section 15.4 presents the QIA work frame, whereas Section 15.5 focuses on describing current imaging systems. Section 15.6 discusses the applicability of QIA in aerobic WWT processes.

15.2 BIOLOGICAL PROCESSES IN WASTEWATER TREATMENT

Physical and chemical processes were the most popular technologies for WWT in the early 20th century, including sedimentation, chemical treatment, and dilution. However, none of these processes offers a high-quality treatment. In this context, the biological (aerobic and anaerobic) WWT processes are capable of producing high-quality clear wastewater. Aerobic processes are well-established biological WWT systems, offering an effective solution for high-quality treated wastewater. Over the past few decades, aerobic biological efforts have generally been based on two distinct principles: (a) suspended growth and (b) attached growth.

The conventional activated sludge process (CAS) is a suspended growth technology widely used in WWTP, comprising an enriched microbial consortia culture in which the contaminants' metabolic conversion promotes an environmentally acceptable final wastewater quality. This system is composed of an aerated tank, where the biochemical phase takes place, and a clarifier, where the settling phase occurs. In the aerated tank, microbial organisms oxidize the pollutants present in the wastewater due to the supply of oxygen. This results in the removal of organic matter and nutrients from the wastewater. On the other hand, the clarifier allows for the settling phase to take place, which is an essential step of the CAS process in biological WWTP. Furthermore, it is associated with sludge thickening and recycling, and effluent clarification, producing final wastewater with low suspended solids. Also, sludge can be removed from the system avoiding biomass overgrowth. The biological nitrogen (nitrification and denitrification) and phosphorous removal stages can also be accomplished by advanced CAS systems with aerobic, anoxic, and anaerobic sections. If the system is completely aerobic, the air supply allows for organic matter and ammonium oxidation. Otherwise, if anoxic zones are added, denitrification takes place, while aerobiosis favors phosphorus removal. However, the performance of the CAS process is affected by a number of factors, among which the aeration efficiency, temperature, pH, sludge settling ability, microbial community profile, and wastewater toxicity should be highlighted (Metcalf & Eddy, 1995; Bitton, 2010).

The control of the microbial aggregates (mainly suspended biomass) structure and function is difficult to maintain in continuous flow systems due to volumetric flow and mass loading variations. Based on this consideration, sequencing batch reactors (SBR) using flocculent activated sludge (AS) have been applied since, for a number of applications, a discontinuous operated system is advantageous. This system is characterized by a series of process phases (e.g., influent feeding, reaction, settling, and effluent withdrawal), each lasting for a defined period of time. Usually, the bioreactors are compact and provide higher treatment efficiencies when compared to CAS systems. Due to its cycle operation flexibility, SBR has been reported as one of the most effective strategies for treating wastewater and has received considerable attention for nutrients removal, namely nitrogen (N) and phosphorus (P) through the use of enhanced biological phosphorus removal (EBPR) systems (Carvalheira et al., 2014) and bioplastics (polyhydroxyalkanoates – PHA) production using complex feedstocks (Oliveira et al., 2017; Shah, 2020, 2021).

Recently, an alternative technology based on the development of granular sludge in SBR is taking its first steps in terms of implementation in full-scale WWTP. This technology is already considered the most important breakthrough emerging from CAS, and during the *IWA World Water Exhibition Congress* in 2014, it was recognized as the new upcoming standard for WWT. Aerobic granular sludge (AGS) technology offers the possibility to design compact WWTP based on simultaneous organic matter, nitrogen and phosphate removal in a single SBR. An additional advantage of the granular sludge is that the required settling time is very short due to its high settling velocity (de Kreuk et al., 2005). AGS has several advantages over CAS, including higher settling abilities and biomass retention; exhibiting a compact and strong microbial structure; the ability to endure high organic loading rates (OLR), high activity, and resistance to toxic compounds (Jiang et al., 2002; Shah, 2021). Up to date, research on AGS technology has been centered on treatment performance and operation conditions, and thus, the assessment of both the AGS structure and the AGS microbial communities is still poorly understood.

Another technology, membrane bioreactors (MBR) integrating CAS with a physicochemical membrane filtration unit, has also attracted attention in water reclamation as a result of the recent technical advances and cost reduction in membranes. MBR has distinctive advantages over CAS processes, including smaller footprint and sludge production; high-quality treated wastewater; operation robustness; and, if designed and maintained appropriately, ease of automation (Ma et al., 2018). MBR is based on biomass-attached growth and uses filtration membranes to provide a solid–liquid barrier, retaining the sludge. This leads to an increased amount of sludge in the reactor and an improved biological reactor operation with high removal efficiency, low sludge production, and resilience to high loads or sudden shock phenomena (Nguyen et al., 2013). However, membrane fouling and energy consumption are still the major challenges to a widespread application of MBR, leading to decreasing permeate flux and increasing operation costs, respectively, with considerable research and development efforts still underway (Krzeminski et al., 2017).

15.3 WASTEWATER BIOMASS AND MICROBIOTA

All the above biological WWT processes are based on microorganisms' consortia. These processes are composed of different types of microorganisms, mostly bacteria, but encompassing also protozoa, metazoa, fungi and algae. In the case of suspended growth biological WWT processes, filamentous and floc-forming bacteria, protozoa, and metazoa are commonly present in abundance. Bacteria are generally prevalent, though imbalances between the different types of bacteria may take place and disturb the WWTP, due to the complex nature of the microbial communities, with significant economic and environmental consequences. A significant protozoa presence is usually associated with good performance, and the species distribution and abundance have been pointed out as indicators of the quality of the effluent exiting an AS plant, providing a useful mechanism to evaluate its performance (Madoni, 1994).

The growth of floc-forming bacteria on a filamentous bacteria backbone forms the basis of AS structures, leaving a clarified supernatant upon settling in the final clarifier. Indeed, sludge flocculation and settling characteristics are crucial to an effective biological treatment, which may be affected by sludge separation problems, such as sludge bulking, foaming, pinpoint sludge, and dispersed growth. The performance of an AS system is dependent on a correct balance between floc-forming and filamentous bacteria. Filamentous proliferation, caused by the excessive growth of filamentous bacteria as well as the overproduction of extracellular polymeric substances (EPS) by floc-forming bacteria, can lead to the deterioration of effluent quality and sludge washout, ultimately collapsing the overall system. The competition between floc-forming and filamentous bacteria is therefore critical to prevent filamentous bulking in full-scale plants. On the other hand, a filamentous bacteria shortage may lead to hardly settleable pin-point flocs (Eikelboom, 2000).

AS malfunctions may be identified by characterizing the aggregated and filamentous biomass structure and contents, and microbiota identification. Weak aggregates may be inferred by large numbers of cells scattered throughout the frontier, whereas in strong aggregates, bacteria are stacked close to one another, resulting in a higher settling velocity. Macroscopically, the aggregates' characteristics include an increased irregularity and a broad distribution of particle sizes.

Filamentous bacteria may be responsible, or take part, in quite different consequences in AS, ranging from filamentous bulking to foaming, among others. Classical methods for the identification of filamentous bacteria have relied on their morphology (Eikelboom, 2000; Jenkins et al., 2003), despite the uncertainty derived from the resemblance between some organisms. As a result, novel identification strategies have been developed relying primarily on the use of immunological or molecular biology techniques combined with the use of fluorescent dyes.

The first step in the research of QIA in wastewater treatment started more than 30 years ago, considering that sludge setting and compaction properties were found to be directly related to the biomass structure, composition, and content. Since then, QIA has become a fundamental tool within the field of environmental science, and the ongoing development has allowed a wide range of applications regarding wastewater biomass and microbiota monitoring.

15.4 QUANTITATIVE IMAGE ANALYSIS

QIA techniques are often decomposed into three main steps: image acquisition (usually through the use of digital cameras), image processing (leading to a final intensity, mask, or binary image), and image analysis (with the calculation of the metrics of interest), next presented.

15.4.1 Image Acquisition

The first step on any QIA system is image acquisition performed, in most cases, by digital (charge-coupled device – CCD or complementary metal–oxide–semiconductor – CMOS) cameras. The signal collected in the camera sensor is then translated into a matrix of picture elements (pixels) proportional to the light intensity received by each color channel. The number of colors in an image is determined by the number of bits assigned to each pixel which, in grayscale images, is usually represented by 8 bits (256 gray levels). Bright-field (without staining), phase-contrast, and even fluorescence (single fluorescent dye or narrow bandpass filters) microscopy traditionally use monochrome images. In fluorescence microscopy, light intensity at a given emission wavelength is of the utmost importance, privileging monochrome cameras with high sensitivity when a single fluorochrome is used (Russ, 2002).

On the other hand, a number of sample preparation methodologies use differential staining to identify microorganisms, enlighten structures or survey for biochemical activity. Indeed, bright-field and fluorescence (using long bandpass filters) staining processes are also widely used, using two or more dyes (or fluorochromes) and producing color images. Both the intensity at a given emission wavelength and the wavelength (color) are important factors in this case. Furthermore, the use of color spaces in color imaging, instead of solely brightness values, allows for the detection of small changes not detectable by the grayscale alone. Consequently, 24-bit RGB (red, green, and blue channels) images are usually acquired, with 8 bits (256 brightness levels) stored for each channel (Russ, 2002).

Fading, bleaching, and background fluorescence issues must all be addressed when using fluorescence microscopy, whereas for bright-field accurate focusing and non-uniform illumination stand out and for phase-contrast the existence of halos and shade-off must also be taken into consideration. Furthermore, color recognition problems may arise, given the limited number of bits assigned to each channel in color imaging, for similar enough colors in the RGB spectrum. For a successful image processing and analysis procedure, these problems should be dealt with and optimized prior to image acquisition (Costa et al., 2013; Mesquita et al., 2013).

15.4.2 Image Processing

Image processing aims at achieving a final grayscale or binary image for a specific application, including a segmentation procedure (in most cases) highlighting the essential information upon the metrics of interest that can be obtained in the image analysis step. Prior to image segmentation, and following image acquisition, monochrome or color images are first pre-processed, which may include background detection and elimination, as well as morphological (opening, closing, top-hat, bottom-hat, etc.) or filter (low pass, high pass, median, edge detection, etc.) image enhancement operations Next, using spatial or frequency domain methodologies, segmentation is employed to distinguish the objects from the background. The most common spatial domain segmentation techniques include manual, histogram-based, first derivative, second derivative and Otsu thresholding, and entropy (texture), watershed, split and merge, clustering and fuzzy (*c*-means or *k*-means) algorithms, among others. On the other hand, frequency domain strategies include the use of wavelets, Fourier, Hough and Radon transforms, among others (Russ, 2002).

The above image processing methods have been extensively used for wastewater treatment images (Costa et al., 2013; Mesquita et al., 2013). More recently, phase-congruency-based methods, built upon the use of Fourier transform and wavelets, have also been proposed for AS images (Khan et al., 2017a). In addition, new methodologies already employ neural network (NN)-based techniques for segmentation, applied to the entire image or local adaptive. Indeed, a growing interest in the use of artificial neural networks (ANN) toward the segmentation of wastewater biomass images is patent on the increasing number of publications with that respect. As an example, techniques such as Cellular Neural Networks (CeNN), including training algorithms such as the Levenberg–Marquardt (LM), Genetic Algorithm (GA) and Artificial Bee Colony (ABC), have been applied to the segmentation of both aggregated and filamentous bacteria images (Boztoprak et al., 2016). On the other hand, Satoh et al. (2021) developed deep learning-based classifiers using a Convolutional Neural Network (CNN), with the pretrained Inception v3 as the CNN architecture, aiming at characterizing floc-forming structures as aggregated or dispersed flocs and establishing the presence, or absence, of filamentous bacteria.

Post-processing of segmented images may also be required for border objects (cut-off by the image's borders) and debris removal and separation of touching objects, among others. Upon the post-processing step, gold approximations of ground truth may be used to assess the resulting images' truthfulness, using metrics such as accuracy, recall, variation of information, F-measure, and Rand index, among others (Khan et al., 2017a, b). Following post-processing, the resulting images may be fed into the image analysis step or serve as mask images to extract intensity images of objects or regions of interest.

15.4.3 Image Analysis

Image analysis data depends primarily on the image type (binary, mask, intensity, etc.) fed to this step and on the intended outcome of the QIA system. One of the most common outcomes of image analysis is the content determination of a given object (or objects) of interest. That is particularly true for binary images, however, for intensity (or mask) images, and especially for fluorescent staining, the aim could be broader (providing information on the biomass physiological status and specific component contents determination) (Costa et al., 2013; Mesquita et al., 2013).

Apart from purely contents data, Euclidian geometry morphological data (Russ, 2002), such as equivalent diameter, area, number, perimeter, length, width, eccentricity, roundness, extent, convexity, compactness, and solidity among others, represent the most common parameters extracted from the resulting images of the image processing step (Costa et al., 2013; Mesquita et al., 2013). However, when objects present self-similarity at various scales, the determination of parameters related to fractal dimensions is often taken into consideration (Mandelbrot, 2004). Indeed, given the complexity of AS aggregates shapes, a growing number of studies have been conducted including fractal dimensions determination (Costa et al., 2013; Mesquita et al., 2013; Smoczyński et al., 2014; Yu et al., 2017; Asensi et al., 2019). On the other hand, the nature of fractal parameters can vary considerably from one application to the next.

15.5 IMAGING IN WASTEWATER TREATMENT

Automated processing and analysis of images acquired by digital cameras enable a near real-time extraction of quantitative information. Among the ever-growing fields of application of this methodology, WWT has already been favored from its implementation, both in lab-scale studies, as in full-scale WWTP (Mesquita et al., 2013). In accordance, grows the number of studies referring to the

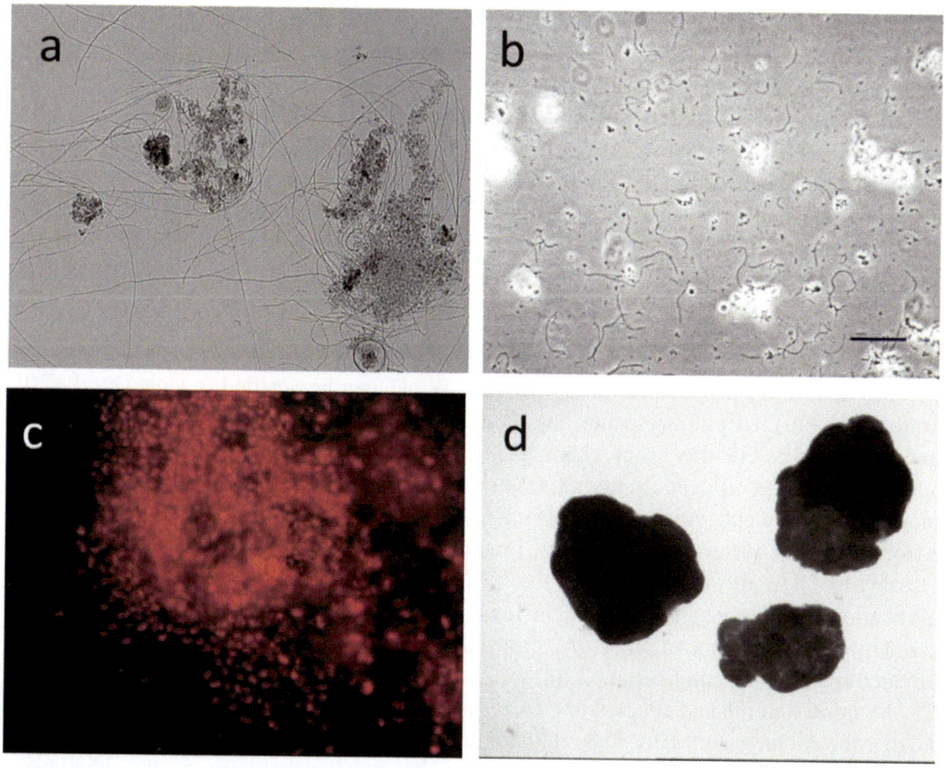

FIGURE 15.1 Example of images acquired for QIA procedures. (a) Bright-field; (b) phase-contrast; (c) epifluorescence; and (d) stereomicroscopy.

development of QIA routines based on samples images acquired using microscopy techniques. The diversity and complexity of the QIA routines are also appreciable and must consider the variety of image sources, such as optical (stereomicroscopy, bright-field, phase-contrast, epifluorescence and confocal laser scanning microscopy - CLSM) or electron microscopy and tomography, and acquisition (two- and three-dimensional frame grabbing, video recording, *in situ*/online) methodologies.

Examples of bright-field, phase-contrast, epifluorescence, and stereomicroscopy are presented in Figure 15.1.

15.5.1 Bright-Field and Phase-Contrast

The relationship between the AS settling ability, filamentous bacteria contents and floc-forming bacteria aggregates structure has been a matter of study, mainly by means of bright-field and/or phase-contrast microscopy, for the last two decades (Costa et al., 2013; Mesquita et al., 2013). Indeed, these methodologies have been preferred from the initial manual filamentous bacteria counting and floc size distribution studies, to the more advanced automated quantification and characterization methods based on QIA allowing to estimate the aggregated biomass contents and structure (both in terms of Euclidean and fractal characterization), as well as the filamentous bacteria contents.

While bright-field microscopy is, undoubtfully, the method of choice for monitoring the aggregated biomass, due to its simplicity and affordability, the transparency of most microbial cells, including filamentous bacteria, renders phase-contrast microscopy as a valuable tool for assessing the filamentous counterpart, not solely for identification purposes, but also for quantitative assessment (Khan et al., 2017a, b). Bright-field microscopy is considered more accurate for determining the aggregates boundaries and assessing short protruding filaments (masked by the aggregates halos in phase-contrast) in AS characterization. However, it may present some limitations, requiring staining procedures for specific microbial communities and intracellular storage compounds identification. On the other hand, phase-contrast microscopy, which contrasts filamentous bacteria without staining, is favored for assessing long protruding filaments and, thus, for filamentous bulking conditions. Furthermore, it is able to distinguish a number of internal structures based on their density (Costa et al., 2013; Mesquita et al., 2013).

Furthermore, when dealing with larger and denser structures, a stereomicroscope is found to be more appropriate than a conventional bright-field/phase-contrast microscope. Indeed, when characterizing both the aerobic and AGS, most studies employ the use of magnifications under 100x and, thus, stereomicroscopes (Costa et al., 2013; Mesquita et al., 2013; Ramos et al., 2017; Leal et al., 2020).

15.5.2 Epifluorescence and CLSM

The widely use of epifluorescence microscopy in WWT can be verified in Costa et al. (2013) and Mesquita et al. (2013). Epifluorescence has been found useful for determining the AS bacteria Gram status, as well as its viability when characterizing induced abnormalities in lab-scale experiments and correlating to the effluent quality parameters Mesquita et al. (2016). Also, Amaral et al. (2017), employed epifluorescence microscopy aiming the determination of PHA inclusions in a mixed culture from an EBPR system. To that effect, images from samples stained with Nile Blue A (NBA), and treated by QIA, allowed to quantify the intracellular bacterial contents on PHA. Similarly, the quantification of polyphosphate (PP) inclusions, within another mixed culture for EBPR, could be assessed by epifluorescence microscopy, upon staining with DAPI (4′, 6-diamidino-2-phenylindole) and treated by QIA (Mesquita et al., 2014).

CLSM, upon staining and coupled to QIA techniques, has been employed to quantify the structure of biological cake layers in MBR, and relate to the spatial change, as well as to determine the contribution of bacterial cell and EPS to the cake layer resistance. For that purpose, the cake layer porosity, bio-volume, and thickness were determined by QIA (Hwang et al., 2012). Again, relating to the enlightenment of biological cake layers in MBR, CLSM upon staining and coupled to QIA techniques has also been employed, alongside optical coherence tomography (OCT), to assess the biovolume and monitor cake layer development by Fortunato et al. (2019). In addition, the abundance of filamentous bacteria in AS samples was determined by Burger et al. (2017) using CLSM, coupled to viability staining, by determining the length of the protruding filamentous bacteria relative to the size of the aggregates. The filamentous abundance was further correlated with the aggregates settling ability, stability and shear strength, allowing to confirm its effects in floc formation solid–liquid separation.

15.5.3 Unconventional Techniques

A number of other microscopy techniques have also been used in WWT studies. Indeed, two-photon excitation laser scanning microscopy (TPE-LSM), coupled to fluorescence microscopy and CLSM, has been found useful for studying the internal structure of aggregates with high cell densities, and laser scanning microscopy (LSM), magnetic resonance imaging (MRI), and scanning transmission X-ray microscopy (STXM) for the analysis of the structure, composition, processes and dynamics of microbial communities (Mesquita et al., 2013; Costa et al., 2013). Also, scanning electron microscopy (SEM) methods have been used to analyze the structure of AS subjected to chemical coagulation (Al and Fe) and electrocoagulation by means of both Euclidean and fractal characterization. The obtained data allowed to further model the phosphate removal and dye sorption by the AS aggregates (Smoczyński et al., 2014).

OCT, a non-destructive and non-invasive technique, has been employed to characterize, *in situ*, the structure and dynamics of biological cake (fouling) layers in an activated sludge membrane bioreactor (AS-MBR), and relate to transmembrane pressure by (Fortunato et al., 2019). To that effect, 2D and 3D OCT QIA techniques were employed, allowing real-time direct imaging of the cake layer, to assess the average thickness, relative roughness and porosity, monitoring the cake layer development.

Considering the complexity of AS microbiota identification, Hu et al. (2021) proposed a QIA methodology intending to replace the use of the protozoa diversity, number, and dominant species. For this purpose, a new procedure was established correlating the microfaunal motion analysis of seven of the most common protozoa found in WWT, in terms of motion trajectory and motion velocity parameters, with AS aggregates Euclidean structure (obtained by principal components analysis [PCA] and related to flocs shape and gaps), MLSS (mixed liquor suspended solids) and SVI (sludge volume index), by the use of video frame imaging.

15.5.4 Color Imaging

The use of multiple color channels in the characterization of aggregated and filamentous bacteria has also been employed, in bright-field, phase-contrast, epifluorescence and CLSM images (Mesquita et al., 2013; Costa et al., 2013). Color images are, for the most part in WWT QIA, represented in Red, Green and Blue (RGB) format. However, such representation is not always ideal for extracting the most relevant information from such images. Indeed, other representations may present an advantage, for specific applications, such as the Hue, Saturation and Value (HSV), Hue, Saturation and Luminosity (HSL), Cyan, Magenta, Yellow and Black (CMYK), and Lightness, green–red and blue–yellow spaces (L*a*b), among others (Russ, 2002).

The use of color is, ultimately, mandatory in applications where different dyes are employed to enlighten a specific attribute of the aggregated or filamentous bacteria under analysis. Such is the case for the study of structural, as Gram status, and physiological aspects, including activity or viability assessment, quantification of specific microorganisms by FISH (*fluorescent in situ hybridization*) and of intracellular inclusions (Costa et al., 2013; Mesquita et al., 2013). In such cases, the use of different color channels in the L*a*b space has already been proven of significance both for determining the aggregated (mainly) and filamentous bacteria Gram status, as well as its viability, when characterizing abnormalities in WWTP and correlating to effluent quality parameters, by using appropriate fluorescent dyes in epifluorescence microscopy (Mesquita et al., 2011, 2016). Also, the use of the HSV format has been well documented in wastewater samples QIA, with each channel segmented separately and compiled to obtain a final segmented (binary) image (Khan et al., 2017b). And, apart from the use of different color spaces than RGB, the use of a specific color channel, instead of the ensemble, may be desirable. Indeed, the blue (RGB) channel intensity in AS macroscopic bright-field images, alongside the use of the Beer-Lambert law, has been found to act as an accurate index for cell concentration measurements (Asgharnejad & Sarrafzadeh, 2020).

In some instances, it is intended to highlight, or quantify, a specific structure within a cell or aggregate. Such is the case of quantifying a number of intracellular storage (nutrient and energy) inclusions, present in the AS, by bright-field microscopy staining, including PHA (polyhydroxyalkanoates) by Sudan Black B (SBB), epifluorescence microscopy staining for PP (polyphosphates) by Methylene Blue (MB) or glycogen by Aniline Blue (AB) (Mesquita et al., 2013, 2014). In addition, fluorescent dyes, coupled to epifluorescence microscopy, have already been used for that purpose, namely NBA for PHA quantification (Amaral et al., 2017). Finally, other uses of color imaging encompass the determination of the RGB mean values of a textile wastewater upon chemical coagulation and sedimentation, in order to assess the removal of suspended solids and color (Yu et al., 2017).

15.5.5 Online/*In Situ* Monitoring

One of the most important aspects in QIA applications in WWT, considered a major turning point in widespread implementation, is the development of online and *in situ* systems for real-time process monitoring and control. In fact, traditional monitoring methods present a delayed response in time, skilled analysts and may not be feasible to implement online, whereas automated QIA procedures have the potential to overcome these drawbacks. Developing online systems to track full-scale WWTPs is, still, troublesome in some cases, that once overcome radically alters the utility of QIA techniques. The main barriers convey the sampling process (capillary or flow cell) and image acquisition (limited depth of focus and poor image quality), both of which require further study and development.

Indeed, examples of online/*in situ* systems already exist, such as in color determination in wastewater monitoring, particles imaging and characterization of bioreactor dispersion (Costa et al., 2013). Also, Koivuranta et al. (2014) developed an optical monitoring method to characterize AS aggregates, and filamentous bacteria, and to study the correlation between the sludge settling ability and aggregates structure. The main advantage of the proposed method relies in its ability to acquire, and analyze, a large number of aggregates in a short timeframe and its *in situ* applicability, with solely sample dilution required as pre-treatment. The main objective of the proposed system is to provide real-time monitoring of the total extended filament length, based on geodesic distance calculation, tackling filamentous bulking and foaming problems. On the other hand, textile wastewater treatment processes, in terms of assessing suspended solids and color removal by chemical coagulation and sedimentation, were the driving force behind the development of an online QIA application by Yu et al. (2017). To that effect, the mean gray and RGB values of the acquired images from lab-scale batch settling tests were determined, alongside particle size, contents, and the fractal dimension.

The most common imaging systems employed for QIA analysis are presented in Table 15.1.

TABLE 15.1
Identification of the Most Employed Imaging Systems for QIA Analysis

Author	Color	Online/*In situ*	Microscopy
Costa et al. (2013), Mesquita et al. (2013)	Dependent	Dependent	Bright-field (most cases), phase contrast, stereomicroscopy, epifluorescence, CLSM, TPE-SLM, LSM, MRI, STXM, color imaging
Amaral et al. (2017)	Yes	No	Epifluorescence microscopy
Asgharnejad and Sarrafzadeh (2020)	Yes	No	–
Burger et al. (2017)	No	No	CLSM
Fortunato et al. (2019)	No	No	CLSM, OCT
Hu et al. (2021)	No	No	Video
Hwang et al. (2012)	No	No	CLSM
Khan et al. (2017a, b)	No/Yes	No	Phase contrast
Koivuranta et al. (2014)	No	Yes	–
Leal et al. (2020)	No	No	Stereomicroscopy
Mesquita et al. (2011)	Yes	No	CLSM
Mesquita et al. (2016)	Yes	No	CLSM
Mesquita et al. (2013)	No	No	Stereomicroscopy
Mesquita et al. (2014)	No	No	Epifluorescence microscopy
Ramos et al. (2017)	No	No	Stereomicroscopy
Smoczyński et al. (2014)	No	No	SEM
Yu et al. (2017)	Yes	Yes	Bright-field

CLSM, confocal laser scanning microscopy; TPE-SLM, two-photon excitation laser scanning microscopy; OCT, optical coherence tomography; SEM, scanning electron microscopy; LSM, laser scanning microscopy; MRI, magnetic resonance imaging; STXM, scanning transmission X-ray microscopy.

15.6 AEROBIC SYSTEMS

15.6.1 CAS Systems

15.6.1.1 Sludge Contents and Settling Ability Studies

QIA techniques have been successfully employed in monitoring AS systems, leading to the establishment of clear relationships between the aggregated bacteria size, structure and contents, the filamentous bacteria contents and the sludge settling ability. In addition, the aggregated bacteria contents have been also found to correlate with the suspended solids. Moreover, direct comparison, regarding the size og aggregates, with conventional particle analyzers corroborated the adequateness of QIA methodologies for such intent. Coagulation and flocculation studies, apart from day-to-day monitoring, have also benefited from the use of QIA techniques, by determining the aggregated biomass characteristics prior and after polyelectrolytes or metallic salts addition. Although most published studies still represent lab-scale experiments, the application of QIA methodologies in the assessment of full-scale WWTP is growing. Furthermore, an advantage that has commonly been endorsed is the ability of QIA to perform timely monitoring of the AS system in order to enlighten the impact of variable operational conditions on the sludge stability and allow for the use of corrective measures to avoid bulking, foaming and other malfunctions (Costa et al., 2013; Mesquita et al., 2013).

One of the most widespread uses of QIA in AS systems regard the quantification and characterization of the aggregated and filamentous bacteria, by means of Euclidean geometry parameters, namely size and shape factors. Such parameters can be later used to monitor the sludge aggregates structure and correlate, or even estimate, the sludge contents and settling ability (mainly in terms of SVI or settling velocity), as well as the overall solid–liquid separation performance. Given that some aggregates generated by AS processes possess self-similitude (fractal) characteristics, fractal dimensions have also been used for morphological characterization (Costa et al., 2013; Mesquita et al., 2013).

One of the most important parameters with respect to the solid–liquid separation in a WWTP clarifier is the sludge settling velocity, which may be determined by the Vesilind equation in the zone of constant fall velocity. Using an optical-based QIA procedure monitoring a series of sedimentation tests on real wastewater, Molina et al. (2020) were able to estimate the settling velocity upon the determination of the number of flocs connected per cluster. In accordance, they were able to determine the sludge sedimentation time, allowing for a timely detection of bulking phenomena and severity. Modeling of the sludge settling ability (in SVI terms), alongside the suspended biomass, was also the goal of the Boztoprak et al. (2016) and Khan et al. (2020) studies which used an optical microscopy approach. Also, Campbell et al. (2019) employed QIA to assess the SVI, by means of determining the filamentous bacteria length by QIA, alongside the size of the aggregates. However, beyond the settling ability, these authors further extended the study to relate the QIA data to specific EPS production, oxygen transfer efficiency and, ultimately, the AS system energy use performance.

WWT systems are prone to be affected by malfunctions such as filamentous or viscous (zoogleal) bulking and pinpoint floc formation, among others. As previously described, QIA techniques may be used to characterize the contents and structure (size and morphology) of aggregated biomass and filamentous bacteria, relating to the above conditions using appropriate chemometric methods. Such was the case of the work of Amaral et al. (2013) who was able to successfully identify all malfunctions by QIA data and estimate the SVI and suspended solids. Also, Oliveira et al. (2018) studied the effect of different operating conditions, including filamentous bulking and washout phenomena, in the structure of sludge aggregates, allowing for a deeper understanding of the aggregates' structure dynamics under process disturbances. QIA was also able to monitor the evolution of AS aggregates morphology from poor to good flocculation properties following a maintenance stoppage at a WWTP (Koivuranta et al., 2014). These authors were also able to diagnose the cause behind the bulking problem.

Different zones within an AS system may present different physicochemical conditions and, thus, diverse microbiota, floc-forming and filamentous bacteria. Such is the case of the existence of anaerobic, or anoxic, zones within the aerated tank. In accordance, it is expected that the sludge within the reactor be affected by the surrounding environment. Indeed, it has already been proven, by QIA methodologies, that oxide-reductive conditions influence the size of the aggregates, leading to differences between the aerobic and the anaerobic zones in a full-scale WWTP (Liwarska-Bizukojc et al., 2015).

One of the parameters determining the operation efficiency of a WWTP is the particle size of the treated effluent. Particle Size Analyzer (PSA), a non-imaging tool, may be used for that purpose; however, it is limited in the amount of information that it presents. On the other hand, light microscopy may be successfully employed for determining colloids size (including sludge aggregates) with the advantage of providing a larger set of data. To that effect, Khanam et al. (2016) analyzed the particle size of WWTP effluents by comparing both techniques. In a related work, Mullins et al. (2018) successfully estimated effluent turbidity due to light scattering and absorption by fine particles in suspension (roughly correlated with the suspended solids) by an automated system based on QIA for a number of different synthetic and real wastewaters.

Coagulation and flocculation experiments are another large field of application for QIA techniques. Electrocoagulation is a process that uses metallic salts to increase phosphates removal from wastewaters. By using QIA, the effect of such process in the aggregates structure may be monitored, as presented in the work of Smoczyński et al. (2014), who assessed sweep flocculation and sludge separation and dehydration by SEM. In an experiment conducted to assess chemical coagulation to remove suspended solids and color from textile wastewater, Yu et al. (2017) successfully used QIA techniques to characterize the size and shape of aggregates. Also, Asensi et al. (2019) employed QIA techniques to assess the effect of ferric chloride addition to improve the settling ability properties of AS toward increased suspended solid removal at the secondary settler. To that intent, the aggregates' size and shape characteristics were determined and correlated to the settling velocity. In all of the above studies, alongside Euclidean geometry parameters, also fractal dimension descriptors were employed.

15.6.1.2 Estimation of Key Quality Parameters

Up to date, few are the published studies that dealt with the estimation of WWT key quality parameters, related to organic matter removal, nitrification, denitrification and phosphorous removal processes, from QIA-based data. Among these, Mesquita et al. (2016) performed a study aiming at evaluating the chemical oxygen demand (COD), ammonium ($N\text{-}NH_4^+$), nitrite ($N\text{-}NO_2^-$) and nitrate ($N\text{-}NO_3^-$) concentrations in the treated effluent. A lab-scale AS system was operated under different experimental conditions and monitored through the use of a QIA procedure allowing to assess the aggregated and filamentous biomass contents and structure. QIA was also used for the microbial community characterization in terms of viability and Gram status. It was found that the use of the morphological and physiological data allowed predicting, to some extent, the effluent studied key quality parameter concentrations based on chemometric techniques.

Also, Tomperi and Leiviskä (2018) have already presented a study in a full-scale WWTP aiming at modeling the effluent COD, biochemical oxygen demand (BOD), total nitrogen (TN), total phosphorous (TP) and SS concentrations, based on QIA and physicochemical data. Apart from the Euclidean-based and fractal parameters characterizing the aggregated and filamentous bacteria, and the studied key quality parameters, physicochemical and operational parameters were also considered. The study also showed the validity of specific optical monitoring variables in modeling a number of key quality parameters.

15.6.1.3 Characterization of the AS Microbiota

The AS microbiota is extremely complex, comprising bacteria, protozoa, metazoa, algae and fungi. The vast majority of the roles that an AS system possesses in wastewater depuration are directly related to bacteria, though protozoa and metazoa also play an important part (mainly in bacteria predation and in bacteria aggregation processes). Therefore, it comes as no surprise that QIA techniques have already been found of use in determining and quantifying the biota present in a given WWTP (Costa et al., 2013; Mesquita et al., 2013).

Regarding bacteria, FISH is, nowadays, a common tool of choice for identification and quantification, using epifluorescence or CLSM methodologies. Beyond FISH, several works have already employed QIA for characterizing the AS biomass physiology, namely regarding viability determination. In addition, the quantification of the aggregated and filamentous bacteria Gram status has already been performed by QIA procedures. In both cases, fluorochromes have been used alongside epifluorescence or CLSM acquisition. Such analyses were further used to elucidate the bacterial distribution within the aggregates and correlated with the settling ability, suspended solids, system malfunctions and even in pharmaceutical active compounds (PhAC) toxicity studies. Furthermore, the uptake activity of nutrients, and other compounds, could also be correlated to the AS phylogenetic groups by means of coupling microautoradiography (MAR) and FISH techniques (Costa et al., 2013; Mesquita et al., 2013).

On the other hand, QIA-based procedures for automatic, or semi-automatic, identification of protozoa and metazoa most important species or groups, have already been successfully implemented and employed for the operation conditions assessment, as well as to relate to bulking events (Costa et al., 2013; Mesquita et al., 2013).

Taking into consideration the role of the bacteria consortia in AS systems for the WWT performance, the number of studies in the literature employing QIA procedures for the identification and quantification of the main bacteria present is rising. For instance, in the study aiming at estimating WWT key quality parameters, Mesquita et al. (2016) employed a QIA procedure allowing to assess the microbial community characterization in terms of viability and Gram status. To that purpose, fluorescence techniques were used, and the resulting stained samples observed, and acquired, through epifluorescence microscopy. Also, aiming at elucidating the bacteria consortia in a WWT system, Abbà et al. (2017) used FISH coupled to an epifluorescence microscope, in samples from a thermophilic aerobic membrane reactor (TAMR). In addition, the membrane reactor sludge was also characterized by QIA in terms of the aggregated and filamentous bacteria contents and structure. Finally, the QIA data was related to the sludge rheological behavior.

Protozoa and metazoa are sensitive to physical, chemical, and operational conditions and, thus, are considered good indicators of treatment quality in AS systems. However, manual identification and quantification of the main groups and species present AS is a skilled task, dependent on experts. Therefore, an automatic QIA-based procedure allowing for the recognition of such organisms has been already the object of study. Indeed, a semi-automatic QIA methodology for protozoa and metazoa recognition has already developed, with the use of chemometrics to analyze the organisms morphological descriptors for identification and classification purposes (Mesquita et al., 2013). The obtained procedure was found adequate for distinguishing most protozoa and metazoa organisms. On a different approach, Hu et al. (2021) analyzed, by means of QIA applied to micro-video frame images, the motion pattern of protozoa species. In that regard, these authors employed both motion trajectory and motion velocity parameters, aiming at providing a methodological foundation for the establishment of a new method for the use of protozoa microfauna as a quality indicator.

Few studies exist in the literature that combine the identification, and quantification, of both bacteria and protozoa species in WWT, standing out the work of Leal et al. (2016) in the enlightenment of foaming events in CAS and oxidation ditch (OD) systems. In accordance, QIA was employed for the identification and quantification of the different filamentous bacteria present, using a specially developed filamentous bacteria index, and for sludge characterization, alongside traditional protozoa and metazoa quantification. This study then proceeded to further enlighten the inner relationships between this set of parameters and relate to the observable bulking events.

15.6.1.4 Extra and Intracellular Compounds Determination

Apart from discriminating whole cells, QIA procedures have also been found useful in quantifying inclusions within filamentous organisms, mainly in AS and EBPR systems, such as intracellular storage inclusions, namely PHA (by means of NBA or SBB staining), PP (MB staining), glycogen (AB staining), among others. Depending on the employed dye (chromogenic or fluorescent) bright-field or fluorescent microscopy (including CLSM) may be required. Furthermore, QIA has been increasingly employed to estimate EPS and used to enlighten the bioflocculation mechanism of aggregated biomass (Costa et al., 2013; Mesquita et al., 2013).

Quantification of AS intracellular storage inclusions, such as PHA, PP or glycogen can be used to monitor the uptake of carbon and phosphate (in EBPR processes) nutrients, in aerobic/anaerobic or famine/feast cycles in SBR. To that effect, bright-field microscopy staining methods, coupled to QIA, have already been developed to assess PHA (by SBB staining), PP (by MB staining) and glycogen (by AB staining) (Mesquita et al., 2013, 2014). In addition, fluorescent dyes, coupled to epifluorescence microscopy, have already been used for that purpose, namely NBA for PHA quantification (Amaral et al., 2017).

On the other hand, Burger et al. (2017), determined, by means of QIA, the floc strength factor (relative change in floc size upon shearing) and studied its correlation with the global and specific (protein: polysaccharide) EPS, alongside filamentous abundance in AS samples from full-scale WWTP.

15.6.2 SBR Systems

Another AS system that has been increasingly studied and applied, mainly for concomitant organic and nutrients (including N and P) removal is the SBR system. As for CAS systems, the ability of QIA to address the specific issues raised by this methodology has also been proven both in lab-scale experiments and in full-scale WWTP. In accordance, a growing number of published works address specifically SBR studies.

Some examples of QIA applications in that regard encompass the study of the correlation between the sludge settling ability and suspended solids with the aggregated and filamentous bacteria contents and structure, aggregated biomass structure and sludge flocculation/deflocculation in saline wastewaters (synthetic and fishery), quantification of AS viability, filamentous bacteria Gram status, and intracellular storage inclusions, effect of surfactants removal in the structure and activity of aggregated biomass, biomass deflocculation in the presence of antibiotics and quantification of antibiotic time-kill activity (Costa et al., 2013; Mesquita et al., 2013).

SBR systems are increasingly studied to perform N and P removal from wastewaters. With that regard, Mesquita et al. (2014) studied the ability of using DAPI staining, coupled to QIA, to quantify PP inclusions in AS biomass from an enhanced biological phosphorous removal (SBR-EBPR) system, operating in aerobic/anaerobic cycles for phosphate uptake from the wastewater. Also, Amaral et al. (2017), monitored in SBR-EBPR systems PHA, PHB (polyhydroxybutyrate) and PHV (polyhydroxyvalerate) inclusions, by means of NBA staining, coupled to QIA, in aerobic dynamic feeding (famine/feast) and aerobic/anaerobic cycles, for increased production of PHA by a mixed microbial cultures (MMC) in WWT.

QIA has also been applied to study a different type of SBR, namely a two-stage settling sequencing batch reactor (TS-SBR) separated by an anoxic stage, operated to improve sludge settleability. A second classical SBR was also operated and the aggregated bacteria structure and filamentous bacteria contents, from both reactors, were monitored and related to the sludge settling ability (Alattabi et al., 2017).

15.6.3 Membrane Bioreactors (MBR)

Membrane bioreactors (MBR) are seen as viable alternatives to traditional AS systems for WWT, mainly in terms of increased potential for sludge retention within the reactor, due to the physical barrier (membrane) that they encompass. It comes as no surprise, thus, that significant research, taking advantage of QIA procedures, has already been performed and published. Bioflocculation monitoring studies are one of the main subjects addressed in published works, being related to filterability, fouling problems and sludge retention times (SRT) (Mesquita et al., 2013).

Hwang et al. (2012) quantitatively assessed membrane biofouling in a lab-scale MBR treating a synthetic wastewater, by QIA combined with CLSM, relating to the sludge EPS contents, the biological cake layer structure and resistance. Another lab-scale activated sludge membrane bioreactor (AS-MBR) was operated with real wastewater by Fortunato et al. (2019), to characterize the structure and dynamics of the cake layers by means of OCT coupled to QIA methodologies during the cake layer development.

Another type of MBR, a TAMR, treating a high-strength wastewater, was studied by QIA procedures to characterize the microbiota, as well as the aggregated and filamentous biomass, and relate to the sludge rheological behavior and, consequently, the hydrodynamics of sludge suspension. To that intent, FISH methodologies were employed for the identification and quantification of main bacterial groups (Abbà et al., 2017).

15.6.4 Aerobic Granular Sludge

Recently, a great deal of attention has been given to the development of small footprint SBR grown aerobic granular sludge (SBR-AGS) for WWT, due to its inherent denser and mechanically stronger structure, higher shock-loading resistance and improved settling ability and solid-effluent separation. However, the granulation process, may not be simple nor fast or straightforward, and should be carefully monitored, to avoid granule disaggregation. To that effect, QIA has already been put into practice, namely to monitor, or enlighten, the aerobic granules structure and stability by means of contents, size and shape parameters (both Euclidean and fractal). Studies have been conducted, and published, relating to the granulation process in SBR and sequential batch airlift reactors (SBAR), among others, treating a variety of wastewaters. Mass transfer limitations is another concern facing AGS systems and CLSM methodologies, coupled to QIA, have already been employed to model mass transfer and characterize cell and EPS distribution inside the granules (Costa et al., 2013).

QIA has also been employed to study the phosphorous removal efficiency of an EBPR process using AGS in a continuous flow reactor (CFR-AGS). The phosphorus removal granular sludge (PRGS) was monitored in terms of phosphorous removal efficiency, granular size and structure while adapting from an SBR inoculum to the continuous flow conditions. QIA proved to be invaluable for monitoring a PRGS-guided operation of continuous flow systems (Li et al., 2016).

AGS has also been characterized, by means of QIA, treating a synthetic wastewater with a high ammonium content and the addition of the recalcitrant 2-fluorophenol (2-FP) and the system enriched with 2-FP degrading bacteria. The granular sludge was monitored throughout the experiment in terms of its structure and the presence of filamentous organisms was also sought (Ramos et al., 2017). Leal et al. (2020), have studied an SBR-based AGS system characterizing, by QIA, the obtained (mature) granular sludge in terms of their contents and structure. In addition, the floccular fraction of the sludge was also characterized in the same terms, alongside the filamentous counterpart. Furthermore, the granular volumetric contents were also determined. Given the objective of avoiding AGS instability (biomass structural and settling properties changes), the QIA data of both fractions was used to predict the sludge SVI as well as the suspended solids (total and volatile).

The most studied AS systems and fields of study in QIA analysis are presented in Table 15.2.

15.7 CHEMOMETRICS

Chemometrics is a term used to describe the use of mathematical and statistical tools to make sense of data collected in biological and chemical processes. The usefulness of chemometric techniques, for process and experiment optimization and development, and information extraction, assessment

TABLE 15.2
Identification of the Most Studied AS Systems and Fields of Study in QIA Analysis

Author	AS System[a]	Studied Fields
Costa et al. (2013), Mesquita et al. (2013)	CAS, SBR, MBR, SBAR, CFR-AGS	Floccular and granular sludge, filamentous bacteria, microbiota (Gram, FISH), viability, protozoa, EPS, storage inclusions (PHA, PP, glycogen), settling ability, effluent quality parameters (COD, BOD, N, P), physicochemical parameters (pH, T, OD, etc.), coagulation/flocculation, toxicity, (bio)fouling, among others.
Abbà et al. (2017)	TAMR	Floccular sludge, filamentous bacteria, microbiota (FISH), physicochemical parameters (pH, T, OD), rheology
Alattabi et al. (2017)	TS-SBR	Floccular sludge, filamentous bacteria, effluent quality parameters (N), settling ability
Amaral et al. (2013)	CAS	Floccular and granular sludge, filamentous bacteria, settling ability, bulking
Amaral et al. (2017)	SBR-EBPR	Storage inclusions (PHA)
Asensi et al. (2019)	–	Floccular sludge, settling ability, EPS, coagulation/flocculation
Boztoprak et al. (2016)	CAS	Floccular sludge, filamentous bacteria, settling ability
Burger et al. (2017)	CAS	Floccular sludge, filamentous bacteria, viability, EPS, rheology
Campbell et al. (2019)	CAS	Floccular sludge, filamentous bacteria, settling ability, EPS contents, oxygen transfer, WWTP energy use
Fortunato et al. (2019)	AS-MBR	(Bio)fouling
Hu et al. (2021)	CAS	Protozoa, settling ability
Hwang et al. (2012)	MBR	EPS, (bio)fouling
Khan et al. (2020)	CAS	Floccular sludge, settling ability
Khanam et al. (2016)	CAS	Floccular sludge, particle size analysis
Koivuranta et al. (2014)	CAS	Floccular sludge, filamentous bacteria, settling ability
Leal et al. (2016)	CAS	Floccular sludge, filamentous bacteria, protozoa, settling ability, foaming
Leal et al. (2020)	SBR-AGS	Granular and floccular sludge, filamentous bacteria, settling ability
Li et al. (2016)	CFR-AGS	Granular sludge, effluent quality parameters (COD, P)
Liwarska-Bizukojc et al. (2015)	CAS	Floccular sludge, effluent quality parameters (BOD), sludge loadings
Mesquita et al. (2016)	CAS	Floccular sludge, filamentous bacteria, microbiota (Gram), viability, effluent quality parameters (COD, N)
Mesquita et al. (2014)	SBR-EBPR	Storage inclusions (PP)
Molina et al. (2020)	CAS	Floccular sludge, settling ability
Mullins et al. (2018)	CAS	Floccular sludge, turbidity
Oliveira et al. (2018)	CAS	Floccular sludge, effluent quality parameters (COD, N)
Ramos et al. (2017)	SBR-AGS	Granular sludge, filamentous bacteria, effluent quality parameters (COD, N), toxicity
Smoczyński et al. (2014)	–	Floccular sludge, coagulation/flocculation
Tomperi and Leiviskä (2018)	CAS	Floccular sludge, filamentous bacteria, effluent quality parameters (BOD, COD, N, P)
Yu et al. (2017)	–	Floccular sludge, density, color, coagulation/flocculation

[a] CAS, conventional activated sludge; SBR, sequencing batch reactor; MBR, membrane bioreactor; SBAR, sequencing batch airlift reactor; CFR-AGS, continuous flow reactor – aggregated granular sludge; TAMR, thermophilic aerobic membrane reactor; TS-SBR, two-stage settling sequencing batch reactor; SBR-EBPR, sequencing batch reactor – enhanced biological phosphorous removal; AS-MBR, activated sludge membrane bioreactor.

and interpretation, further increases as the volume of collected data grows. That is even more true regarding environmental studies and applications, referring to three main areas of interest: quantitative (bio)chemical analysis, environmental quality monitoring, and toxicological effects modeling and prediction. Regarding WWT, timely detecting disturbances in the system is critical for performance monitoring, due to ever more stringent effluent quality legislations (Mesquita et al., 2013).

Given the vast amount of data gathered by QIA techniques, chemometrics stands out as an invaluable tool in organizing, translating, reducing and labeling the resulting datasets. Most of the chemometric techniques employed in QIA data originated from WWT possess one (or more) of the following objectives: clustering, classification, modeling or prediction. Furthermore, some processes are so complex that they are better dealt with by black box approaches such as ANN.

15.7.1 Clustering and Classification

Clustering and classification methods allow for information organization, assembling, comparison and labeling. For that intend, a number of methodologies exist, being the cluster analysis (CA), PCA, discriminant analysis (DA) and decision trees (DT) among the most widely used. PCA and DA are extremely feasible methods for extracting relevant information from large datasets. They act by projecting the original variables into latent variables (LVs) reducing high-dimensional and strongly correlated data to an uncorrelated, easily interpreted, smaller dataset able for interrelationships analysis (Einax et al., 2004).

Applications wise, PCA has been employed in WWT for QIA data selection, determination of wastewater composition, AS systems disturbances diagnosis, sludge settling ability monitoring, toxicity assessment, microbiota Gram status and viability evaluation, nitrification and phosphate removal processes efficiency and identification of operating conditions (Mesquita et al., 2013). On the other hand, CA, DA and PCA have been employed to identify the factors affecting domestic sewage composition, similarity or dissimilarity, and spatial and temporal variations, whereas DA and DT have been employed for protozoa and metazoa automatic identification by QIA procedures, among others (Mesquita et al., 2013). This procedure allowed to select the most appropriate set of variables (LVs with respect to DA and original values in DT) to classify a given protozoa (or metazoa) in the corresponding group or species.

AS systems disturbances, encompassing viscous and filamentous bulking, and pinpoint phenomena have already been characterized by the use of PCA techniques. The use of LVs, originated from QIA data of aggregated and filamentous biomass contents and structure, allowed to clearly distinguish between the studied malfunctions and normal operating conditions. Furthermore, DT was also used for the identification of each studied condition, and definition of the key parameters limits (Amaral et al., 2013). In an analog perspective, foaming events in full-scale WWTP (CAS and OD) have already been successfully monitored by QIA data of aggregated and filamentous biomass, as well as protozoa and metazoa, with mild and severe events being correlated, by means of PCA, cross-correlation (CC) and DT to assess the foaming occurrences, and enlighten the inner relationships (Leal et al., 2016). PCA was also employed in the assessment of intracellular PHA inclusions by Amaral et al. (2017) for enlightening the fundamental differences between the QIA datasets of aerobic and anaerobic stages on one hand and famine/feast cycles, on the other, thus leading to independent model determination for each case. Regarding a correlation analysis of microfauna motion velocity with the AS structure, Hu et al. (2021) classified and reduced into two comprehensive indexes the set of original morphological parameters of the sludge aggregated biomass sludge by PCA.

15.7.2 Modeling and Prediction

Modeling and prediction allow for a given parameter to be assessed by a set of related variables, presenting an advantage in the case of tedious, labor-intensive, time-consuming or leading to the use or production of harmful chemical species in its determination. Two of the most employed methods are the ordinary least squares (OLS), in terms of multilinear regression (MLR), and partial least squares (PLS). In both cases, it is intended to select, and weight, a reduced subset of uncorrelated original input variables in order to model (or predict) a known output parameter. PLS is a linear multivariate statistical method determining a set of new uncorrelated components (latent vectors), from the original input dataset, relevant for the output parameter(s) model (Einax et al., 2004).

A significant number of published studies, in the field of WWT, already take advantage of least squares methodologies to model (or predict) a given parameter, providing an alternative method to analytical off-line measurements. Literature shows the use of PLS approaches in lab-scale and full-scale WWTP, based (entirely or in part) on AS QIA data, to build predictive models for monitoring WWTP performance, diagnose WWTP processes and predict (or isolate) disturbances (washout, bulking and other phenomena), predict operational and effluent quality parameters, sludge settling ability properties and even intracellular storage compounds (Mesquita et al., 2013).

Tomperi and Leiviskä (2018), in their study to model effluent key quality parameters, based on QIA and physicochemical data employed stepwise selection (SS) as one of the nine different approaches (including correlation-based selection, forward selection, GA and a successive projections algorithm (SPA) combined with a GA) for image segmentation of the phase-contrast images of AS samples. The study also showed the validity of specific optical monitoring variables in modeling a number of key quality parameters. Also, following a selection procedure of QIA-based AS structure parameters, with the intent of predicting WWTP biomass contents and solid–liquid separation, stepwise regression (SR) has been used for feature selection and model generalization regarding the monitoring of WWTP to accommodate variable sludge settling ability and suspended solids contents (Khan et al., 2020).

PLS has also been extensively used in WWT studies. Amaral et al. (2013) employed PLS to predict the sludge settling ability (in terms of SVI) and suspended solids in a lab-scale AS system operated under four different operational conditions (filamentous bulking, viscous bulking, pinpoint flocs formation, and normal conditions). Mesquita et al. (2016) also studied AS systems, operated under different experimental conditions, and employed PLS models to correlate QIA data with wastewater key quality parameters, validating the use of the AS structural and physiological data with the effluent COD, $N\text{-}NH_4^+$ and $N\text{-}NO_3^-$ concentrations. With a different objective, the intracellular PHA concentrations, in AS of an EBPR with MMC, were successfully predicted using PLS models fed with QIA data obtained from epifluorescence microscopy upon NBA staining (Mesquita et al., 2013; Amaral et al., 2017). Also, Asensi et al. (2019) in their study to assess the effect of ferric chloride addition to improve the settling ability properties of AS employed PLS regression, applied to the collected data, revealed the sludge volumetric fraction, suspended solid concentration and aggregated biomass fractal dimension to be the variables most correlated with the settling velocity.

A different approach from the above was envisaged by Leal et al. (2020) who employed MLR to obtain linear model fits for suspended solids determination (both for the suspended and granular biomass fractions) and sludge settling ability from a QIA based of the aggregated (floccular and granular) and filamentous biomass from an AGS.

15.7.3 Other Supervised Learning Methods

ANN are particularly good at capturing non-linear interactions between variables in complex systems, and able to be used where there isn't enough process information to build a predefined model of the system. On the other hand, support vector machines (SVM) are learning algorithms processing data for classification and regression purposes. being most effective in high-dimensional spaces and in cases where the number of dimensions is greater than the number of samples (Cherkassky & Mulier, 2007).

One of the most employed supervised learning methods for treating the datasets originated (fully or partially) from QIA applications in WWT is ANN, used for the classification of AS systems protozoa and metazoa images (alongside DA and DT) and to estimate influent COD loads to a WWTP, allowing the differentiation between situations of low, normal and high influent loads (Mesquita et al., 2013). Also, Yu et al. (2017) successfully used a backpropagation ANN for the prediction of suspended solids and color removal efficiencies in textile wastewater, upon chemical coagulation and sedimentation, based on an online QIA application characterizing the AS system aggregates size and shape. Alongside DA and DT, have employed a backpropagation ANN, has also been employed for protozoa and metazoa automatic identification by QIA procedures, by weighting to a set of variables, and establishing the most relevant node connections to classify a given protozoa (or metazoa) in the corresponding group or species (Mesquita et al., 2013).

As previously referred, Boztoprak et al. (2016) have already employed CeNN techniques, including the LM, GA and ABC training algorithms, to segment both aggregated and filamentous bacteria images, with the objective of modeling of the sludge settling ability and the suspended biomass of an AS system. With a different objective, but also regarding image processing, Satoh et al. (2021) developed deep learning-based classifiers using an Inception v3 CNN architecture, for the characterization of aggregated and dispersed flocs and establishing the presence, or absence, of filamentous bacteria.

SVM have already been used to select WWTP AS structure parameters, derived from a bright-field imaging QIA procedure, aiming at predicting biomass contents and solid–liquid separation. SVM was applied throughout the QIA process in image acquisition, aggregates partitioning into size ranges, morphological features extraction, and SVI and suspended solids modeling (Khan et al., 2020).

The combination of different chemometric techniques (clustering and classification, modeling and prediction and other supervised learning methods) is also commonly employed, as reflected by the works described previously on Mesquita et al. (2013), Amaral et al (2013) and Khan et al. (2020).

The main chemometric tools applied in QIA analysis studies are presented in Table 15.3.

TABLE 15.3
Identification of the Main Chemometric Tools Applied in QIA Analysis Studies

Author	Chemometric Tool[a]
Mesquita et al. (2013)	CC, CA, PCA, DA, DT, OLS (MLR), PLS, ANN
Amaral et al. (2017)	PCA. PLS
Amaral et al. (2013)	DT. PLS
Asensi et al. (2019)	PLS
Boztoprak et al. (2016)	CeNN (LM, GA, ABC)
Hu et al. (2021)	PCA
Khan et al. (2020)	SR, SVM
Leal et al. (2020)	OLS (MLR)
Leal et al. (2016)	CC, PCA, DT
Mesquita et al. (2016)	PLS
Satoh et al. (2021)	CNN (Inception v3)
Tomperi and Leiviskä (2018)	SS, GA, SPA
Yu et al. (2017)	ANN

[a] CC, cross-correlation; CA, cluster analysis; PCA, principal components analysis; DA, discriminant analysis; DT, decision trees; OLS (MLR), ordinary least squares (multilinear regression); PLS, partial least squares; ANN, artificial neural networks; CeNN, cellular neural networks; LM, Levenberg–Marquardt; GA, genetic algorithm; ABC, artificial bee colony; SR, stepwise regression; SVM, support vector machine; CNN, convolutional neural network; SPA, successive projections algorithm; SS, stepwise selection.

15.8 CHALLENGES AND FUTURE TRENDS

As research tools in WWT, imaging techniques combined with staining procedures are of particular interest, providing meaningful information on biomass characterization, viability, and composition in near real time. The use of QIA in the monitoring of WWT processes is well established in this context and regarded as a powerful technology with significant potential for development. The ability to detect potential problems at an early stage, allowing to perform corrective actions, is the main benefit of QIA over traditional analytical procedures. Some noteworthy results have already been obtained; however, further research is still needed to widespread and broaden its applications.

The number of QIA studies for the characterization of AS has increased over time, with the goal of producing cleaner and safer effluents by combining operational parameters, biomass characterization and sludge structural data. A prominent issue in AS systems concerns the settling ability, with key relationships already established using QIA data, allowing the recognition of potential bulking phenomena at an earlier stage. Albeit no universal equation has been agreed upon, the use of chemometric techniques, neural networks and expert systems is being explored with promising results. QIA procedures can also be used to circumvent the current lack of data on microbial structure, distribution, and dynamics, particularly in EBPR and AGS systems, and directly related to the imaging of stained samples. FISH, Gram, Neisser and other staining protocols are, thus, cornerstones for enlightening the biomass viability, composition and intracellular storage compounds.

Two of the major developing fields of QIA application are the study of the granulation/degranulation processes, and granule stability, in AS systems treating, or in the presence of, toxics, recalcitrant or PhAC as well as for AGS (in lab-scale and, more pronouncedly in full-scale WWTP). Although a few studies have already been published, a great deal of work characterizing the biomass structure and microbiota of these systems are undergoing, at the time being (AGS stability in long-term operations, EPS characterization in full-scale WWTP, effect of PhAC in AGS structure and settling ability, microbiota characterization, among others), or remains to be performed.

The ultimate objective is the development of online/*in situ* QIA methodologies for real-time process monitoring and control, believed to be a critical step toward the widespread implementation of these procedures in WWT processes. The key constraint to this implementation is not the QIA procedure by itself, but rather the technical challenges of collecting undamaged, homogeneous, and representative samples from real WWTP, equipment cost, and poor image quality. Therefore, the development of such procedures, alongside the acquisition of well-focused and contrasted images remains a challenge.

In conclusion, the use of QIA, alongside multivariate statistical analysis, as a monitoring tool has been found to significantly contribute to survey AS in full-scale WWTP, minimizing effluent composition differences, occurrence of plant malfunctions and helping to propose appropriate correction measures.

ACKNOWLEDGMENTS

The authors thank the Portuguese Foundation for Science and Technology (FCT) under the scope of the strategic funding of UIDB/ BIO/04469/2020 unit. Daniela P. Mesquita and Cristina Quintelas thanks FCT for funding through program "DL 57/2016 – Norma transitória".

REFERENCES

Abbà, A., Collivignarelli, M.C., Manenti, S., Pedrazzani, R., Todeschini, S., & Bertanza, G. (2017). Rheology and microbiology of sludge from a thermophilic aerobic membrane reactor. *Journal of Chemistry, 2017*, Article ID 8764510. DOI: 10.1155/2017/8764510.

Alattabi, A.W., Harris, C.B., Alkhaddar, R.M., Hashim, K.S., Ortoneda-Pedrola, M., & Phipps, D. (2017). Improving sludge settleability by introducing an innovative, two-stage settling sequencing batch reactor. *Journal of Water Process Engineering, 20*(November), 207–216. DOI: 10.1016/j.jwpe.2017.11.004.

Amaral, A.L., Abreu, H., Leal, C., Mesquita, D.P., Castro, L.M., & Ferreira, E.C. (2017). Quantitative image analysis of polyhydroxyalkanoates inclusions from microbial mixed cultures under different SBR operation strategies. *Environmental Science and Pollution Research*, *24*(17), 15148–15156. DOI: 10.1007/s11356-017-9132-0.

Amaral, A.L., Mesquita, D.P., & Ferreira, E.C. (2013). Automatic identification of activated sludge disturbances and assessment of operational parameters. *Chemosphere*, *91*(5), 705–710. DOI: 10.1016/j.chemosphere.2012.12.066.

Asensi, E., Zambrano, D., Alemany, E., & Aguado, D. (2019). Effect of the addition of precipitated ferric chloride on the morphology and settling characteristics of activated sludge flocs. *Separation and Purification Technology*, *227*(June), 115711. DOI: 10.1016/j.seppur.2019.115711.

Asgharnejad, H., & Sarrafzadeh, M.H. (2020). Development of digital image processing as an innovative method for activated sludge biomass quantification. *Frontiers in Microbiology*, *11*, 574966. DOI: 10.3389/fmicb.2020.574966.

Bitton, G. (2010). *Wastewater Microbiology*. Wiley-Blackwell. DOI: 10.1002/9780470901243.

Boztoprak, H., Özbay, Y., Güçlü, D., & Küçükhemek, M. (2016). Prediction of sludge volume index bulking using image analysis and neural network at a full-scale activated sludge plant. *Desalination and Water Treatment*, *57*(37), 17195–17205. DOI: 10.1080/19443994.2015.1085909.

Burger, W., Krysiak-Baltyn, K., Scales, P.J., Martin, G.J.O., Stickland, A.D., & Gras, S.L. (2017). The influence of protruding filamentous bacteria on floc stability and solid-liquid separation in the activated sludge process. *Water Research*, *123*, 578–585. DOI: 10.1016/j.watres.2017.06.063.

Campbell, K., Wang, J., & Daniels, M. (2019). Assessing activated sludge morphology and oxygen transfer performance using image analysis. *Chemosphere*, *223*, 694–703. DOI: 10.1016/j.chemosphere.2019.02.088.

Carvalheira, M., Oehmen, A., Carvalho, G., & Reis, M.A.M. (2014). The effect of substrate competition on the metabolism of polyphosphate accumulating organisms (PAOs). *Water Research*, *64*, 149–159. DOI: 10.1016/j.watres.2014.07.004.

Cherkassky, V., & Mulier, F.M. (2007). *Learning* from *Data: Concepts, Theory, and Methods*. Wiley-IEEE Press.

Costa, J.C., Mesquita, D.P., Amaral, A.L., Alves, M.M., & Ferreira, E.C. (2013). Quantitative image analysis for the characterization of microbial aggregates in biological wastewater treatment: A review. *Environmental Science and Pollution Research*, *20*(9), 5887–5912. DOI: 10.1007/s11356-013-1824-5.

de Kreuk, M.K., Heijnen, J.J., & Van Loosdrecht, M.C.M. (2005). Simultaneous COD, nitrogen, and phosphate removal by aerobic granular sludge. *Biotechnology and Bioengineering*, *90*(6), 761–769. DOI: 10.1002/bit.20470.

Eikelboom, D.H. (2000). *Process Control of Activated Sludge Plants by Microscopic Investigation*. IWA, London.

Einax, J.W., Zwanziger, H.W., & Geiss, S. (2004). *Chemometrics in Environmental Analysis*. VCH, Weinheim.

Fortunato, L., Li, M., Cheng, T., Rehman, Z.U., Heidrich, W., & Leiknes, T.O. (2019). Cake layer characterization in activated sludge membrane bioreactors: Real-time analysis. *Journal of Membrane Science*, *578*, 163–171. DOI: 10.1016/j.memsci.2019.02.026.

Hu, X., Wang, K., Chang, J., Zhang, L., Zhong, M., & Nie, Y. (2021). Establishment of a comprehensive analysis method for the microfaunal movement in activated sludge. *Environmental Science and Pollution Research International*, *28*(14), 17084–17097. DOI: 10.1007/s11356-020-12090-3.

Hwang, B.K., Lee, C.H., Chang, I.S., Drews, A., & Field, R. (2012). Membrane bioreactor: TMP rise and characterization of bio-cake structure using CLSM-image analysis. *Journal of Membrane Science*, *419–420*, 33–41. DOI: 10.1016/j.memsci.2012.06.031.

Jeison, D., & Chamy, R. (1998). Novel technique for measuring the size distribution of granules from anaerobic reactors for wastewater treatment. *Biotechnology Techniques*, 12, 659–662.

Jiang, H.L., Tay, J.H., & Tay, S.T.L. (2002). Aggregation of immobilized activated sludge cells into aerobically grown microbial granules for the aerobic biodegradation of phenol. *Letters in Applied Microbiology*, *35*(5), 439–445. DOI: 10.1046/j.1472-765X.2002.01217.x.

Khan, M.B., Nisar, H., & Ng, C.A. (2017). Image processing and analysis of phase-contrast microscopic images of activated sludge to monitor the wastewater treatment plants. *IEEE Access*, *6*, 1778–1791. DOI: 10.1109/ACCESS.2017.2780249.

Khan, M.B., Nisar, H., & Ng, C.A. (2020). Generalized modeling of the sludge volume index and suspended solids to monitor activated sludge bioflocculation using image analysis. *Environment Protection Engineering*, *46*(3), 17–37. DOI: 10.37190/epe200302.

Khan, M.B., Nisar, H., Ng, C.A., Yeap, K.H., & Lai, K.C. (2017b). Segmentation approach towards phase-contrast microscopic images of activated sludge to monitor the wastewater treatment. *Microscopy and Microanalysis*, *23*(6), 1130–1142. DOI: 10.1017/S1431927617012673.

Khanam, T., Syuhada Wan Ata, W.N., & Rashedi, A. (2016). Particle size measurement in waste water influent and effluent using particle size analyzer and quantitative image analysis technique. *Advanced Materials Research, 1133*, 571–575. DOI: 10.4028/www.scientific.net/amr.1133.571.

Koivuranta, E., Keskitalo, J., Stoor, T., Hattuniemi, J., Sarén, M., & Niinimäki, J. (2014). A comparison between floc morphology and the effluent clarity at a full-scale activated sludge plant using optical monitoring. *Environmental Technology (United Kingdom), 35*(13), 1605–1610. DOI: 10.1080/09593330.2013.875065.

Krzeminski, P., Leverette, L., Malamis, S., & Katsou, E. (2017). Membrane bioreactors – A review on recent developments in energy reduction, fouling control, novel configurations, LCA and market prospects. *Journal of Membrane Science, 527*, 207–227. DOI: 10.1016/j.memsci.2016.12.010.

Leal, C., Amaral, A.L., & Costa, M.de L. (2016). Microbial-based evaluation of foaming events in full-scale wastewater treatment plants by microscopy survey and quantitative image analysis. *Environmental Science and Pollution Research, 23*(15), 15638–15650. DOI: 10.1007/s11356-016-6750-x.

Leal, C., Val del Río, A., Mesquita, D.P., Amaral, A.L., Castro, P.M.L., & Ferreira, E.C. (2020). Sludge volume index and suspended solids estimation of mature aerobic granular sludge by quantitative image analysis and chemometric tools. *Separation and Purification Technology, 234*(116049). DOI: 10.1016/j.seppur.2019.116049.

Li, D., Lv, Y., Zeng, H. & Zhang, J. (2016). Enhanced biological phosphorus removal using granules in continuous-flow reactor. *Chemical Engineering Journal, 298*, 107–116. DOI: 10.1016/j.cej.2016.03.152.

Liwarska-Bizukojc, E., Klepacz-Smółka, A., & Andrzejczak, O. (2015). Variations of morphology of activated sludge flocs studied at full-scale wastewater treatment plants. *Environmental Technology (United Kingdom), 36*(9), 1123–1131. DOI: 10.1080/09593330.2014.982717.

Ma, J., Dai, R., Chen, M., Khan, S.J., & Wang, Z. (2018). Applications of membrane bioreactors for water reclamation: Micropollutant removal, mechanisms and perspectives. *Bioresource Technology, 269*, 532–543. DOI: 10.1016/j.biortech.2018.08.121.

Madoni, P. (1994). A sludge biotic index (SBI) for the evaluation of the biological performance of activated sludge plants based on the microfauna analysis. *Water Research, 28*, 67–75.

Mandelbrot, B.B. (2004). *Fractals and Chaos*. Springer-Verlag, New York. DOI: 10.1007/978-1-4757-4017-2.

Mesquita, D.P., Amaral, A.L., & Ferreira, E.C. (2011). Identifying different types of bulking in an activated sludge system through quantitative image analysis. *Chemosphere, 85*(4), 643–652. DOI: 10.1016/j.chemosphere.2011.07.012.

Mesquita, D.P., Amaral, A.L., & Ferreira, E.C. (2013). Activated sludge characterization through microscopy: A review on quantitative image analysis and chemometric techniques. *Analytica Chimica Acta, 802*, 14–28. DOI: 10.1016/j.aca.2013.09.016.

Mesquita, D.P., Amaral, A.L., & Ferreira, E.C. (2016). Estimation of effluent quality parameters from an activated sludge system using quantitative image analysis. *Chemical Engineering Journal, 285*, 349–357. DOI: 10.1016/j.cej.2015.09.110.

Mesquita, D.P., Amaral, A.L., Leal, C., Carvalheira, M., Cunha, J.R., Oehmen, A., Reis, M.A.M., & Ferreira, E.C. (2014). Monitoring intracellular polyphosphate accumulation in enhanced biological phosphorus removal systems by quantitative image analysis. *Water Science and Technology, 69*(11), 2315–2323. DOI: 10.2166/wst.2014.146.

Metcalf, & Eddy. (2003). *Wastewater Engineering: Treatment and Reuse*. McGraw-Hill, Boston.

Molina, M.A., Pérez, C.A.A., & Leiva, C.A. (2020). Characterization of filamentous flocs to predict sedimentation parameters using image analysis. *Journal of Sensors, 2020*, 5248509. DOI: 10.1155/2020/5248509.

Mullins, D., Jones, E., Glavin, M., Coburn, D., Hannon, L., & Clifford, E. (2018). A novel image processing-based system for turbidity measurement in domestic and industrial wastewater. *Water Science and Technology, 77*(5), 1469–1482. DOI: 10.2166/wst.2018.030.

Nguyen, T.T., Ngo, H.H., & Guo, W. (2013). Pilot scale study on a new membrane bioreactor hybrid system in municipal wastewater treatment. *Bioresource Technology, 141*, 8–12. DOI: 10.1016/j.biortech.2013.03.125.

Oliveira, P., Alliet, M., Coufort-Saudejaud, C., & Frances, C. (2018). New insights in morphological analysis for managing activated sludge systems. *Water Science and Technology, 77*(10), 2415–2425. DOI: 10.2166/wst.2018.189.

Oliveira, C.S.S., Silva, C.E., Carvalho, G., & Reis, M.A. (2017). Strategies for efficiently selecting PHA producing mixed microbial cultures using complex feedstocks: Feast and famine regime and uncoupled carbon and nitrogen availabilities. *New Biotechnology, 37*, 69–79. DOI: 10.1016/j.nbt.2016.10.0.

Ramos, C., Amorim, C.L., Mesquita, D.P., Ferreira, E.C., Carrera, J., & Castro, P.M.L. (2017). Simultaneous partial nitrification and 2-fluorophenol biodegradation with aerobic granular biomass: Reactor performance and microbial communities. *Bioresource Technology, 238*, 232–240. DOI: 10.1016/j.biortech.2017.03.173.

Russ, J.C. (2002). *The Image Processing Handbook*, 4th Ed. CRC Press, Boca Raton.

Satoh, H., Kashimoto, Y., Takahashi, N. & Tsujimura, T. (2021). Deep learning-based morphology classification of activated sludge flocs in wastewater treatment plants. *Environmental Science: Water Research & Technology*, *7*, 298–305.

Shah, M.P. (2020). *Advanced Oxidation Processes for Effluent Treatment Plants*. Elsevier.

Shah, M.P. (2020). *Microbial Bioremediation and Biodegradation*. Springer.

Shah, M.P. (2021). *Removal of Emerging Contaminants through Microbial Processes*. Springer.

Smoczyński, L., Ratnaweera, H., Kosobucka, M., Kvaal, K., & Smoczyński, M. (2014). Image analysis of sludge aggregates obtained at preliminary treatment of sewage. *Water Science and Technology*, *70*(6), 1048–1055. DOI: 10.2166/wst.2014.332.

Tomperi, J., & Leiviskä, K. (2018). Comparison of modelling accuracy with and without exploiting automated optical monitoring information in predicting the treated wastewater quality. *Environmental Technology (United Kingdom)*, *39*(11), 1442–1449. DOI: 10.1080/09593330.2017.1331267.

Yu, R.F., Chen, H.W., Cheng, W.P., & Huang, H.D. (2017). Applying online image analysis to simultaneously evaluate the removals of suspended solids and color from textile wastewater in chemical flocculated sedimentation. *Journal of Environmental Informatics*, *29*(1), 29–38. DOI: 10.3808/jei.201500303.